丝路国际产能合作促进中心
Belt and Road International Production Capacity Cooperation Promotion Center (BRICC)
"一带一路"建筑设计系列丛书
BRICC A "Belt and Road" Series for Architecture Design

建筑设计技术指南
——助力"一带一路"建设（中英文对照）

Architecture Design Technical Guide:
For "Belt and Road" Initiative Construction（Chinese and English）

深圳市联合创艺建筑设计有限公司　编
Shenzhen United Architecture Design Co., Ltd.

中国建筑工业出版社
CHINA ARCHITECTURE & BUILDING PRESS

图书在版编目（CIP）数据

建筑设计技术指南：助力"一带一路"建设：汉英对照 / 深圳市联合创艺建筑设计有限公司编 . — 北京：中国建筑工业出版社，2019.9
（"一带一路"建筑设计系列丛书）
ISBN 978-7-112-23881-1

Ⅰ.①建… Ⅱ.①深… Ⅲ.①民用建筑 — 建筑设计 — 指南 — 汉、英 Ⅳ.① TU24-62

中国版本图书馆 CIP 数据核字（2019）第 118501 号

责任编辑：武晓涛　朱晓瑜
责任校对：李欣慰

"一带一路"建筑设计系列丛书
建筑设计技术指南
——助力"一带一路"建设（中英文对照）
深圳市联合创艺建筑设计有限公司　编

*

中国建筑工业出版社出版、发行（北京海淀三里河路 9 号）
各地新华书店、建筑书店经销
北京点击世代文化传媒有限公司制版
北京中科印刷有限公司印刷

*

开本：787×1092 毫米　1/16　印张：23¾　字数：488 千字
2019 年 11 月第一版　2019 年 11 月第一次印刷
定价：82.00 元
ISBN 978-7-112-23881-1
（34182）

版权所有　翻印必究
如有印装质量问题，可寄本社退换
（邮政编码 100037）

本书编委会成员及其他成员组成

The members and other members of the editorial board of this book

专家委员会主任：廖志斌
Director of the Committee of Experts: Liao Zhibin

审定：丘志强
Validation: Qiu Zhiqiang

校对：张伯林
Proofreading: Zhang Bolin

专家委员会委员：张伯林　张　伟　秦　谱　李建平　江潮树　陶　健　孟庆君
Member of the Committee of Experts: Zhang Bolin, Zhang Wei, Qin Pu, Li Jianping, Jiang Chaoshu, Tao Jian, Meng Qingjun

主编：廖志斌
Editor-in-chief: Liao Zhibin

副主编：张伯林　丘志强　陈　强
Deputy editor-in-chief: Zhang Bolin, Qiu Zhiqiang, Chen Qiang

参编单位：深圳市骏业建筑科技有限公司
Participating units: Shenzhen Junye Construction Technology Co., Ltd.

1 总则	编写：张伯林	校对：张　伟
1 General Principles	Preparation: Zhang Bolin	Proofreading: Zhang Wei
2 术语	编写：张伯林	校对：张　伟
2 Terminology	Preparation: Zhang Bolin	Proofreading: Zhang Wei
3 基本规定	编写：张　伟	校对：张伯林
3 Basic Provisions	Preparation: Zhang Wei	Proofreading: Zhang Bolin
4 城市规划对建筑的限定	编写：秦　谱	校对：张伯林
4 Limits to Building in Urban Planning	Preparation: Qin Pu	Proofreading: Zhang Bolin
5 总平面设计	编写：张伯林　韩广智	校对：张　伟
5 Site Planning	Preparation: Zhang Bolin　Han Guangzhi	Proofreading: Zhang Wei
6 建筑防火一般规定	编写：江潮树	校对：李建平
6 Fire Protection Design for Buildings	Preparation: Jiang Chaoshu	Proofreading: Li Jianping

7	地下室和半地下室	编写：李建平　校对：江潮树
7	Basements and Semi-basements	Preparation: Li Jianping　Proofreading: Jiang Chaoshu
8	墙体	编写：孟庆君　校对：张伯林
8	Walls	Preparation: Meng Qingjun　Proofreading: Zhang Bolin
9	幕墙、采光顶	编写：孟庆君　校对：张伯林
9	Curtain Wall, Lighting Roof	Preparation: Meng Qingjun　Proofreading: Zhang Bolin
10	室内装修工程设计	编写：孟庆君　校对：张伯林
10	Interior Decoration Engineering Design	Preparation: Meng Qingjun　Proofreading: Zhang Bolin
11	屋面和吊顶	编写：张伯林　校对：李建平
11	Roof and Suspended Ceiling	Preparation: Zhang Bolin　Proofreading: Li Jianping
12	楼梯、台阶、坡道	编写：江潮树　校对：李建平
12	Staircases, Steps and Ramps	Preparation: Jiang Chaoshu　Proofreading: Li Jianping
13	电梯、自动扶梯、自动人行道	编写：江潮树　校对：李建平
13	Elevators, Escalators and Moving Walkways	Preparation: Jiang Chaoshu　Proofreading: Li Jianping
14	门窗	编写：张伯林　校对：江潮树
14	Doors and Windows	Preparation: Zhang Bolin　Proofreading: Jiang Chaoshu
15	建筑物无障碍设计	编写：李建平　校对：江潮树
15	Accessibility Design for Buildings	Preparation: Li Jianping　Proofreading: Jiang Chaoshu
16	厨房	编写：张伯林　校对：李建平
16	Kitchen	Preparation: Zhang Bolin　Proofreading: Li Jianping
17	卫生间	编写：李建平　校对：江潮树
17	Toilets	Preparation: Li Jianping　Proofreading: Jiang Chaoshu
18	设备用房	编写：张伯林　校对：江潮树
18	Equipment Rooms	Preparation: Zhang Bolin　Proofreading: Jiang Chaoshu
19	电缆井、管道井、烟道、通风道和垃圾道	编写：张伯林　校对：江潮树
19	Cable Wells, Pipeline Shafts, Flues, Ventilation Channels and Refuse Chutes	Preparation: Zhang Bolin　Proofreading: Jiang Chaoshu
20	景观设计	编写：陶　健　校对：张伯林
20	Design of Landscape	Preparation: Tao Jian　Proofreading: Zhang Bolin

深圳市联合创艺建筑设计有限公司

Shenzhen United Architecture Design Co., Ltd.

2018 年 8 月

August 2018

序 言
Preface

  中国建筑自改革开放以来 40 年积淀的先进的技术水平、管理能力和大量项目实战经验，加上"一带一路"国际市场的锤炼，已经成为名副其实的全球基建强国。我们不仅承揽了全球最多的基建工程项目，而且建成了许多最大的、最难的、全球瞩目的标志性项目。由此，中国的建筑企业也得到了快速成长，这从美国《工程新闻记录》(ENR)250强中中资建筑企业数量和排名的快速进步就可得到印证。

  Through 40 years accumulation of advanced technology, administrative ability and plenty of practical experience in projects since the Reform and Opening-up, added with the "Belt and Road" international market tempering, China's construction has become an indeed superpower in the world's field of infrastructure. We have undertaken not only most of the global infrastructure engineering projects, but also finished a lot of biggest landmark projects with global attention under utmost difficulties. Therefore, Chinese construction enterprises have grown rapidly, which can be proven from the fast progress for the number and ranking of Chinese construction enterprises recorded in the Top 250 International General Contractors of the *US Engineering News-Record* (ENR).

  随着中国建筑企业的走出去，尤其是随着中国境外项目投资的增大，随着建筑企业境外 EPC 总包业务的增多，中国的建筑设计也在国际市场上崭露头角。

  Along with the "Going-Global" of Chinese construction enterprise, especially with the investment increase of China's overseas projects and increase of overseas EPC general contract business by construction enterprises, China's architectural design is also emerging in the international market.

  但令人遗憾的是，设计业务的份额增长还不足。这其中有多方面的原因，有中国设计标准与国际标准本身存在差异的问题，也有各方对中国标准的理解不够的问题。而后者的缺憾需要通过优化中国标准的制定思维，通过出版外文版的中国标准，通过提高中国设计师的外语沟通能力等来加以弥补。

  Yet unfortunately, the growth of design business' market shares still has a long way to go. There are reasons from several aspects, such as difference of Chinese design standards to those of international and lack of understanding to Chinese standards by world people. For the solution to the later, we need to optimize the thinking of Chinese standard formulation, to publish foreign versions of Chinese standards, and to enhance Chinese designer's communication abilities in foreign language.

另外，现在也有越来越多的外国设计企业进入中国，他们也面临着尽快熟悉中国设计标准，按中国标准来开展业务的问题。因此，为适应建设市场的快速发展，规范国内建筑市场，增强境外建筑设计竞争力，加快与国际市场的接轨，十分需要中英文结合的建筑设计标准方面的指导书。

Moreover, there are more and more foreign design companies entering China presently, who are facing the same issues of getting familiar with Chinese design standards and carrying out business on Chinese standards. Therefore, to adapt to the fast development of construction market, to regulate domestic construction market, to improve the competitiveness of overseas architectural design and to facilitate the integration with the international market, we need an instruction book urgently on architectural design standards bilingually in Chinese and in English.

本书就是这样的一本工具书。它采用中英文对照的方式，针对一般建筑工程项目，集成了国家各种规范和规定的主要设计内容，可以指导设计师快速查找规范和规定的相关条文。

This book is the one. It's written both in Chinese and English, focusing on general construction engineering projects, combing with major design content of various of national regulations and rules, so that designers can be guided to spot out quickly items of regulations and rules related.

此书作为建筑工程设计行业的工具书，可以成为在"一带一路"沿线国家进行建筑工程设计技术交流时的一个有效载体。通过它可以让在国内从事建筑设计的外国专家以及工程技术人员了解到中国的技术规范与规程，也可为"走出去"从事建筑设计的人员将中国标准与外国标准进行有效对接，让中国先进的建筑施工技术在中国设计的保障下在项目所在国推广应用。另外，此书也可以作为大、中专院校，工程类学生的课外辅导参考书，帮助学生熟悉工程专业术语，为今后的工作打下坚实的基础。

As a reference book of architectural engineering design industry, this book can be an effective carrier for the exchange of architectural engineering design technology in countries along "the Belt and Road". Foreign experts and engineering technicians who are involving in the business of Chinese architectural design can understand Chinese technical regulations and rules through the book; Chinese "Going-Global" people of architectural design can effectively docking Chinese standard with that of other countries under the help of the book, and enable Chinese advanced construction engineering technology to be applied in the world under the escort of China Design. Moreover, this book can also be an extra-curricular reference book for students of colleges and universities majoring in engineering, to help them to get familiar with engineering terminology, paving the way for the work after graduation.

本套丛书以建筑工程设计标准为准则，针对各类工程新技术，以深入浅出的语言，图文并茂的方式进行呈现，可以方便读者理解。

This book based on construction engineering design standard, showing various of new technologies of engineering in simple language and abundant illustrations to serve for better understanding.

值得肯定的是，本书是由深圳市联合创艺建筑设计有限公司组织编写的。作为一家民营的设计企业，不仅业务上令人瞩目，还有强烈的社会责任感，为了"一带一路"建设的高质量发展，专门组织行业内的专家教授，历时一年，编写了此书。这是值得赞许，也是令人欣慰的。

It is remarkable that this book was written by Shenzhen Lianhe Chuangyi Architectural Design Co., Ltd., which is an private design enterprise with both impressive achievement in the field and strong sense of social responsibility. For sake of quality development of the "Belt and Road" construction, the company endeavored in organizing professors and experts within the industry and spent a whole year for the completion of the book, which is praise-worthy and gratifying.

作为丝路国际产能合作促进中心，我们也肩负着协助中国企业"走出去"的使命，出版实战化的"走出去"丛书也是我们工作的一部分。本书内容具体而实用，相信将十分有利于建筑企业的境外业务开拓，让中国的建筑设计也加快走向世界。

Belt and Road International Production Capacity Cooperation Promotion Center (BRICC) shoulders the mission of assisting Chinese enterprises "Going-Global", and publishing practical "Going Out" series is also a part of our job. With specific and practical content, the book is believed to be very conducive to the development of overseas business for construction enterprises and will help Chinese architectural design "Going-Global" as well.

此为序。
As conclusion of Preface.

Xie Yangjun

丝路国际产能合作促进中心　主任

Belt and Road International Production Capacity Cooperation Promotion Center (BRICC) Director

2019 年 9 月

September, 2019

前　言
Foreword

　　本书是深圳市联合创艺建筑设计有限公司编写的第一部中英文版统一技术措施，是以现行国家规范和全国技术标准以及行业标准为准则编制的。

　　This book *Architecture Design Tehnical Guide* is the first unified technical measures in Chinese and English version prepared by Shenzhen United Architectural Design Co., Ltd.. It is based on existing national norms and national technical standards, as well as industry standards.

　　本书编制内容和编排方式，力图从设计需要出发通过对原规范、标准条文的总结和归纳、提炼，尽量表格化、简图化，以达到简单方便、查找快捷的使用目的。同时采用中英文是为了适应设计市场国际化的发展需要。

　　The contents and arrangement of this book starts from design needs, proceeds through the original norms, standards of the provisions by summary, induction and refining, as well as tries to use tabular, sketch to achieve simple and convenient, find quickly. Moreover, Chinese and English are used to meet the development needs of the internationalization of the design market.

　　本书在编写过程中得到有关领导、专家、同行的大力支持、帮助、指导。在此表示衷心的感谢。

　　Strong support, help, guidance have been given to this book in the preparation process by the relevant leadership, experts and peers. We would like to express my heartfelt thanks.

　　由于本书内容广、工作量大、时间和水平有限，一定还存在很多问题，敬请批评指正，以便在今后不断修正和完善。

　　Due to the wide contents of this Book, heavy workload and limited time, there must still be a lot of problems, please criticize and correct, we will continuously revise and improve the Book in the future.

深圳市联合创艺建筑设计有限公司
Shenzhen United Architecture Design Co., Ltd.
2018 年 8 月
August, 2018

目 录
Contents

1	总则	1
1	General Principles	1
2	术语	7
2	Terminology	7
3	基本规定	17
3	Basic Provisions	17
	3.1 民用建筑分类	18
	3.1 Classification of Civil Buildings	18
	3.2 建筑高度、层数、净高	20
	3.2 Height, Storey and Net Height of Building	20
	3.3 主要技术经济指标	24
	3.3 Main Technical and Economic Indicators	24
	3.4 建筑气候分区对建筑基本要求	27
	3.4 Basic Requirements of the Climate Zoning of Buildings	27
	3.5 建筑与环境	30
	3.5 Buildings and Environment	30
4	城市规划对建筑的限定	33
4	Limits to Building in Urban Planning	33
	4.1 【通则】4.1条：建筑基地	34
	4.1 [Code for design of civil buildings] 4.1: Construction Base	34
	4.2 建筑突出物	38
	4.2 Building Protrusions	38
	4.3 建筑高度控制	39
	4.3 Height Control of Building	39
	4.4 建筑密度、容积率和绿地率	40
	4.4 Building Density, Plot Ratio and Greening Rate	40
5	总平面设计	43
5	Site Planning	43
	5.1 一般规定	44
	5.1 General Requirements	44
	5.2 竖向设计	48
	5.2 Vertical Design	48
	5.3 居住区道路及停车场	53
	5.3 Roads and Parking Areas in Residential Areas	53
	5.4 广场、商业步行区及室外活动运动场	66

	5.4	Squares, Commercial Walking Areas and Outdoor Activity Venues	66
	5.5	管线综合	69
	5.5	Integrated Layout of Pipelines	69
6		建筑防火一般规定	71
6		Fire Protection Design for Buildings	71
	6.1	建筑分类和耐火等级	72
	6.1	Building Classification and Refractory Grade	72
	6.2	防火分区	74
	6.2	Fire Compartments	74
	6.3	安全疏散距离	77
	6.3	Safety Evacuation Distance	77
	6.4	疏散走道	81
	6.4	Evacuation Corridor	81
7		地下室和半地下室	87
7		Basements and Semi-basements	87
	7.1	术语	88
	7.1	Terminology	88
	7.2	一般规定	91
	7.2	General Provisions	91
	7.3	地下汽车库	93
	7.3	Underground Garages	93
	7.4	地下商业	112
	7.4	Underground Commercial Buildings	112
	7.5	地下室防水	119
	7.5	Waterproof of Basements	119
8		墙体	131
8		Walls	131
	8.1	墙体类型与材料	132
	8.1	Types and Materials of Walls	132
	8.2	墙体防潮、防水、隔汽	135
	8.2	Moisture Proofing, Waterproof and Vapor Proofing of Walls	135
	8.3	墙体防火	138
	8.3	Fireproofing of Walls	138
	8.4	【隔声规】4条：墙体隔声	147
	8.4	[Code for design of sound insulation of civil buildings] 4: Wall Sound Insulation	147
9		幕墙、采光顶	151
9		Curtain Wall, Lighting Roof	151
	9.1	【幕墙规】2条：常用类型	152
	9.1	[Technical code for metal and stone curtain walls engineering] 2: Commonly Used Type	152
	9.2	【幕墙规】3条：主要材料及选用要点	153
	9.2	[Technical code for metal and stone curtain walls engineering] 3:	

	Main Materials and Selection Points ··· 153

10 室内装修工程设计 ·· 157
10 Interior Decoration Engineering Design ··· 157

10.1 【通则】6.15 条：一般规定 ··· 158
10.1 [Code for design of civil buildings] 6.15: General Provisions ······················· 158
10.2 【装修防火规范】3 条：装修材料的分类和分级 ····································· 159
10.2 [Code for fire prevention in design of interior decoration of buildings] 3: Classification and Grading of Decoration Materials ································· 159
10.3 【装修防火规范】4 条：特别场所 ·· 161
10.3 [Code for fire prevention in design of interior decoration of buildings] 4: Special Places ········ 161
10.4 【装修防火规范】5 条：民用建筑 ·· 166
10.4 [Code for fire prevention in design of interior decoration of buildings] 5: Civil Buildings ······· 166

11 屋面和吊顶 ··· 169
11 Roof and Suspended Ceiling ·· 169

11.1 【屋面技规】3.0.2 条：屋面类型及基本构造层次 ······································ 170
11.1 [Technical code for roof engineering] 3.0.2: Roof Types and Basic Structure Levels ··········· 170
11.2 【屋面技规】3.0.5 条：屋面防水等级和设防要求 ······································ 171
11.2 [Technical code for roof engineering] 3.0.5: Waterproof Level and Fortification Requirements of Roofs ·· 171
11.3 屋面的找坡层和找平层 ·· 172
11.3 Sloping Layer and Leveling Layer of Roof ··· 172
11.4 【屋面技规】4.2 条：屋面的排水设计 ·· 173
11.4 [Technical code for roof engineering] 4.2: Drainage Design for Roofs ············· 173
11.5 保温层和隔热层设计 ·· 175
11.5 Design of Insulation Layers and Heat Shielding Layers ································ 175
11.6 卷材及涂膜防水层设计 ·· 179
11.6 Design of Coiled Materials and Coated Film Waterproof Layers ····················· 179
11.7 【屋面技规】4.7 条：保护层和隔离层设计 ·· 184
11.7 [Technical code for roof engineering] 4.7: Design of Protective Layers and Isolation Layers ······· 184
11.8 瓦屋面 ·· 185
11.8 Tile Roofs ·· 185
11.9 金属屋面设计 ·· 190
11.9 Design for Metal Board Roofs ··· 190
11.10 玻璃采光顶设计 ·· 194
11.10 Design for Glass Daylighting Roofs ··· 194

12 楼梯、台阶、坡道 ·· 195
12 Staircases, Steps and Ramps ··· 195

12.1 一般规定 ·· 196
12.1 General Provisions ·· 196
12.2 楼梯、楼梯间设计 ·· 198
12.2 Design of Staircase and Staircase Room ··· 198

		12.3	楼梯、楼梯间防火设计	203
		12.3	Fire Protection Design for Stairs and Staircases	203
		12.4	台阶、坡道	218
		12.4	Step and Ramp	218
13	电梯、自动扶梯、自动人行道			223
13	Elevators, Escalators and Moving Walkways			223
		13.1	一般规定	224
		13.1	General Provisions	224
		13.2	电梯	227
		13.2	Elevators	227
		13.3	自动扶梯	239
		13.3	Escalators	239
		13.4	自动行人道	243
		13.4	Moving Sidewalk	243
		13.5	防火设计要点	245
		13.5	Key Points for Fire Protection Design	245
		13.6	局部布置及构造	249
		13.6	Local Arrangement and Construction	249
14	门窗			257
14	Doors and Windows			257
		14.1	一般规定	258
		14.1	General Provisions	258
		14.2	门、窗设计	260
		14.2	Design of Door and Window	260
15	建筑物无障碍设计			267
15	Accessibility Design for Buildings			267
		15.1	无障碍设计范围	268
		15.1	Accessibility Design Scope	268
		15.2	无障碍设施设计要求	268
		15.2	Design Requirements for Accessibility Facilities	268
16	厨房			287
16	Kitchen			287
		16.1	一般规定	288
		16.1	General Provisions	288
		16.2	住宅厨房	289
		16.2	Kitchen of the Residential Buildings	289
		16.3	公用厨房	293
		16.3	Communal Kitchen	293
17	卫生间			299
17	Toilets			299
		17.1	一般规定		300

	17.1	General Provisions	300
	17.2	公共卫生间	302
	17.2	Public Toilets	302
	17.3	住宅卫生间	314
	17.3	Toilets of Residential Buildings	314
	17.4	卫生间防水	317
	17.4	Waterproofing in Toilets	317
18	设备用房		321
18	Equipment Rooms		321
	18.1	给水和排水	322
	18.1	Water Supply and Drainage	322
	18.2	【通则】8.3 条：建筑电气	323
	18.2	[Code for design of civil buildings] 8.3: Building Electricity	323
	18.3	智能化系统机房设计要求	327
	18.3	Design Requirements for Intelligent System Room	327
	18.4	【通则】8.3 条：暖通和空调	328
	18.4	[Code for design of civil building] 8.3: Heating, Ventilation and Air Conditioning	328
	18.5	暖通空调设备用房一览表	329
	18.5	Schedule of HVAC Equipment Rooms	329
	18.6	管道井	330
	18.6	Pipeline Shafts	330
19	电缆井、管道井、烟道、通风道和垃圾道		331
19	Cable Wells, Pipeline Shafts, Flues, Ventilation Channels and Refuse Chutes		331
20	景观设计		335
20	Design of Landscape		335
	20.1	一般规定	336
	20.1	General Provisions	336
	20.2	平面布局	337
	20.2	Plane Layout	337
	20.3	竖向设计	340
	20.3	Vertical Design	340
	20.4	园路及铺装场地	341
	20.4	Garden Paths and Pavement Sites	341
	20.5	景观小品	343
	20.5	Landscape Accessories	343
	20.6	【09 措施】2.6.1 条：水景	350
	20.6	[Measures 2009] 2.6.1: Waterscape	350
	20.7	【09 措施】2.7 条：种植设计	353
	20.7	[Measures 2009] 2.7: Planting Design	353

1

总则
General Principles

1.0.1 本指南的编制严格遵守国家现行规划法律、法规、标准和规范，目的是进一步提高建筑工程质量和设计效率，对于民用建筑设计中的共性问题所编制的民用建筑设计技术指南。

1.0.1 In the establishment of the Guide, the current national planning laws, regulations, standards and specifications shall be strictly followed, to further improve the quality of building engineering and design efficiency. The Guide are the unified technical measures for civil buildings, targeting for the common problems in civil building design.

1.0.2 本指南适用于新建、改建和扩建的民用建筑设计。

1.0.2 The unified technical Guide for techniques are applicable to the design for civil buildings in construction, alteration and extension.

1.0.3 使用本指南时，除应遵守国家方针政策外，还应遵守所在省、直辖市、自治区等地方政策与法规。

1.0.3 Any application of the Guide, shall be in appliance with the local policies and regulations of provinces, municipalities or autonomous regions, in addition to the State guide lines and policies.

1.0.4 本指南中国家标准、行业标准系最低要求，凡是地方标准严于国家标准、行业标准者应按地方标准执行，凡本指南严于地方标准者应按本指南执行。除了更严格的部分外，本指南如有与国家标准、行业标准及地方标准法规相矛盾时，应按国家标准、行业标准及地方法规执行。

1.0.4 The national standards and industrial standards in the Guide shall be the minimum requirements. If the local standards are more stringent than the national standards and industrial standards, the local standards shall be followed in the implementation. If the Guide are more stringent than the local standards, the Guide shall be followed in the implementation. Except for the more strict parts, if the Guide contradicts with the national standards and industrial standards and local standards and regulations, the national standards and industrial standards and local regulations shall be followed during implementation.

1.0.5 本指南原文引用的现行国家规范、技术标准、技术规程的条目均加以"【规范名称的缩写】规范条目"表示。简写规则见表1.0.5。

1.0.5 The current national norms, technical standards, and technical specifications cited in the original text of this Guide are added with "abbreviation of the specification+ specification entries". The abbreviated rules see Tab.1.0.5.

相关国家规范、技术标准、技术规程简写 表 1.0.5

The Abbreviated Rules of the Current National Norms, Technical Standards, and technical Specification　　Tab.1.0.5

规范名称 Specification name	编号 Code	本指南简写方式 Short description of this measure
民用建筑设计通则 Code for design of civil buildings	GB 50352—2005	【通则】 [Code for design of civil buildings]
建筑设计防火规范 Code for fire protection design of buildings	GB 50016—2014（2018年版）	【防火规】 [Code for fire protection design of buildings]
办公建筑设计规范 Design code for office building	JGJ 67—2006	【办规】 [Design code for office building]
中小学校设计规范 Code for design of school	GB 50099—2011	【校规】 [Code for design of school]
托儿所、幼儿园建筑设计规范 Code for design of nursery and kindergarten buildings	JGJ 39—2016	【托规】 [Code for design of nursery and kindergarten buildings]
旅馆建筑设计规范 Code for design of hotel building	JGJ 62—2014	【旅规】 [Code for design of hotel building]
综合医院建筑设计规范 Code for design of general hospital	GB 51039—2014	【医规】 [Code for design of general hospital]
图书馆建筑设计规范 Code for design of library buildings	JGJ 38—2015	【图规】 [Code for design of library buildings]
档案馆建筑设计规范 Code for design of archives buildings	JGJ 25—2010	【档规】 [Code for design of archives buildings]
博物馆建筑设计规范 Code for design of museum building	JGJ 66—2015	【博规】 [Code for design of museum building]
人民防空地下室设计规范 Code for design of civil air defence basement	GB 50038—2005	【人防规】 [Code for design of civil air defence basement]
绿色建筑评价标准 Assessment standard for green building	GB/T 50378—2014	【绿建评】 [Assessment standard for green building]
民用建筑隔声设计规范 Code for design of sound insulation of civil buildings	GB 50118—2010	【隔声规】 [Code for design of sound insulation of civil buildings]
金属与石材幕墙工程技术规范 Technical code for metal and stone curtain walls engineering	JGJ 133—2001	【幕墙规】 [Technical code for metal and stone curtain walls engineering]
玻璃幕墙工程技术规范 Technical code for glass curtain wall engineering	JGJ 102—2003	【玻幕规】 [Technical code for glass curtain wall engineering]
公共建筑节能设计标准 Designstandardfor energy efficiency of public buildings	GB 50189—2015	【公共节能】 [Design standard for energy efficiency of public buildings]
夏热冬暖地区居住建筑节能设计标准 Design standard for energy efficiency of residential buildings in hot summer and warm winter zone	JGJ 75—2012	【夏冬规】 [Design standard for energy efficiency of residential buildings in hot summer and warm winter zone]

续表

规范名称 Specification name	编号 Code	本指南简写方式 Short description of this measure
建筑玻璃应用技术规程 Technical specification for application of architectural glass	JGJ 113—2015	【玻技规】 [Technical specification for application of architectural glass]
公园设计规划 Code for design of park	GB 51192—2016	【公园规】 [Code for design of park]
全国民用建筑工程设计技术措施：规划·建筑·景观（2009年版） National Technical Measures for Design of Civil Construction Planning·Architecture·Landscape	2009JSCS—1	【09措施】 [Measures 2009]
建筑内部装修设计防火规范 code for fire prevention in design of interior decoration of buildings	GB 50222—2017	【装修防火规范】 [Code for fire prevention in design of interior decoration of buildings]
住宅设计规范 Code for design of Residential Buildings	GB 50096—2011	【住设规】 [Design code of residential buildings]
住宅建筑规范 Residential building code	GB 50368—2005	—
城市公共厕所设计标准 Standard for design of urban public toilets	CJJ 14—2016	【公厕设计标准】 [Standard for design of urban public toilets]
地下工程防水技术规范 Technical code for waterproofing of underground works	GB 50108—2008	【地下防水规】 [Technical code for waterproofing of underground works]
建筑室内防水工程技术规程 Technical specification for interior waterproof of residential buildings	CECS 196：2006	【室内防水规】 [Technical specification for interior waterproof of residential buildings]
汽车库、修车库、停车场设计防火规范 Code for fire protection design of garage, motor repair shop and parking area	GB 50067—2014	【汽防规】 [Code for fire protection design of garage, motor repair shop and parking area]
车库建筑设计规范 Code for design of parking garage building	JGJ 100—2015	【车库建规】 [Code for design of parking garage building]
商店建筑设计规范 Code for design of store buildings	JGJ 48—2014	【商店建规】 [Code for design of store buildings]
城市居住区规划设计标准 Standard for urban residential areas planning &design	GB 50180—2018	【居规划标】 [Standard for urban residential areas planning and design]
深圳市建筑设计规则 Shenzhen Architectural Design Regulation	2019年版	【深圳建规则】 [Shenzhen architectural design regulation]
砌体结构设计规范 Code for design of masonry structures	GB 50003—2011	【砌体结构规范】 [Code for design of masonry structures]
民用建筑工程室内环境污染控制规范 Code for indoor environmental pollution control of civil building engineering	GB 50325—2010	【环控规】 [Code for indoor environmental pollution control of civil building engineering]

续表

规范名称 Specification name	编号 Code	本指南简写方式 Short description of this measure
屋面工程技术规范 Technical code for roof engineering	GB 50345—2012	【屋面技规】 [Technical code for roof engineering]
种植屋面技术规程 Technical specification for planting roof engineering	JGJ 155—2013	【种植规程】 [Technical specification for planting roof engineering]
建筑结构荷载规范 Load code for the design of building structures	GB 50009—2012	【建结规】 [Load code for the design of building structures]
老年人照料设施建筑设计标准 Standard for design of care facilities for the aged	JGJ 450—2018	【老年人照料设施标】 [Standard for design of care facilities for the aged]
城乡建设用地竖向规划规范 Code for vertical planning on urban and rural development land	CJJ 83—2016	【城竖规】 [Code for urban vertical planning]
建筑地面设计规范 Code for design of building ground	GB 50037—2013	【地面规】 [Code for design of building ground]
城市道路交通规划设计规范 Code for transport planning on urban road	GB 50220—95	【城道规】 [Code for transport planning on urban road]
城市工程管线综合规划规范 Code for urban engineering pipelines comprehensive planning	GB 50289—2016	【城管线规】 [Code for design of urban pipelines]
无障碍设计规范 Codes for accessibility design	GB 50763—2012	【无规】 [Codes for accessibility design]
饮食建筑设计标准 Standard for design of dietetic buildings	JGJ 64—2017	【饮食标】 [Standard for design of dietetic buildings]
国家建筑标准设计图集——无障碍设计 Atlas of National Building Standard Design-Barrier-Free design	12J926	【无障碍设计 12J926】 [Design of accessibility facilities 12J926]
疗养院建筑设计标准 Architectural Design Code for Sanatorium	JGJ 40	【疗养院规】 [Code for design of sanatoriums]
住宅设计规范 Design Code of Residential Buildings	GB 50096—2011	【住设规】 [Design code of residential buildings]
住宅建筑规范 Residential building code	GB 50368—2005	【住建规】 [Residential building code]
城市公共厕所设计标准 Standard for design of urban public toilets	CJJ 14—2016	【公厕设计标准】 [Standard for design of urban public toilets]
地下工程防水技术规范 Technical code for waterproofing of underground works	GB 50108—2008	【地下防水规】 [Technical code for waterproofing of underground works]
建筑室内防水工程技术规程 Technical specification for interior waterproof engineering of building	CECS 196：2006	【室内防水规】 [Technical specification for interior waterproof engineering of building]

续表

规范名称 Specification name	编号 Code	本指南简写方式 Short description of this measure
宿舍建筑设计规范 Code for design of dormitory building	JGJ 36—2016	【宿舍规】 [Code for design of dormitory building]
建筑玻璃采光顶技术要求 Technical requirements of building glass skylight system	JG/T 231—2018	【采光顶】 [Technical requirement of building glass skylight system]
住宅性能评定技术标准 Technical standard for performance assessment of residential buildings	GB/T 50362—2005	【住宅性能标准】 Technical standard for performance assessment of residential buildings
建筑外门窗气密、水密、抗风压性能分级及检测方法 Graduations and test methods of air permeability, watertightness, wind load resistance performance for building external windows and doors	GB/T 7106—2008	【建筑外门窗气密、水密、抗风压性能分级及检测方法】 Graduations and test methods of air permeability, watertightness, wind load resistance performance for building external windows and doors

2
术语
Terminology

2.0.1 【通则】2.0.1 条：民用建筑

2.0.1 [Code for design of civil buildings] 2.0.1: Civil Building

供人们居住和进行公共活动的建筑的总称。

Generic terms for any building in which people reside or conduct public activities.

2.0.2 【通则】2.0.2 条：居住建筑

2.0.2 [Code for design of civil buildings] 2.0.2: Residential Building

供人们居住使用的建筑。

Buildings for people to live in and use.

2.0.3 【通则】2.0.3 条：公共建筑

2.0.3 [Code for design of civil buildings] 2.0.3: Public Building

供人们进行各种公共活动的建筑。

Buildings for people to conduct various public activities.

2.0.4 【通则】2.0.4 条：无障碍设施

2.0.4 [Code for design of civil buildings] 2.0.4: Barrier-free Facility

方便残疾人、老年人等行动不便或有视力障碍者使用的安全设施。

Safety facilities to facilitate the use of people with mobility problems such as the disabled, the aged etc., or those with visual impairments.

2.0.5 【通则】2.0.6 条：建筑地基

2.0.5 [Code for design of civil buildings] 2.0.6: Construction Site

根据用地性质和使用权属确定的建筑工程项目的使用场地。

Field designated to any construction project determined as per the use natures and property rights.

2.0.6 【通则】2.0.7 条：道路红线

2.0.6 [Code for design of civil buildings] 2.0.7: Boundary Lines for Road

规划的城市道路（含居住区级道路）用地的边界线。

Boundary lines of planned lands for urban roads (including roads of residential district level).

2.0.7 【通则】2.0.8 条：用地红线

2.0.7 [Code for design of civil buildings] 2.0.8: Boundary Lines for Land

各类建筑工程项目用地的使用权属范围的边界线。

Boundary lines of the use rights of various construction project lands.

2.0.8 【通则】2.0.9 条：建筑控制线

2.0.8 [Code for design of civil buildings] 2.0.9: Building Control Line

有关法规或详细规划确定的建筑物、构筑物的基底位置不得超出的界线。

Boundary lines that shall not be exceeded within which the positions of building and structure foundations are confirmed in the relevant regulations or detailed planning.

2.0.9 【通则】2.0.10 条：建筑密度

2.0.9 [Code for design of civil buildings] 2.0.10: Building Density

在一定范围内，建筑物的基底面积总和占用地面积的比例（％）。

Proportion（％）of the total area of the building foundation to the land area within a certain range.

2.0.10 【通则】2.0.11 条：容积率

2.0.10 [Code for design of civil buildings] 2.0.11: Plot Ratio

在一定范围内，建筑面积总和与用地面积的比值。

Ratio of the total building area to the land area within a certain range.

2.0.11 【通则】2.0.12 条：绿地率

2.0.11 [Code for design of civil buildings] 2.0.12: Greening Rate

一定地区内，各类绿地总面积占该地区总面积的比例（％）。

Proportion（％）of the total area of various green lands to the total area of the region in a certain area.

2.0.12 【通则】2.0.13 条：日照标准

2.0.12 [Code for design of civil buildings] 2.0.13: Insolation Standard

根据建筑物所处的气候区、城市大小和建筑物的使用性质确定的，在规定的日照标准日（冬至日或大寒日）的有效日照时间范围内，以底层窗台面为计算起点的建筑外窗获得的日照时间。

Sunshine time obtained from the exterior windows determined as per the climate zone, city size and use nature of the building within the effective sunshine time range of the specified standard sunshine day（winter solstice or great cold day），with the low-storey windowsill as the starting point for the calculation.

2.0.13 【防火规】附录 A：建筑高度和建筑层数的计算方法

2.0.13 [Code for fire protection design of buildingss] Appendix A: Calculation Methods of Building Height and Storeys

A.0.1 建筑高度的计算应符合下列规定：

A.0.1 The height of the building shall be calculated in accordance with the following provisions:

1.建筑屋面为坡屋面时，建筑高度应为建筑室外设计地面至其檐口与屋脊的平均高度。

1. If the roof of the building is a sloping roof, the building height should be the average

height of the building outdoor design ground to its eaves and roof.

2. 建筑屋面为平屋面（包括有女儿墙的平屋面）时，建筑高度应为建筑室外设计地面至其屋面面层的高度。

2. When the building roofs are flat roofs (including flat roofs with parapets), the height of the building should be the height of the building outdoor design ground to the surface of its roof.

3. 同一座建筑有多种形式的屋面时，建筑高度应按上述方法分别计算后，取其中最大值。

3. When the same building has many forms of roofs, the height of the building should be calculated separately according to the above method, taking the maximum value.

4. 对于台阶式地坪，当位于不同高程地坪上的同一建筑之间有防火墙分隔，各自有符合规范规定的安全出口，且可沿建筑的两个长边设置贯通式或尽头式消防车道时，可分别计算各自的建筑高度。否则，应按其中建筑高度最大者确定该建筑的建筑高度。

4. For stair flooring, the respective building height can be calculated separately if the same building located on different elevation floor is separated by a firewall, each has a safe exit in accordance with the specifications, and can be set through or end-type fire lanes along the two long edges of the building. Otherwise, the building height of the building should be determined according to the maximum height of the building.

5. 局部突出屋顶的瞭望塔、冷却塔、水箱间、微波天线间或设施、电梯机房、排风和排烟机房以及楼梯出口小间等辅助用房占屋面面积不大于1/4者，可不计入建筑高度。

5. If the lookout tower, cooling tower, water tank room, microwave antenna or facilities, elevator room, exhaust and smoke extraction room, as well as staircase exit small room and other auxiliary rooms have local protruding roofs, whose roof area is not more than 1/4 of the room area, the heights of auxiliary rooms can not be counted into the building height.

6. 对于住宅建筑，设置在底部且室内高度不大于2.2m的自行车库、储藏室、敞开空间，室内外高差或建筑的地下或半地下室的顶板面高出室外设计地面的高度不大于1.5m的部分，可不计入建筑高度。

6. For residential buildings, bicycle library, storage room and open space set at the bottom and indoor height is not greater than 2.2m, and the parts (if indoor and outdoor height difference or the height of roof surface of buildings' underground or semi-basement higher than the outdoor design floor is not greater than 1.5m), their height can not be counted into the building height.

A.0.2 建筑层数应按建筑的自然层数计算，下列空间可不计入建筑层数：

A.0.2 The number of building floors shall be calculated on the basis of the natural floors of a building, and the following spaces may not be counted in the number of building floors:

1. 室内顶板面高出室外设计地面的高度不大于1.5m的地下或半地下室；

1. Buildings underground or semi-basement, with height of roof surface not greater 1.5m than the outdoor design floor;

2. 设置在建筑底部且室内高度不大于2.2m的自行车库、储藏室、敞开空间；

2. Bicycle library, storage room and open space set at the bottom, with indoor height not greater than 2.2m；

3. 建筑屋顶上突出的局部设备用房、出屋面的楼梯间等。

3. Protruding local equipment housing on the building roof, and the stairwell out of roofs etc.

2.0.14 【通则】2.0.14条：层高

2.0.14 [Code for design of civil buildings] 2.0.14: Storey Height

建筑物各层之间以楼、地面面层（完成面）计算的垂直距离，屋顶层由该层楼面面层（完成面）至平屋面的结构面层或至坡顶的结构面层与外墙外皮延长线的交点计算的垂直距离。

Vertical distance between the surfaces of floors, or the ground surface（finished）, or the vertical distance between the top floor surface（finished）and the structural layer of the flat roof, or the intersection point of the structural layer and the extension line of the exterior wall surface in the event of pitched roof.

2.0.15 【通则】2.0.15条：室内净高

2.0.15 [Code for design of civil buildings] 2.0.15: Ceiling Height

从楼、地面面层（完成面）至吊顶或楼盖、屋盖底面之间的有效使用空间的垂直距离。

The vertical distances of the effective use spaces from floor surfaces and ground surfaces (finished surfaces) to suspended ceilings, or bottom surfaces of floors or roofs.

2.0.16 【防火规】2.1.1条：高层建筑

2.0.16 [Code for fire protection design of buildings] 2.1.1: High-Rise Building

建筑高度大于27m的住宅建筑和建筑高度大于24m的非单层厂房、仓库和其他民用建筑。

Residential buildings with the building height greater than 27m and non-single-storey plant buildings, warehouses and other civil buildings with the building height greater than 24m.

2.0.17 【防火规】2.1.2条：裙房

2.0.17 [Code for fire protection design of buildings] 2.1.2: Podium

在高层建筑主体投影范围外，与建筑主体相连且建筑高度不大于24m的附属建筑。

Subsidiary buildings outside the main projection range of the high-rise building, connected with the main body of the building, with the building height not greater than 24m.

2.0.18 【防火规】2.1.3 条：重要公共建筑

2.0.18 [Code for fire protection design of buildings] 2.1.3: Important Public Building

发生火灾可能造成重大人员伤亡、财产损失和严重社会影响的公共建筑。

Public buildings where heavy casualties, property losses and serious social impact may occur in case of a fire.

2.0.19 【防火规】2.1.4 条：商业服务网点

2.0.19 [Code for fire protection design of buildings] 2.1.4: Commercial Service Station

设置在住宅建筑的首层或首层及二层，每个分隔单元建筑面积不大于 300m² 的商店、邮政所、储蓄所、理发店等小型营业性用房。

Small-sized business premises on the first floor or first and second floors of the residential building, such as stores, post offices, saving agencies, or barber shops, whose the separation unit covers less than 300m².

2.0.20 【防火规】2.1.5 条：高架仓库

2.0.20 [Code for fire protection design of buildings] 2.1.5: High-rack Storage

货架高度大于 7m 且采用机械化操作或自动化控制的货架仓库。

Shelf warehouses with the shelf height greater than 7m, in mechanized operation or automatic control.

2.0.21 【防火规】2.1.6 条：半地下室

2.0.21 [Code for fire protection design of buildings] 2.1.6: Semibasement

房间地面低于室外设计地面的平均高度大于该房间平均净高 1/3，且不大于 1/2 者。

Rooms of which the average height of the floor below the outdoor design ground is between 1/3 and 1/2 of the net average room height.

2.0.22 【防火规】2.1.7 条：地下室

2.0.22 [Code for fire protection design of buildings] 2.1.7: Basement

房间地面低于室外设计地面的平均高度大于该房间平均净高 1/2 者。

Rooms where the average height of the floor below the outdoor design ground is above 1/2 of the net average height of rooms.

2.0.23 【防火规】2.1.10 条：耐火极限

2.0.23 [Code for fire protection design of buildings] 2.1.10: Fire Endurance

在标准耐火试验条件下，建筑构件、配件或结构从受到火的作用时起，至失去承载能力、完整性或隔热性时止所用时间，用小时表示。

Time duration (in hours) in which, under the standard fire resistance test, building components, accessories or structures endure from the action of fire to the loss of load-bearing capacity, integrity or thermal insulation.

2.0.24 【防火规】2.1.11 条：防火隔墙
2.0.24 [Code for fire protection design of buildings] 2.1.11: Fire Partition
建筑内防止火灾蔓延至相邻区域且耐火极限不低于规定要求的不燃性墙体。
Non-combustible walls with the fire endurance not lower than the requirements of the Code to prevent fire from spreading to the adjacent areas.

2.0.25 【防火规】2.1.12 条：防火墙
2.0.25 [Code for fire protection design of buildings] 2.1.12: Fire Wall
防止火灾蔓延至相邻建筑或相邻水平防火分区且耐火极限不低于3.00h 的不燃性墙体。
Non-combustible walls with the fire endurance not less than 3.00h for the prevention of fire from spreading to adjacent buildings or adjacent horizontal fire compartments.

2.0.26 【防火规】2.1.13 条：避难层（间）
2.0.26 [Code for fire protection design of buildings] 2.1.13: Refuge Floor（Room）
建筑内用于人员暂时躲避火灾及其烟气危害的楼层（房间）。
Floors（rooms）for personnel to temporarily avoid the harm from fire and smoke in buildings.

2.0.27 【防火规】2.1.17 条：避难走道
2.0.27 [Code for fire protection design of buildings] 2.1.17: Refuge Corridor
采取防烟措施且两侧设置耐火极限不低于3.00h 的防火隔墙，用于人员安全通行至室外的走道。
Corridors for safe passage of personnel to outdoors in which smoke preventive measures are adopted and the fire partitions with the fire endurance not less than 3.00 h are arranged on both sides.

2.0.28 【防火规】2.1.14 条：安全出口
2.0.28 [Code for fire protection design of buildings] 2.1.14: Safety Exit
供人员安全疏散用的楼梯间和室外楼梯的出入口或直通室内外安全区域的出口。
Staircases, and entrances and exits of the outdoor staircases, or exits directly to the outdoor safe areas, used for safety evacuation.

2.0.29 【防火规】2.1.15 条：封闭楼梯间
2.0.29 [Code for fire protection design of buildings] 2.1.15: Enclosed Staircase
在楼梯间入口处设置门，以防止火灾的烟和热气进入的楼梯间。
Staircases with a door at the staircase entrance to prevent fire smoke and hot air from traveling the staircase in the event of fire.

2.0.30 【防火规】2.1.16 条：防烟楼梯间
2.0.30 [Code for fire protection design of buildings] 2.1.16: Smokeproof Staircase

在楼梯间入口处设置防烟的前室、开敞式阳台或凹廊（统称前室）等设施，且通向前室和楼梯间的门均为防火门，以防止火灾的烟和热气进入的楼梯间。

Staircases with the facilities of smoke-proof anteroom, open balcony or concave corridor (collectively referred to as the anteroom), etc, arranged at the staircase entrance, where fireproof doors shall be designed at the entrance to the staircase or the anterooms to prevent the entry of fire smoke and hot air into the staircase.

2.0.31 【防火规】2.1.21 条：防火间距

2.0.31 [Code for fire protection design of buildings] 2.1.21: Fire Separation Distance

防止着火建筑在一定时间内引燃相邻建筑，便于消防扑救的间隔距离。

Separation distance for preventing fire buildings from igniting their adjacent buildings within a certain period of time and facilitating fire fighting.

2.0.32 【防火规】附录 B.0.1：防火间距的计算方法

2.0.32 [Code for fire protection design of buildings] Appendix B.0.1: Calculation Methods for the Fire Separation Distance

建筑物之间的防火间距应按相邻建筑外墙的最近水平距离计算，当外墙有凸出的可燃或难燃构件时，应从其凸出部分外缘算起。

The fire separation distance between buildings shall be calculated as per the minimum horizontal distance between the exterior walls of the adjacent buildings, and when there are protruding combustible or flame-retardant components on the exterior walls, the distance shall be calculated from the outer edges of the protruding parts.

建筑物与储罐、堆场的防火间距，应为建筑外墙至储罐外壁或堆场中相邻堆垛外缘的最近水平距离。

The fire separation distance between buildings and storage tanks or storage yards shall be the minimum horizontal distance from the exterior wall of the building to the outer wall of the storage tank or the outer edge of the adjacent stack at the stockyard.

2.0.33 【防火规】2.1.22 条：防火分区

2.0.33 [Code for fire protection design of buildings] 2.1.22: Fire Compartment

在建筑内部采用防火墙、楼板及其他防火分隔设施分隔而成，能在一定时间内防止火灾向同一建筑的其余部分蔓延的局部空间。

Local space inside a building partitioned by firewalls, floor slabs and other fireproof partition facilities, which is able to prevent fire from spreading to the rest parts of the building within a certain period of time.

2.0.34 【通则】2.0.26 条：建筑幕墙

2.0.34 [Code for design of civil buildings] 2.0.26: Building Curtain Wall

由金属构架与板材组成的,不承担主体结构荷载与作用的建筑外围护结构。

Enclosure structures of buildings consisting of metal frames and panels which bear no load and effects of the main structures.

2.0.35 【通则】2.0.27 条:吊顶

2.0.35 [Code for design of civil buildings] 2.0.27: Suspended Ceiling

悬吊在房屋屋顶或楼板结构下的顶棚。

Ceilings suspended from roofs or floor slabs.

2.0.36 【通则】2.0.31 条:装修

2.0.36 [Code for design of civil buildings] 2.0.31: Decoration

以建筑物主体结构为依托,对建筑内、外空间进行的细部加工和艺术处理。

Detail processing and aesthetic treatment on the interior and/or exterior spaces of buildings, based on the main structures of the buildings.

3 基本规定
Basic Provisions

3.1 民用建筑分类

3.1 Classification of Civil Buildings

3.1.1 【公共节能】条文说明 1.0.2：按功能分类

3.1.1 [Design standard for energy efficiency of public buildings] Explanation1.0.2: Classification as per functions

民用建筑分为居住建筑和公共建筑两大类，其中居住建筑包括住宅建筑。

Civil buildings are divided into residential buildings and public buildings, in which residential buildings include residential buildings.

公共建筑包含办公建筑（包括写字楼、政府办公楼等），商场建筑（如商场、金融建筑等），旅游建筑（如旅馆饭店、娱乐场所等），科教文卫建筑（包括文化、教育、科研、医疗卫生、体育建筑等），通信建筑（如邮电、通信、广播用房等）以及交通运输用房（如机场、车站建筑等）。

Public buildings include office buildings (including office building, government office building, etc.), shopping malls (such as shopping mall, financial building, etc.) , tourist buildings (such as hotel hotels, entertainment venues, etc.) , science and education buildings (including culture, education, scientific research, health care, sports construction, etc.), communications buildings (such as housing for post and telecommunications, communications, broadcasting), and transportation housing (such as airports, station buildings, etc.).

【防火规】5.1.1 条：宿舍、公寓等非住宅类为公共建筑的规定。

[Code for fire protection design of buildings] 5.1.1: Regulation of dormitories, apartments and other non-residential categories belonging to public buildings.

【防火规】5.1.1 条：建筑高度大于54m 的住宅建筑（包括设置商业服务网点的住宅建筑）。

[Code for fire protection design of buildings] 5.1.1: Residential buildings with a higher than 54m (including residential buildings with commercial service outlets) .

3.1.2 【防火规】5.1.1 条：按建筑高度、功能、火灾危险性和扑救难易程度分类：民用建筑根据其建筑高度和层数可分为单、多层民用建筑和高层民用建筑。高层民用建筑根据其建筑高度、使用功能和楼层的建筑面积可分为一类和二类。民用建筑的分类应符合表3.1.2 的规定。

3.1.2 [Code for fire protection design of buildings] 5.1.1: Classify according to the building height, function, fire hazard and the degree of difficulty in fighting: civil buildings can be divided into single-story, multistoried and high-rise civil buildings according to its building height and number of floors. High-rise civil buildings can be divided into the first category and the secondary category according to their building height, use function and floor building area. The classification of civil buildings shall conform to the provisions of Tab. 3.1.2.

【防火规】表 5.1.1：民用建筑的分类 表 3.1.2
[Code for fire protection design of buildings] Tab. 5.1.1: Classification of Civil Buildings Tab. 3.1.2

名称 Type	高层民用建筑 High-rise civil buildings		单、多层民用建筑 Single-storey and multi-storey civil buildings
	一类 Class I	二类 Class II	
住宅建筑 Residential buildings	建筑高度大于 54m 的住宅建筑（包括设置商业网点的住宅建筑） Residential buildings with the building height greater than 54 m (including residential buildings with commercial network)	建筑高度大于27m，但不大于54m的住宅建筑（包括设置商业网点的住宅建筑） Residential buildings with the building height greater than 27m but not greater than 54m (including residential buildings with commercial network)	建筑高度不大于27m 的住宅建筑（包括设置商业网点的住宅建筑） Residential buildings with the building height not greater than 27m(including residential buildings with commercial network)
公共建筑 Public buildings	1. 建筑高度大于 50m 的公共建筑； 1. Public buildings with the building height greater than 50m; 2. 建筑高度 24m 以上任一楼层建筑面积大于 1000m² 的商店、展览、电信、邮政、财贸金融建筑和其他多种功能组合的建筑； 2. Stores, exhibition buildings, telecommunication buildings, postal buildings, financial buildings and other multi-functional buildings with the building height above 24m and the building area of any floor above 1000m²； 3. 医疗建筑、重要公共建筑、独立建造的老年照料设施； 3. Medical buildings, important public buildings, and facilities independently built for caring for the aged; 4. 省级及以上的广播电视指挥调度建筑、网局级和省级电力调度建筑； 4. Broadcast and television commanding and dispatching buildings of the provincial level or above, and power dispatching buildings of the grid bureau level and provincial level; 5. 藏书超过 100 万册的图书馆、书库 5. Libraries and and book warehouses with the numb er of books more than 1 million	除一类高层公共建筑外的其他高层公共建筑 Other high-rise public buildings except class I	1. 建筑高度大于 24m 的单层公共建筑； 1. Single-storey public buildings with the building height greater than 24 m; 2. 建筑高度不大于 24m 的其他公共建筑 2. Other public buildings with the building height not greater than 24m

注：1. 表中未列入的建筑，其类别应根据本表类比确定。
Notes:1. For the buildings unlisted in the table, the category shall be determined by analogy with this table.
2. 除本规范另有规定外，宿舍、公寓等非住宅类居住建筑的防火要求，应符合本规范有关公共建筑的规定。
2. Unless otherwise specified in the Code, the fire protection requirements of non-residential habitational buildings of dormitories, apartments, etc., shall conform to the provisions relevant to public buildings in the Code.
3. 除本规范另有规定外，裙房的防火要求应符合本规范有关高层民用建筑的规定。
3. Unless otherwise specified in the Code, the fire protection requirements of podiums shall conform to the provisions relevant to high-rise civil buildings in the Code.
注意：宿舍和公寓按公建消防要求。
Note: For dormitories and apartments, the fire protection requirements for public buildings shall be followed.

3.2 建筑高度、层数、净高

3.2 Height, Storey and Net Height of Building

3.2.1 控制建筑高度根据【通则】4.3.1条：建筑高度不应危害公共空间安全、卫生和景观，下列地区应实行建筑高度控制。

3.2.1 For the control of building height, in accordance with [Code for design of civil buildings] 4.3.1, the building height shall not endanger the safety, hygiene and landscapes of public space, and in the following regions, the building height shall be controlled.

1. 对建筑高度有特别要求的地区，应按城市规划要求控制建筑高度；

1. In the regions with special requirements for the building height, the building height shall be controlled as per the urban planning requirements;

2. 沿城市道路的建筑物，应根据道路的宽度控制建筑裙楼和主体塔楼的高度；

2. The height of podiums and main towers of buildings along urban roads shall be controlled in accordance with the width of the road;

3. 机场、电台、电信、微波通信、气象台、卫星地面站、军事要塞工程等周围的建筑，当其处在各种技术作业控制区范围内时，应按净空要求控制建筑高度；

3. The height of buildings around airports, radio stations, telecommunications, microwave communications, meteorological observatories, satellite earth stations, military fortresses, and the like, within the scope of various technical operation control areas shall be controlled on the basis of the clearance requirements;

4. 在国家或地方公布的各级历史文化名城、历史文化保护区、文物保护单位和风景名胜区的各项建设，应按国家或地方制定的保护规划和有关条例进行。

4. The construction of famous historical and cultural cities, historical and cultural protection areas, cultural relics protection units and scenic spots at various level released by the state or local authorities shall be carried out in accordance with the protection plans and relevant regulations formulated by the state or local authorities.

注：建筑高度控制尚应符合当地城市规划行政主管部门和有关专业部门的规定。
Note: The control of building height shall conform to the provisions of the local competent department of city planning administration and relevant professional departments.

3.2.2 建筑层数计算

3.2.2 Calculation of Building Storeys

建筑层数应按建筑的自然层数计算，下列空间可不计入建筑层数：

The building storeys shall be calculated as per the natural storeys of buildings, but the following spaces shall be not included into the building storeys:

1. 室内顶板面高出室外设计地面的高度不大于1.5m的地下或半地下室；

1. Basements and semi-basements with indoor top boards 1.5m higher than outdoor design floors;

2. 设置在建筑底部且室内高度不大于2.2m的自行车库、储藏室、敞开空间；

2. Bicycle garages, storage rooms and open space arranged at the bottoms of buildings and with the height indoor not greater than 2.2m;

3. 建筑屋顶上突出的局部设备用房、出屋面的楼梯间等。

3. Local equipment rooms protruding above the rooftop, or staircase rooms above the roof.

3.2.3 【住设规】4.0.5条：建筑层高和净高，住宅楼的层数计算应符合下列规定。

3.2.3 [Design code of residential buildings] 4.0.5: Building storey height and net height, the calculation of the number of floors of residential buildings shall comply with the following requirements.

1. 当住宅楼的所有楼层的层高不大于3.00m时，层数应按自然层数计。

1. When the floor height of all floors of a residential building is not greater than 3.00m, the number of floors should be counted according to the natural floor.

2. 当住宅和其他功能空间处于同一建筑物内时，应将住宅部分的层数与其他功能空间的层数叠加计算建筑层数。当建筑中有一层或若干层的层高大于3.00m时，应对大于3.00m的所有楼层按其高度总和除以3.00m进行层数折算，余数小于1.50m时，多出部分不应计入建筑层数，余数大于或等于1.50m时，多出部分应按1层计算。

2. When the residential and other functional spaces are in the same building, the number of floors of the residential parts should be superimposed with the number of floors in other functional spaces to be the number of floors of the building. When there is one or more floors in the building, with storey height greater than 3.00m, all floors responding to the storey-height greater than 3.00m are divided by 3 by their height sum. The extra part should not be counted as the number of building floors if the remainder is less than 1.50m; or else, the remainder is calculated as 1 floor.

3. 层高小于2.20m的架空层和设备层不应计入自然层数。

3. The overhead floor and equipment floor with storey-height less than 2.20m should not be counted into the natural floors.

4. 高出室外设计地面小于2.20m的半地下室不应计入地上自然层数。

4. Semi-basement with ground less than 2.20m above outdoor design should not be included

in the number of natural floors above ground.

各类建筑不同房间的净高要求见表 3.2.2-1、表 3.2.2-2。

The requirements of the net height of different rooms in various buildings are shown in Tab. 3.2.2-1, Tab. 3.2.2-2.

层高和净高控制
Control of Storey Height and Net height
表 3.2.2-1 Tab. 3.2.2-1

类别 Category	房间部位 Room area	最小净高（m） Minimum net height（m）		备注 Note
		无空调 No air-conditioned and air-conditioned	有空调 Air-conditioned and air-conditioned	
住宅 Residential	卧室、起居室（厅） Bedroom, living room（hall）	2.4	—	局部净高不应低于2.10m，且局部净高的室内面积不应大于室内使用面积的1/3 Local net height should not be less than 2.10m, and the local net height of the indoor area should not be greater than 1/3 of the indoor areas of usage
	厨房、卫生间 Kitchen, toilet	2.2	—	
	利用坡屋顶内空间作起居室(厅)、卧室时 When using the space in the sloping roof as a living room（hall）and bedroom	1/2 使用面积≥ 2.1 One-second usage area ≥ 2.1		—
	走道 Aisle	2.2		
中小学校 School		小学 Primary	初中 Junior	高中 Senior	
	普通教室、史地、美术、音乐教室 General classroom, steve, art, music classroom	3.00	3.05	3.10	
	舞蹈教室 Dance classroom	4.50		
	科学教室、实验室、计算机教室、劳动教室、技术教室、合班教室 Science classrooms, laboratories, computer classrooms, labor classrooms, technical classrooms, classroom classes	3.10		
	阶梯教室 Staircase classroom	最后一排（楼地面最高处）距顶棚或上方突出物最小距离为 2.20m The minimum distance of the last row（the highest ground floor）is 2.20m from the ceiling or above the protruding object		

续表

类别 Category	房间部位 Room area			最小净高（m） Minimum net height（m）			备注 Note
				无空调 No air-conditioned and air-conditioned	有空调 Air-conditioned and air-conditioned		
中小学各类体育场地【校规】7.2.1 条 Various sports venues in primary and middle schools [Code for design of School] 7.2.1	田径 Athletics	篮球 Basketball	排球 Volleyball	羽毛球 Badminton	乒乓球 Table tennis	体操 Gymnastics	田径场地可减少部分项目降低净高 Track and field sites can reduce some items to reduce net height
	9	7	7	9	4	6	
托幼【托规】4.1.17 条 Childcare [Code for design of nursery and kindergarten buildings] 4.1.17	房间名称 Room name			最小净高 Minimum net height			
	活动室、寝室、乳儿室 Activity room, dormitory, milk			3.0			
	多功能活动室 Multifunctional activity room			3.9			
旅馆【旅规】4.2.9 Hostel [Code for design of hotel building] 4.2.9	客房 Guest room			2.6	2.4		
	利用坡屋顶内空间作客房 Use of sloping roof space for guest rooms			2.4	—		
	卫生间 Bathroom			2.1	—		
	客房层公共走道及客房内走道 Guest room floor public walkway and in-room walkway			2.1			
医院【医规】5.1.9 Hospital [Code for design of general hospital] 5.1.9	诊查室 Diagnostic room			2.6			
	病房 Ward			2.8			
	公共走道 Public walkways			2.3			

【商店建规】表 4.2.3：商业营业厅的净高   表 3.2.2-2
[Code for design of store buildings] Tab. 4.2.3: Net height of Business Halls   Tab. 3.2.2-2

通风方式 Ventilation mode	自然通风 Natural ventilation			机械开窗和自然通风相结合 Combination of mechanical window and natural ventilation	空气调节系统 Air conditioning system
	单面开窗 Single-opened Window	前面开敞 Open in the front	前后开窗 Open in the front and back		
最大进深与净高比 Ratio of maximum length to clear height	2∶1	2.5∶1	4∶1	5∶1	—
最小净高（m） Minimum net height（m）	3.20	3.20	3.50	3.50	3.00

3.3 主要技术经济指标

3.3 Main Technical and Economic Indicators

3.3.1 常用技术经济指标及释义（表 3.3.1）

3.3.1 Common Technical and Economic Indicators and Definitions（Tab.3.3.1）

【09 措施】表 2.5.11 条：主要技术经济指标表　　　　　　　　表 3.3.1
[Measures 2009] Tab. 2.5.11: Table of Main Technical and Economic Indicators　　Tab. 3.3.1

序号 No	分项名称 Item name		单位 Unit	数量 Quantity	备注 Remarks
1	总用地面积 Total land area				含待征地和非建设用地 Land to be expropriated and non-construction land
2	建设用地面积 Area of construction lands		hm²		即规划净用地 i.e., the planned clear lands
3	总建筑面积 Total building area		m²		
其中 Where	计容建筑面积 Capacity building area		m²		
	其中 Where	商业建筑面积 Area of commercial buildings	m²		按主要的不同功能分列 Divided as per different main functions
		住宅建筑面积 Area of residential buildings	m²		
	不计容面积 Non-capacity building area		m²		
	其中 Where	地下车库面积 Area of underground garages	m²		
		架空层建筑面积 Building area of overhead floors	m²		
4	建筑占地面积 Floor area		m²		
5	建筑密度 Building density		%		又称建筑覆盖率 Also known as building coverage rate
6	容积率 Plot ratio		%		
7	绿地率 Greening rate		%		

续表

序号 No	分项名称 Item name		单位 Unit	数量 Quantity	备注 Remarks
8	机动车位数量 Number of parking spaces for motorized vehicles		个 piece		客车、货车可分列，规划要求时注明地上车位比率 Passenger cars and trucks can be divided into lines, and the ground parking space ratio shall be indicated if required in planning
	其中 Where	地上停车位 Ground parking spaces	个 piece		
		地下停车位 Underground parking spaces	个 piece		
9	非机动车位 Non-motorized vehicle parking spaces		个 piece		

3.3.2 居住区规划技术经济指标
Technical and Economic Indicators of Residential District Planning

【居规划标】4.0.10 条：居住区规划设计应汇总重要的技术指标，并应符合本标准附录 A 第 A.0.3 条的规定。

[Standard for urban residential areas planning and design] 4.0.10: Important technical indicators shall be summarized in the planning and design of resident areas, and shall conform to the provisions of Appendix A, Paragraph A.0.3.

【居规划标】附录 A.0.3：居住区综合技术指标应符合表 A.0.3 的要求。

[Standard for urban residential areas planning and design] Appendix A.0.3: Comprehensive technical indicators of resident areas shall meet the requirements of Table A.0.3.

【居规划标】附表 A.0.3：居住区综合技术指标　　　　表 3.3.2

[Standard for urban residential areas planning and design] Appendix A.0.3: Comprehensive Technical Indicators of Resident Areas　　Tab. 3.3.2

项目 Item			计量单位 Unit	数值 Value	所占比例（%） Proportion（%）	人均面积指标（m^2/人） Area per capita（m^2/person）
各级生活圈居住区指标 Indicators of living areas at all levels	居住区用地	总用地面积 Total land area	hm^2	▲	100	▲
		其中 Where 住宅用地 Residential land	hm^2	▲	▲	▲
		配套设施用地 Land for supporting facilities	hm^2	▲	▲	▲
		公共绿化 Public greening	hm^2	▲	▲	▲

续表

项目 Item				计量单位 Unit	数值 Value	所占比例（%） Proportion (%)	人均面积指标（m²/人） Area per capita (m²/person)
各级生活圈居住区指标 Indicators of living areas at all levels		城市道路用地 Urban road land		hm²	▲	▲	—
	居住总人口 Total resident population			人 person	▲	—	—
	居住总套（户）数 Total residential units			套 set	▲	—	—
	住宅建筑总面积 Total area of residential buildings			万 m² 10000m²	▲	—	—
居住街坊指标 Residential neighborhood index	用地面积 Land area			hm²	▲	—	▲
	容积率 Plot ratio			—	▲	—	—
	地上建筑面积 Ground floor area	总建筑面积 Gross building area		万 m² 10000m²	▲	100	—
		其中 Including	住宅建筑 Residential building	万 m² 10000m²	▲	▲	—
			便民服务设施 Facilities	万 m² 10000m²	▲	▲	—
	地下总建筑面积 Underground floor area			万 m² 10000m²	▲	▲	—
	绿地率 Greening rate			%	▲	—	—
	集中绿地面积 Intensive green area			m²	▲	—	▲
	住宅套（户）数 Number of residential units			套 set	▲	—	—
	住宅套均面积 Average area of residential units			m²/套 m²/set	▲	—	—
	居住人数 Number of residents			人 person	▲	—	—
	住宅建筑密度 Residential building density			%	▲	—	—
	住宅建筑平均层数 Average story of residential building			层 story	▲	—	—
	住宅建筑高度控制最大值 Maximum of residential building height control			m	▲	—	—
	停车位 Parking spaces	总停车位 Total parking spaces		辆	▲	—	—

项目 Item			计量 单位 Unit	数值 Value	所占比例(%) Proportion (%)	人均面积指标 (m²/人) Area per capita (m²/person)
居住街坊 指标 Residential neighborhood index	停车位 Parking spaces	其中 (among)	地上停车位 Ground parking spaces	辆	▲	—	—
			地下停车位 Underground parking spaces	辆	▲	—	—
	地面停车位 Ground parking spaces		辆	▲	—	—

3.4 建筑气候分区对建筑基本要求

3.4 Basic Requirements of the Climate Zoning of Buildings

3.4.1 【公共节能】3.1.2 条：建筑气候分区及代表城市（表 3.4.1）

3.4.1 [Design standard for energy efficiency of public buildings] 3.1.2: The Climate Zoning of Buildings and Representative Cities (Tab. 3.4.1)

【公共节能】表 3.1.2：代表城市建筑热工设计分区　　　　表 3.4.1
[Design standard for energy efficiency of public buildings] Tab. 3.1.2:
Thermal Engineering Design Zoning of Representative Cities　　　Tab. 3.4.1

气候分区及气候子区 Climate region and subregion		代表城市 Representative city
严寒地区 Severely cold areas	严寒 A 区 Severely cold regions A	博克图、伊春、呼玛、海拉尔、满洲里、阿尔山、玛多、黑河、嫩江、海伦、齐齐哈尔、富锦、哈尔滨、牡丹江、大庆、安达、佳木斯、二连浩特、多伦、大柴旦、阿勒泰、那曲 Bugt, Yichun, Humar, Hailar, Manchuria, Arxan, Maduo, Heihe, Nenjiang, Hailun, Tsitsihar, Fujin, Harbin, Mudanjiang, Daqing, Anda, Kiamusze, Erenhot, Duolun, Da Qaidam, Altay and Naqu
	严寒 B 区 Severely cold regions B	
	严寒 C 区 Severely cold region C	长春、通化、延吉、通辽、四平、抚顺、阜新、沈阳、本溪、鞍山、呼和浩特、包头、鄂尔多斯、赤峰、额济纳旗、大同、乌鲁木齐、克拉玛依、酒泉、西宁、日喀则、甘孜、康定 Changchun, Tonghua, Yanji, Tongliao, Siping, Fushun, Fuxin, Shenyang, Benxi, Anshan, Hohhot, Baotou, Erdos, Chifeng, Ejin Banner, Datong, Urumchi, Karamay, Jiuquan, Xining, Shigatse, Garze and Kangding
寒冷地区 Cold regions	寒冷 A 区 Chillness A distinguishes	丹东、大连、张家口、承德、唐山、青岛、洛阳、太原、阳泉、晋城、天水、榆林、延安、宝鸡、银川、平凉、兰州、喀什、伊宁、阿坝、拉萨、林芝、北京、天津、石家庄、保定、邢台、济南、德州、兖州、郑州、安阳、徐州、运城、西安、咸阳、吐鲁番、库尔勒、哈密 Dandong, Dalian, Zhangjiakou, Chengde, Tangshan, Qingdao, Luoyang, Taiyuan, Yangquan, Jincheng, Tianshui, Yulin, Yan'an, Baoji, Yinchuan, Pingliang, Lanzhou, Kashgar, Yining, Aba, Lhasa, Linzhi, Beijing, Tianjin, Shijiazhuang, Baoding, Xingtai, Jinan, Dezhou, Yanzhou, Zhengzhou, Anyang, Xuzhou, Yuncheng, Xi'an, Xianyang, Turpan, Korla and Hami
	寒冷 B 区 Cold regions B	

续表

气候分区及气候子区 Climate region and subregion		代表城市 Representative city
夏热冬冷地区 Regions hot in summer and cold in winter	夏热冬冷 A 区 Hot-summer and cold-winter area A	南京、蚌埠、盐城、南通、合肥、安庆、九江、武汉、黄石、岳阳、汉中、安康、上海、杭州、宁波、温州、宜昌、长沙、南昌、株洲、永州、赣州、韶关、桂林、重庆、达州、万州、涪陵、南充、宜宾、成都、遵义、凯里、绵阳、南平 Nanjing, Bengbu, Yancheng, Nantong, Hefei, Anqing, Jiujiang, Wuhan, Huangshi, Yueyang, Hanzhong, Ankang, Shanghai, Hangzhou, Ningbo, Wenzhou, Yichang, Changsha, Nanchang, Zhuzhou, Yongzhou, Ganzhou, Shaoguan, Guilin, Chongqing, Dazhou, Wanzhou, Fuling, Nanchong, Yibin, Chengdu, Zunyi, Kaili, Mianyang and Nanping
	夏热冬冷 B 区 Regions B hot in summer and cold in winter	
夏热冬暖地区 Areas hot in summer and warm in winter	夏热冬暖 A 区 Area A hot in summer and warm in winter	福州、莆田、龙岩、梅州、兴宁、英德、河池、柳州、贺州、泉州、厦门、广州、深圳、湛江、汕头、南宁、北海、梧州、海口、三亚 Fuzhou, Putian, Longyan, Meizhou, Xingning, Yingde, Hechi, Liuzhou, Hezhou, Quanzhou, Xiamen, Guangzhou, Shenzhen, Zhanjiang, Shantou, Nanning, Beihai, Wuzhou, Haikou and Sanya
	夏热冬暖 B 区 Area B hot in summer and warm in winter	
温和地区 Moderate climate area	温和 A 区 Area A with moderate climate	昆明、贵阳、丽江、会泽、腾冲、保山、大理、楚雄、曲靖、泸西、屏边、广南、兴义、独山 Kunming, Guiyang, Lijiang, Huize, Tengchong, Baoshan, Dali, Chuxiong, Qujing, Luxi, Pingbian, Guangnan, Xingyi and Dushan
	温和 B 区 Area B with moderate climate	瑞丽、耿马、临沧、澜沧、思茅、江城、蒙自 Ruili, Gengma, Lincang, Lancang, Simao, Jiangcheng and Mengzi

3.4.2 建筑规划设计适应气候的基本要求（表 3.4.2）

3.4.2 Basic Requirements for Adaptation of Architectural Planning and Design to Climate (Tab.3.4.2)

【通则】3.2.1 条：气候分区及各分区对建筑基本要求　　　　表 3.4.2
[Code for design of civil buildings] 3.2.1: Climate Zoning and Basic Requirements of Each Zone for Buildings　　　　Tab. 3.4.2

分区代号 Region code	分区名称 Region name	气候主要指标 Main climate indicators	各区辖行政区范围 Administrative area of each zone	建筑基本要求 Basic requirements for buildings
I	IA IB IC ID	严寒地区 Severely cold areas	1 月平均气温 ≤ -10℃ Average temperature in January ≤ -10℃ 7 月平均气温 ≤ 25℃ Average temperature in July ≤ 25℃ 7 月平均相对湿度 ≥ 50% Average relative humidity in July ≥ 50%	黑龙江、吉林全境；辽宁大部；内蒙古中、北部及陕西、山西、河北、北京北部的部分地区 Heilongjiang and entire region of Jilin; most parts of Liaoning; central and northern parts of Inner Mongolia and some parts in north of Shaanxi, Shanxi, Hebei and Beijing	1. 建筑物必须满足冬季保温、防寒、防冻等要求 1. For the buildings, the requirements of insulation, cold-resistant, freeze protection, etc., in winter must be met 2. IA、IB 区应防止冻土、积雪对建筑物的危害 2. Measures must be taken to prevent congealed ground and snow in Areas IA and IB 3. IB、IC、ID 区西部，建筑物应防冰雹、防风沙 3. Hail and sandstorm shall be prevented for buildings in the west of Areas IB, IC and ID

续表

分区代号 Region code		分区名称 Region name	气候主要指标 Main climate indicators	各区辖行政区范围 Administrative area of each zone	建筑基本要求 Basic requirements for buildings
II	IIA IIB	寒冷地区 Cold regions	1月平均气温 –10～0℃ Average temperature in January –10-0℃ 7月平均气温 18～28℃ Average temperature in July 18-28℃	天津、山东、宁夏全境；北京、河北、山西、陕西大部；辽宁南部；甘肃中、东部以及河南、安徽、江苏北部的部分地区 Tianjin, Shandong and whole territory of Ningxia; Beijing, Hebei, Shanxi and the most parts of Shaanxi; south of Liaoning; Middle and eastern Gansu, Henan, Anhui,and some parts of northern Jiangsu	1. 建筑物应满足冬季保温、防寒、防冻等要求，夏季部分地区应兼顾防热 1. For the buildings, the requirements of insulation, cold-resistant, freeze protection, etc., in winter shall be met, and heat protection shall be carried out in some parts in summer 2. IIA建筑物应防热、防潮、防暴风雨、沿海地带应防盐雾侵蚀 2. For buildings in Area IIA, heatstroke, moisture and rainstorm shall be prevented, and salt fog erosion shall be prevented in coastal areas
III	IIIA IIIB IIIC	夏热冬冷地区 Regions hot in summer and cold in winter	1月平均气温 0～10℃ Average temperature in January 0-10℃ 7月平均气温 25～30℃ Average temperature in July: 25-30℃	上海、浙江、江西、湖北、湖南全境；江苏、安徽、四川大部；陕西、河南南部；贵州东部；福建、广东、广西北部和甘肃南部的部分地区 Shanghai, Zhejiang, Jiangxi, Hubei and whole territory of Hunan; Jiangsu, Anhui and most parts of Sichuan; Shaanxi and southern Henan; eastern Guizhou; Fujian, Guangdong, Northern Guangxi and some parts of southern Gansu	1. 建筑物必须满足夏季防热、遮阳、通风降温要求，冬季应兼顾防寒 1. Buildings must meet the requirements of heat protection, shading, ventilation and cooling in summer and shall be cold resistant in winter 2. 建筑物应防雨、防潮、防洪、防雷电 2. Buildings shall be preventive against rain, moisture, flood and lighting. 3.IIIA区应防台风、暴雨袭击及盐雾侵蚀 3. Typhoon, rainstorm attack and salt fog erosion shall be prevented in Regions IIIA
IV	IVA IVB	夏热冬暖地区 Areas hot in summer and warm in winter	1月平均气温 >10℃ Average temperature in January ＞ 10℃ 7月平均气温 25～29℃ Average temperature in July: 25-29℃	海南、台湾全境；福建南部；广东、广西大部以及云南西南部和元江河谷地区 Hainan and whole territory of Taiwan; southern Fujian; Guangdong, the most parts of Guangxi, southwestern Yunnan and the river valley of Yuanjiang River	1. 建筑物必须满足夏季防热、遮阳、通风、防雨要求 1. Buildings must meet the requirements of heat protection, shading, ventilation and rain protection in summer 2. 建筑物应防暴雨、防潮、防洪、防雷电 2. Buildings shall be preventive against rainstorm, moisture, flood and lighting 3.IVA应防台风暴雨袭击及盐雾侵蚀 3. Typhoon, rainstorm attack and salt fog erosion shall be prevented in Regions IVA
V	VA VB	温和地区 Moderate climate area	1月平均气温 0～13℃ 7月平均气温 18～25℃ Average temperature in July 18-25 ℃	云南大部、贵州、四川西南部、西藏南部一小部分地区 Most parts of Yunnan, Guizhou, southwestern Sichuan and a small part of the regions of southern Tibet	1. 建筑物应满足防雨和通风要求 1. Buildings must meet the requirements of rain-proof and ventilation 2.VA区建筑物应注意防寒，VB区建筑物应特别注意防雷电 2. Buildings in VA regions shall be cold-proof, and buildings in VB regions shall be lighting-proof

续表

分区代号 Region code	分区名称 Region name	气候主要指标 Main climate indicators	各区辖行政区范围 Administrative area of each zone	建筑基本要求 Basic requirements for buildings
VI	VIA VIB VIC 严寒地区 Severely cold areas 寒冷地区 Cold regions	1月平均气温 −22~0℃ Average temperature in January −22-0℃ 7月平均气温 <18℃ Average temperature in July <18℃	青海全境；西藏大部；四川西部、甘肃西南部；新疆南部部分地 Whole territory of Qinghai; most parts of Tibet; west of Sichuan and southwest of Gansu; some parts of southern Xinjiang	建筑热工设计应符合严寒和寒冷区相关要求 Thermal performance design of buildings shall conform to the relevant requirements of the severely cold and cold regions
VII	VIIA VIIB VIIC VIID 严寒地区 Severely cold areas	1月平均气温 −20~−5℃ Average temperature in January: −20- −5℃ 7月平均气温 ≥ 18℃ Average temperature in July ≥ 18℃ 7月平均相对湿度 <50% Average relative humidity in July <50%	新疆大部；甘肃北部；内蒙古西部 Most parts of Xinjiang; northern Gansu; western part of Inner Mongolia	建筑热工设计应符合严寒和寒冷地区相关要求 Thermal performance design of buildings shall conform to the relevant requirements of the severely cold and cold regions

3.5 建筑与环境

3.5 Buildings and Environment

3.5.1 建筑与环境的关系

3.5.1 Relationship between Buildings and Environment

【通则】3.4.1 条：建筑与环境的关系应符合下列要求。

[Code for design of civil buildings] 3.4.1: The relationship between buildings and environment shall conform to the requirements in the following.

1. 建筑基地应选择在无地质灾害或洪水淹没等危险的安全地段；

1. The construction base shall be selected in the safe areas without the dangers of geological disasters, floods, etc.;

2. 建筑总体布局应结合当地的自然与地理环境特征，不应破坏自然生态环境；

2. The overall layout of the building shall not destroy the natural ecological environment in combination with the characteristics of the local natural and geographical environment;

3. 建筑物周围应具有能获得日照、天然采光、自然通风等的卫生条件；

3. The conditions of access to sunshine, natural daylighting, and natural ventilation shall be

provided around the building;

4. 建筑物周围环境的空气、土壤、水体等不应构成对人体的危害，确保卫生安全的环境；

4. The air, soil, and water in the environment of buildings shall not constitute a hazard to human bodies and ensure a hygienic and safe environment;

5. 对建筑物使用过程中产生的垃圾、废气、废水等废弃物应进行处理，并应对噪声、眩光等进行有效的控制，不应引起公害；

5. The wastes of garbage, exhaust gas, waste water, and the like. generated in the use of buildings shall be treated, and the noise, glare, and the like. shall be effectively controlled and shall not cause public hazards;

6. 建筑整体造型与色彩处理应与周围环境协调；

6. The overall modeling and color treatment of buildings shall be coordinated with the surrounding environment;

7. 建筑基地应做绿化、美化环境设计，完善室外环境设施。

7. The construction base shall be greened and landscaped, and the outdoor environmental facilities shall be improved.

4

城市规划对建筑的限定
Limits to Building in Urban Planning

4.1 【通则】4.1条：建筑基地

4.1 [Code for design of civil buildings] 4.1: Construction Base

4.1.1 【通则】4.1.1条：基地总平面设计应以所在城市总体规划、分区规划、控制性详细规划及当地主管部门提供的规划条件为依据。

4.1.1 [Code for design of civil buildings] 4.1.1: The general layout design for bases shall be on the basis of the overall planning, district planning, detailed control planning of cities where the bases are located in and the planning conditions provided by the local competent authorities.

4.1.2 【通则】4.1.2条：基地总平面设计应结合工程特点、使用要求，注重节地、节能、节水、节材、保护环境和减少污染，为人们提供健康适用的空间，以适应建设发展的需要。

4.1.2 [Code for design of civil buildings] 4.1.2: The general layout design for bases shall provide people with healthy and applicable space in combination with the engineering characteristics and application requirements, and attaching importance to land, energy, water and material saving, environment protection and pollution reduction, to adapt to the requirements of construction and development.

4.1.3 【通则】4.1.3条：基地总平面设计应结合当地气候条件、自然地形、周围环境、地域文化和建筑环境，因地制宜地确定规划指导思想。

4.1.3 [Code for design of civil buildings] 4.1.3: The guiding thoughts of planning for the general layout design for bases shall be determined in combination with the local climatic conditions and natural terrains, surroundings, regional cultures, and building environment in accordance with the local conditions.

4.1.4 【通则】4.1.4条：基地总平面设计应保护自然植被、自然水域、水系，保护生态环境。

4.1.4 [Code for design of civil buildings] 4.1.4: In the general layout design for bases, the natural vegetation, natural water areas, water systems and ecological environment shall be protected.

4.1.5 【通则】4.1条：建筑基地各条规定如下。

4.1.5 [Code for design of civil buildings] 4.1: The provisions for construction bases are specified as follows.

4.1.5.1 【通则】4.1.1条：基地内建筑使用性质应符合城市规划确定的用地性质。

4.1.5.1 [Code for design of civil buildings] 4.1.1: The application properties of buildings in the base shall conform to the determined land natures of urban planning.

4.1.5.2 【通则】4.1.2条：基地应与道路红线相邻接，否则应设基地道路与道路红线所划定的城市道路相连接。基地内建筑面积小于或等于3000m² 时，基地道路的宽度不应小于4m，基地内建筑面积大于3000m²且只有一条基地道路与城市道路相连接时，基地道路的宽度不应小于7m，若有两条以上基地道路与城市道路相连接时，基地道路的宽度不应小于4m。

4.1.5.2 [Code for design of civil buildings] 4.1.2: The base shall be adjacent to the boundary lines of roads, otherwise the roads in the base shall be arranged and connected to the urban roads demarcated with boundary lines of roads. When the building area in the base is less than or equal to 3000m², the width of roads in the base shall be not less than 4m; when the building area in the base is more than 3000m² and there is one road of the base connected to the urban roads, the width of roads in the base shall be not less than 7m, and if there are more than two roads in the base connected to the urban roads, the width of roads in the base shall be not less than 4m.

4.1.5.3 【通则】4.1.3条：基地地面高程应符合下列规定。

4.1.5.3 [Code for design of civil buildings] 4.1.3: The ground elevation of bases shall conform to the following provisions.

1. 基地地面高程应按城市规划确定的控制标高设计；

1. The ground elevation of bases shall be designed as per the determined control elevation of urban planning;

2. 基地地面高程应与相邻基地标高协调，不妨碍相邻各方的排水；

2. The ground elevation of bases shall be coordinated with the elevation of the adjacent base and shall not impede the drainage of the adjacent parties;

3. 基地地面最低处高程宜高于相邻城市道路最低高程，否则应有排除地面水的措施。

3. The elevation of the lowest grounds of bases shall be higher than the lowest elevation of the adjacent urban roads. Otherwise, measures shall be taken to drain surface water.

4.1.5.4 【通则】4.1.4条：相邻基地的关系应符合下列规定。

4.1.5.4 [Code for design of civil buildings] 4.1.4: The relationship between adjacent bases shall conform to the following provisions.

1. 建筑物与相邻基地之间应按建筑防火等要求留出空地和道路。当建筑前后各自留有空地或道路，并符合防火规范有关规定时，则相邻基地边界两边的建筑可毗连建造。

1. The open spaces and roads shall be reserved between the building and the adjacent base as per the requirements for the fireproofing of buildings. When open spaces or roads are reserved in front of and at the back of the building, and the relevant provisions in the code for

fire protection design of buildings are met, the buildings on both sides of the boundaries of the adjacent bases can be built in an adjacent way.

2. 本基地内建筑物和构筑物均不得影响本基地或其他用地内建筑物的日照标准和采光标准。

2. The buildings and structures in the bases shall not affect the standards for sunshine and standards for daylighting of buildings in the bases or other lands.

3. 除城市规划确定的永久性空地外，紧贴基地用地红线建造的建筑物不得向相邻基地方向设洞口、门、外平开窗、阳台、挑檐、空调室外机、废气排出口及排泄雨水。

3. Except for the determined permanent open spaces of urban planning, for the determined permanent open spaces of urban planning, the openings, doors, external casement windows, balconies, overhanging eaves, outdoor units of air conditioners, exhaust gas outlets and rainwater drainage outlets shall not be arranged in the direction of the adjacent base.

4.1.5.5 【通则】4.1.5 条：基地机动车出入口位置应符合下列规定。

4.1.5.5 [Code for design of civil buildings] 4.1.5: The positions of exits and entrances of motorized vehicles of the base shall conform to the following provisions.

1. 与大中城市主干道交叉口的距离，自道路红线交叉点量起不应小于 70m；

1. The distance to the crossings of main roads in large and medium-sized cities shall be not less than 70m from the crossing points of boundary lines of the roads;

2. 与人行横道线、人行过街天桥、人行地道（包括引道、引桥）的最边缘线不应小于 5m；

2. The distance to the outermost edge lines of pedestrian crossings, pedestrian overpasses and pedestrian underpasses (including approach roads and approach bridges) shall be not less than 5m;

3. 距地铁出入口、公共交通站台边缘不应小于 15m；

3. The distances to the entrances and exits of subways and edges of the public transport platforms shall be not less than 15m;

4. 距公园、学校、儿童及残疾人使用建筑的出入口不应小于 20m；

4. The distances to the entrances and exits of parks, schools and buildings for children and the disabled people shall be not less than 20m;

5. 当基地道路坡度大于 8% 时，应设缓冲段与城市道路连接；

5. When the gradient of roads in the base is more than 8%, the buffer section shall be arranged and connected with the urban roads;

6. 与立体交叉口的距离或其他特殊情况，应符合当地城市规划行政主管部门的规定。

6. The distance to the grade-separated junction or other special matters shall conform to the provisions of the local competent department of city planning administration.

4.1.5.6 【通则】4.1.6 条：大型、特大型的文化娱乐、商业服务、体育、交通等人员密集建筑的基地应符合下列规定。

4.1.5.6 [Code for design of civil buildings] 4.1.6: The bases of densely populated buildings for large and extra large cultural entertainment, business service, sports, transportation, and the like shall conform to the following provisions.

1. 基地应至少有一面直接邻接城市道路，该城市道路应有足够的宽度，以减少人员疏散时对城市正常交通的影响；

1. For the base, at least one side shall be directly adjacent to the urban road, and this urban road shall be wide enough to reduce the impact on normal urban traffic during personnel evacuation;

2. 基地沿城市道路的长度应按建筑规模或疏散人数确定，并至少不小于基地周长的 1/6；

2. The length of the base along the urban road shall be determined as per the building scale or the number of evacuees number, and at least not less than 1/6 of the perimeter of the base;

3. 基地应至少有两个或两个以上不同方向通向城市道路的（包括以基地道路连接的）出口；

3. At least two or more exits in different directions leading to urban roads (including the case that the connection is made through the base roads) shall be arranged for the base;

4. 基地或建筑物的主要出入口，不得和快速道路直接连接，也不得直对城市主要干道的交叉口；

4. The main entrances and exits of bases or buildings shall not be directly connected with the high speed roads, and shall not have a direct access to the intersections of trunk roads;

5. 建筑物主要出入口前应有供人员集散用的空地，其面积和长宽尺寸应根据使用性质和人数确定；

5. Open spaces for people gathering and evacuation shall be provided in front of the main entrances and exits of buildings, the area, length and width shall be determined as per the use nature and the number of people;

6. 绿化和停车场布置不应影响集散空地的使用，并不宜设置围墙、大门等障碍物。

6. The layout of greening and parking areas shall not affect the use of open spaces for collection and distribution, and the obstructions such as enclosure walls and gates, and the like shall not be arranged.

4.2 建筑突出物

4.2 Building Protrusions

4.2.1 【通则】4.2.1 条：建筑突出物规定如下。

4.2.1 [Code for design of civil buil.dings] 4.2.1: The provisions for building protrusions are specified as follows.

建筑物及附属设施不得突出道路红线和用地红线建造，不得突出的建筑突出物为：

The buildings and ancillary facilities shall not be constructed beyond the boundary lines for roads and boundary lines for lands, and these building protrusions are listed as follows:

——地下建筑物及附属设施，包括结构挡土桩、挡土墙、地下室、地下室底板及其基础、化粪池等；

——Underground buildings and ancillary facilities, including the structural retaining piles, retaining walls, basements, bottom boards of basements and the foundations, septic tanks, etc.;

——地上建筑物及附属设施，包括门廊、连廊、阳台、室外楼梯、台阶、坡道、花池、围墙、平台、散水明沟、地下室进排风口、地下室出入口、集水井、采光井等；

——Ground and ground buildings, and ancillary facilities, including the porches, connecting corridors, balconies, outdoor staircases, steps, ramps, flower beds, enclosure walls, platforms, apron open ditches, air inlets and outlets of basements, entrances and exits of basements, water-collecting wells, light wells, etc.;

——除基地内连接城市的管线、隧道、天桥等市政公共设施外的其他设施。

——Facilities except those connected to the municipal public facilities of pipelines, tunnels, overpasses, etc., in the base.

4.3 建筑高度控制

4.3 Height Control of Building

4.3.1 【通则】4.3.1 条：建筑高度不应危害公共空间安全、卫生和景观，下列地区应实行建筑高度控制。

4.3.1 [Code for design of civil buildings] 4.3.1: The building height shall not endanger the safety, hygiene and landscapes of public space, and the building height shall be controlled in the following regions.

1. 对建筑高度有特别要求的地区，应按城市规划要求控制建筑高度；

1. In the regions with special requirements for the building height, the building height shall be controlled as per the urban planning requirements;

2. 沿城市道路的建筑物，应根据道路的宽度控制建筑裙楼和主体塔楼的高度；

2. The height of podiums and main towers of the buildings along urban roads shall be controlled in accordance with the width of the road;

3. 机场、电台、电信、微波通信、气象台、卫星地面站、军事要塞工程等周围的建筑，当其处在各种技术作业控制区范围内时，应按净空要求控制建筑高度；

3. The height of buildings around airports, radio stations, telecommunications, microwave communications, meteorological observatories, satellite earth stations, military fortresses, and the like, within the scope of various technical operation control areas shall be controlled on the basis of the clearance requirements;

4. 当建筑处在本通则第 1 章第 1.0.3 条第 8 款所指的保护规划区内。

4. The height of buildings in the protection planning areas specified in Item 8 of Clause 1.0.3, Chapter 1 in this code shall be controlled.

注：建筑高度控制尚应符合当地城市规划行政主管部门和有关专业部门的规定。

Note: The control of building height shall conform to the provisions specified by the local competent department of urban planning administration and relevant professional departments.

4.3.2 【通则】4.3.2 条：建筑高度控制的计算应符合下列规定。

4.3.2 [Code for design of civil buildings] 4.3.2: The calculation of the building height control shall conform to the following provisions.

1. 第 4.3.1 条 3、4 款控制区内建筑高度，应按建筑物室外地面至建筑物和构筑物最高点的高度计算。

1. The building height in the control areas of Items 3 and 4, Clause 4.3.1, shall be calculated as per the height from outdoor floors of buildings to the highest points of the buildings and structures.

2. 非第4.3.1条3、4款控制区内建筑高度：平屋顶应按建筑物室外地面至其屋面面层或女儿墙顶点的高度计算；坡屋顶应按建筑物室外地面至屋檐和屋脊的平均高度计算。下列突出物不计入建筑高度内：

2. In the control areas, the height of buildings not included in Items 3 and 4 of Clause 4.3.1: for the flat roofs, the height shall be calculated as per the height from outdoor floors of buildings to the roof floors or the top points of the parapet walls; and calculated for the pitched roofs as per the average height from outdoor floors of buildings to the eaves and roof ridges; the following protrusions shall not be included into the building height:

1）局部突出屋面的楼梯间、电梯机房、水箱间等辅助用房占屋顶平面面积不超过1/4者；

1）If the proportion of areas of the auxiliary rooms of staircases, elevator rooms, water tank rooms, and the like which partially protrude out of the roofs, to areas of roof planes does not exceed 1/4;

2）突出屋面的通风道、烟囱、装饰构件、花架、通信设施等；

2）Ventilation ducts, chimneys, decorative components, pergolas, communication facilities, etc., protruding out of roofs;

3）空调冷却塔等设备。

3）Equipments like air-conditioning cooling towers, and so forth.

4.4 建筑密度、容积率和绿地率

4.4 Building Density, Plot Ratio and Greening Rate

4.4.1 【通则】4.4.1条：建筑密度（也称建筑覆盖率）

4.4.1 [Code for design of civil buildings] 4.4.1: The building density（is also known as building coverage rate）

4.4.1.1 【居规划标】4.0.2条注2：建筑密度是居住街坊内，住宅建筑及其便民服务设施建筑基底面积与该居住街坊用地面积的比率（%）。

4.4.1.1 [Standard for urban residential areas planning and design] 4.0.2 Note 2: The building density is the ratio（%）of the floor area of residential buildings and facilities building in a residential neighborhood to the area of the residential neighborhood.

4.4.1.2 【深圳建规则】3.3.1.2条：深圳地区分类。

4.4.1.2 [Shenzhen architectural design regulation] 3.3.1.2: Classification in Shenzhen.

1. 一级建筑覆盖率＝建筑基地面积/建设用地面积×100%

1. Coverage rate of Grade I buildings = area of the construction base/area of the construction land × 100%

2. 二级建筑覆盖率＝塔楼建筑基底面积/建设用地面积×100%

2. Grade II building coverage rate = area of the building foundations of towers/area of the construction land × 100%

4.4.1.3 建筑基底面积：建筑物接触地面自然层建筑外墙或结构外围水平投影面积。

4.4.1.3 Area of the building foundation: horizontal projection area of the exterior wall contacting the natural ground layer or the outer periphery of the structure.

4.4.2 容积率

4.4.2 Plot Ratio

4.4.2.1 【通则】4.4.1条：建筑设计应符合法定规划控制的建筑密度、容积率和绿地率的要求。

4.4.2.1 [Code for design of civil buildings] 4.4.1: The architectural design shall conform to the requirements of the building density, plot ratio and greening rate controlled as per the legal planning.

4.4.2.2 【通则】4.4.2条：当建设单位在建筑设计中为城市提供永久性的建筑开放空间，无条件地为公众使用时，该用地的既定建筑密度和容积率可给予适当提高，且应符合当地城市规划行政主管部门有关规定。

4.4.2.2 [Code for design of civil buildings] 4.4.2: When the construction organization provides permanent open building spaces for the city in the building design, to be used by the public without any conditions, the determined building density and plot ratio for the land can be appropriately increased, but the relevant provisions of the local urban planning authorities shall be met.

4.4.3 绿地率

4.4.3 Greening rate

4.4.3.1 【居规划标】4.0.2条注3：绿地率是居住街坊内绿地面积之和与该居住街坊用地面积的比率（%）。

4.4.3.1 [Standard for urban residential areas planning and design] 4.0.2 Note 2: Greening rate is the ratio（%）of the total green space in a residential neighborhood to the land area of the residential neighborhood.

4.4.3.2 【居规划标】4.0.4 条：新建各级生活圈居住区应配套规划建设公共绿地，并应集中设置具有一定规模，且能开展休闲、体育活动的居住区公园；公共绿地控制指标应符合表 4.4.3.1 的规定。

4.4.3.2 [Standard for urban residential areas planning and design] 4.0.4: Green space should be planned and constructed for the newly built residential areas at all levels and residencel parks with a certain scale where leisure and sports activities can be carried out should be set up centrally; the control index of public green space should comply with the provisions of Tab. 4.4.3.1.

【居规划标】表 4.0.4：公共绿地控制指标　　　　表 4.4.3.1

[Standard for urban residential area planning and design] Tab. 4.0.4: Control Index of Public Green Space　　　Tab. 4.4.3.1

类别 Category	人均公共绿地面积（m²/人） Area of public green space per capita（m²/person）	居住区公园 Residence park		备注 Note
		最小规模 Minimum scale（hm²）	最小宽度 Minimum width（m）	
15 分钟生活圈居住区 15 minutes living quarters	2.0	5.0	80	不含 10 分钟生活圈居住区及以下级居住区的公共绿地控制指标 The control index of public green space excluding 10 minutes living quarters and below
10 分钟生活圈居住区 10 minutes living quarters	1.0	1.0	50	不含 5 分钟生活圈居住区及以下级居住区的公共绿地控制指标 The control index of public green space excluding 5 minutes living quarters and below
5 分钟生活圈居住区 5 minutes living quarters	1.0	0.4	30	不含居住街坊的绿地指标 The index of green space excluding residential neiborhood

注：1. 居住区公园应设置 10%～15% 的体育活动场地。
Note：1. 10%～15% sport space should be set in residence parks.
2. 居住区配套设施设置（即 15 分钟、10 分钟及 5 分钟生活圈配套设施设置）详见【居规划标】附录 B 居住区配套设施设置规定。
2. Supporting facilities construction in the residential area (that is, construction of supporting facilities in 15 minutes, 10 minutes and 5minutes living quarters) can be found in Regulations of construction of supporting facilities in the residential area in the Appendix B of [Standard for urban residential areas planning and design].

5

总平面设计
Site Planning

5.1 一般规定

5.1 General Requirements

5.1.1 【通则】5.1.1 条：民用建筑应根据城市规划条件和任务要求，按照建筑与环境关系的原则，对建筑布局、道路、竖向、绿化及工程管线等进行综合性的场地设计。

5.1.1 [Code for design of civil buildings] 5.1.1: The comprehensive site design shall be carried out for the layout, roads, verticality, greening, engineering pipelines, etc., of the civil building in accordance with the urban planning conditions and task demands on the basis of the principles for the relationship between architectures and environments.

5.1.2 【通则】5.1.2 条：建筑布局应符合下列规定。

5.1.2 [Code for design of civil buildings] 5.1.2: The building layout shall conform to the following provisions.

1. 建筑间距应符合防火规范要求；

1. The spacing of buildings shall conform to the requirements in code for fire protection design of buildings;

2. 建筑间距应满足建筑用房天然采光（本规范 7.1 采光）的要求，并应防止视线干扰；

2. The spacing of buildings shall meet the natural lighting requirements for building rooms (clause 7.1 of this code), and the interference of sightline shall be prevented;

3. 有日照要求的建筑应符合日照标准的要求，并应执行当地城市规划行政主管部门制定的相应的建筑间距规定；

3. The buildings with requirements for sunshine shall conform to the requirements for sunshine standards, and the corresponding provisions of building spacing formulated by the local competent department of urban planning administration shall be implemented;

4. 对有地震等自然灾害地区，建筑布局应符合有关安全标准的规定；

4. In the area suffering from natural disasters such as earthquakes, etc., the building layout shall conform to the provisions for relevant safety standards;

5. 建筑布局应使建筑基地内的人流、车流与物流合理分流，防止干扰，并有利于消防、停车和人员集散；

5. In the building layout, reasonable arrangement shall be made to diverse the flow of

people, vehicle and logistics in the building base, to prevent any interference, and facilitate fire protection, parking and the assembly and evacuation of people;

6. 建筑布局应根据地域气候特征，防止和抵御寒冷、暑热、疾风、暴雨、积雪和沙尘等灾害侵袭，并应利用自然气流组织好通风，防止不良小气候产生；

6. In the building layout, the attack of disasters such as coldness, summer heat, high wind, rainstorm, accumulated snow, sand-dust storm, etc., shall be prevented and resisted as per the climate characteristics of the region, and ventilation shall be well organized through natural airflow, to prevent the occurrence of undesirable microclimates;

7. 根据噪声源的位置、方向和强度，应在建筑功能分区、道路布置、建筑朝向、距离以及地形、绿化和建筑物的屏障作用等方面采取综合措施，以防止或减少环境噪声；

7. In accordance with the positions, directions and strength of noise sources, comprehensive measures shall be adopted in the aspects of functional zoning of buildings, layout of roads, building directions, distances and terrain, barrier action of greening and buildings, etc., to prevent or reduce environmental noise;

8. 建筑物与各种污染源的卫生距离，应符合有关卫生标准的规定。

8. The sanitary distances between buildings and various pollution sources shall conform to the relevant hygienic standards.

5.1.3 建筑间距

5.1.3 Spacing of Buildings

5.1.3.1 【通则】2.0.13 条：日照标准

5.1.3.1 [Code for design of civil buildings] 2.0.13: Sunshine Standard

根据建筑物所处的气候区、城市大小和建筑物的使用性质确定的，在规定的日照标准日（冬至日或大寒日）的有效日照时间范围内，以底层窗台面为计算起点的建筑外窗获得的日照时间。

Sunshine time of the exterior windows, determined as per the climate zone, city size and use nature of the building within the effective sunshine time range of the specified standard sunshine day (winter solstice or great cold day), with the surfaces of windowsills on the ground floor as the starting point for calculation.

5.1.3.2 住宅建筑日照标准

5.1.3.2 Sunshine Stargard of Residential Building

1.【居规划标】表 4.0.9：住宅建筑日照标准（表 5.1.3.2）

1. [Standard for urban residential area planning and design] Tab. 4.0.9:Sunshine standard of residential building（Tab. 5.1.3.2）

住宅建筑日照标准　　　　　　　　　　　　　　　　　　　表 5.1.3.2
Sunshine Standard for Residential Building　　　　　　Tab. 5.1.3.2

建筑气候区划 Building climate division	I、II、III、VII 气候区 I, II, III and VII climate region		IV 气候区 IV climate region		V、VI 气候区 V and VI climate region
城区常住人口（万人） Urban resident population（10000 Persons）	≥ 50	< 50	≥ 50	< 50	无限定 Unlimited
日照标准日 Standard sunshine day	大寒日 Great cold				冬至日 Winter solstice
日照时数（h） Sunshine hours（h）	≥ 2		≥ 3		≥ 1
有效日照时间带（当地真太阳时） Effective sunshine time slot（local true solar time）	8 时~ 16 时 8am to 4 pm				9 时~ 15 时 9am to 3 pm
计算起点 Calculating start	底层窗台面 Surfaces of windowsills on the ground floor				

注：底层窗台面是指距室内地坪 0.9m 高的外墙位置。
Note: Surfaces of windowsills on the ground floor refers to the place of external wall of 0.9m high from indoor floor.

5.1.4　建筑之间的防火间距
5.1.4　Fire Separation between Buildings

5.1.4.1　民用建筑之间的防火间距见表 5.1.4.1。

5.1.4.1　Fire Separation between Civil Buildings see Tab. 5.1.4.1.

【防火规】表 5.2.2：民用建筑之间的防火间距（m）　　　　　表 5.1.4.1
[Code for fire protection design of buildings] Tab. 5.2.2:
Fire Separation between Civil Buildings（m）　　　　　　Tab. 5.1.4.1

建筑类别 Building category		高层民用建筑 High-rise civil buildings	裙房和其他民用建筑 Podiums and other civil buildings		
		一、二级 Grade I and Grade II	一、二级 Grade I and Grade II	三级 Grade III	四级 Grade IV
高层民用建筑 High-rise civil buildings	一、二级 Grade I and Grade II	13	9	11	14
裙房和其他民用建筑 Podiums and other civil buildings	一、二级 Grade I and Grade II	9	6	7	9
	三级 Grade III	11	7	8	10
	四级 Grade IV	14	9	10	12

注：1. 相邻两座单、多层建筑，当相邻外墙为不燃性墙体且无外露的可燃性屋檐，每面外墙上无防火保护的门、窗、洞口不正对开设且该门、窗、洞口的面积之和不大于外墙面积的 5% 时，其防火间距可按本表的规定减少 25%。

Notes: 1. For two adjacent single-storey and multi-storey buildings, when the adjacent exterior walls are non-combustible walls and non-exposed flammable eaves, doors, windows and openings without fire protection on each exterior wall are not arranged in an opposite way, and the sum of the areas of the doors, windows and openings is not greater than 5% of the area of the exterior wall, the fire protection separation can be reduced by 25% as per the provisions in the table.

2. 两座建筑相邻较高一面外墙为防火墙，或高出相邻较低一座一、二级耐火等级建筑的屋面15m及以下范围内的外墙为防火墙时，其防火间距不限。

2. When the higher exterior walls of two adjacent buildings are fire walls, or the external walls higher than the roof of the lower building in the adjacent buildings with the fire resistance ratings of Grade I and Grade II by 15 m or below are fire walls, the fire separation shall be not limited.

3. 相邻两座高度相同的一、二级耐火等级建筑中相邻任一侧外墙为防火墙，屋顶的耐火极限不低于1.00h时，其防火间距不限。

3. When the exterior walls on any side of two adjacent buildings with Grade I or Grade II fire resistance rating and of the same height are fire walls, and the fire endurance of the roofs is not less than 1.00 hour, the fire separation shall be not limited.

4. 相邻两座建筑中较低一座建筑的耐火等级不低于二级，相邻较低一面外墙为防火墙且屋顶无天窗，屋顶的耐火极限不低于1.00h时，其防火间距不应小于3.5m；对于高层建筑，不应小于4m。

4. The fire resistance rating of the lower of two adjacent buildings shall be not less than Grade II, and when the exterior wall on the adjacent lower side is a fire wall, no skylight is arranged in the roof, and the fire endurance of the roof is not less than 1.00 h, the fire separation shall be not less than 3.5m; For high-rise buildings, not less than 4m.

5. 相邻两座建筑中较低一座建筑的耐火等级不低于二级且屋顶无天窗，相邻较高一面外墙高出较低一座建筑的屋面15m及以下范围内的开口部位设置甲级防火门、窗，或设置符合现行国家标准《自动喷水灭火系统设计规范》GB 50084规定的防火分隔水幕或本规范第6.5.3条规定的防火卷帘时，其防火间距不应小于3.5m；对于高层建筑，不应小于4m。

5. When the fire resistance rating of the lower of two adjacent buildings shall be not less than Grade II, and no skylight is not arranged in the roof, Grade A fire doors and windows shall be arranged at the opening position in the higher exterior wall in the adjacent walls with the height above the roof of the lower building by 15 m and below, or the fire compartment drencher sprinkler system conforming to the provisions in the current national standard specified in the *Code for Design of Sprinkler Systems* GB 50084 or fireproof shutters specified in Clause 6.5.3 of this Code are arranged, the fire separation shall be not less than 3.5 m; for high-rise buildings, not less than 4 m.

6. 相邻建筑通过连廊、天桥或底部的建筑物等连接时，其间距不应小于本表的规定。

6. When the adjacent buildings are connected through corridors, overbridges or buildings at the bottoms, the spacing shall be not less than the provisions in the table.

7. 耐火等级低于四级的既有建筑，其耐火等级可按四级确定。

7. For the existing buildings with the fire resistance rating lower than Grade IV, the fire resistance rating shall be determined as per Grade IV.

5.1.4.2 【防火规】5.2.4条：除高层民用建筑外，数座一、二级耐火等级的住宅建筑或办公建筑，当建筑物的占地面积总和不大于250m² 时，可成组布置，但组内建筑物之间的间距不宜小于4m。组与组或组与相邻建筑物的防火间距不应小于本规范第5.2.2条的规定。

5.1.4.2 [Code for fire protection design of buildings] 5.2.4: Except high-rise civil buildings, when the total floor area is not greater than 250m², several residential buildings or office buildings of fire resistance rating of Grade I or Grade II can be arranged in groups, but the spacing between buildings in a group shall be appropriately not less than 4 m. The fire separation between groups or between a group and the adjacent buildings shall be not less than the provisions in clause 5.2.2 of the code.

5.1.4.3 【防火规】5.2.6 条：建筑高度大于 100m 的民用建筑与相邻建筑的防火间距，当符合本规范第 3.4.5 条、第 3.5.3 条、第 4.2.1 条和第 5.2.2 条允许减小的条件时，仍不应减小。

5.1.4.3 [Code for fire protection design of buildings] 5.2.6: Although the buildings conform to the conditions allowed to be reduced in clauses 3.4.5, 3.5.3, 4.2.1 and 5.2.2 of the Code, the fire separation between civil buildings with the building height greater than 100m, the fire separation between adjacent buildings shall not be reduced.

5.2 竖向设计

5.2 Vertical Design

5.2.1 基本原则

5.2.1 Basic Principles

1. 【城竖规】3.0.7 条：同一城市的用地竖向规划应采用统一的坐标和高程系统。

1. [Code for urban vertical planning] 3.0.7: Uniform coordinate and elevation system shall be adopted in the vertical land planning of the same city.

2. 【城竖规】3.0.7 条附录表 1：水准高程系统换算。

[Code for urban vertical planning] Appendix 1 of 3.0.7: Standard elevation system conversion.

水准高程系统换算　　　　　表 5.2.4
Standard elevation system conversion　　Tab 5.2.4

被转换者 Datum to be converted	转换者 Datum			
	56 黄海高程基准 Yellow sea elevation Datum (1956)	85 高程基准 Elevation datum (1985)	吴淞高程基准 Wusong elevation datum	珠江高程基准 Zhujiang elevation datum
56 黄海高程基准 Yellow sea elevation datum (1956)	—	+0.029	−1.688	+0.586
85 高程基准 Elevation datum (1985)	−0.029	—	−1.717	+0.557
吴淞高程基准 Wusong elevation datum	−1.688	+1.717	—	+2.274
珠江高程基准 Zhujiang elevation datum	−0.586	−0.577	−2.274	—

备注：高程基准之间的差值为各地区精密水准点之间的差值的平均数
Remarks: the difference between elevation data is the mean value of the difference values between the precision datum points of all the regions

5.2.2 【09措施】3.1.1条：竖向设计的内容

5.2.2 [Measures 2009] 3.1.1: Contents of Vertical Design

1. 制定利用与改造地形的方案，合理选择、设计场地的地面形式；依据不同的自然地形坡度，场地的地面形式可分别处理成平坡式、台阶式和混合式。

1. The programs for using and changing terrains shall be established, and the ground forms of the sites shall be reasonably selected and designed; On the basis of the gradients of different natural terrains, the ground forms of the sites can be treated into flat-type, step-type and mixed-type respectively.

2. 确定场地坡度、控制点高程、地面形式。

2. The gradient of the site, elevation at the control point and ground form shall be determined.

3. 制定合理利用、储存和收集雨水的方案。在干旱、贫水地区，竖向设计应做到使雨水就地渗入地下，或使雨水便于收集储存和利用。

3. The program for reasonable use, storage and collection of rainwater shall be established. In arid and water-poor regions, vertical design shall achieve the effect that rainwater is allowed to infiltrate underground, or rainwater can be easy to collect, store and use.

4. 制定合理排除地面和路面雨水的方案。在降雨量大、洪涝多发地区，为减少排放至市政管网及江、河、湖、海的雨水量，竖向设计可考虑雨水就地收集与利用，以利于排洪调蓄。

4. The reasonable program for the drainage of rainwater on the ground and pavement shall be established. In the areas with heavy rainfall and floods, in order to reduce the amount of rainwater drained to the municipal pipe network, rivers, lakes and seas, in vertical design, the local rainwater collection and utilization can be taken into consideration, to facilitate flood discharge and storage regulation.

5. 合理组织场地的土石方工程和防护工程。

5. The earthwork and stonework engineering and protection engineering in the site shall be reasonably organized.

6. 结合道路设计和景观设计，提出合理的竖向设计条件与要求。

6. In combination of the road and landscape design, reasonable conditions and requirements for the vertical design shall be proposed.

7. 有利于保护和改善建设场地及周围场地的环境景观。

7. The environmental landscapes shall be favorable to the protection and improvement of the construction site and surrounding sites.

5.2.3 【居规划标】6.0.4条：居住街坊内附属道路的规划设计应满足消防、救护、搬

家等车辆的通达要求，并应符合下列规定

5.2.3 [Standard for urban residential areas planning and design] 6.0.4: The planning and design of roads in a residential neighborhood should met the accessibility requirements of fire, ambulance, moving vehicles, and shall comply with the following provisions

1. 主要附属道路至少应有两个车行出入口连接城市道路，其路面宽度不应小于 4.0m；其他附属道路的路面宽度不宜小于 2.5m。

1. The main road should have at least two vehicular entrances and exits connetcting urban roads, and the road width should not be less than 4.0m; and the width of other affiliated road should not be less than 2.5m.

2. 人行出入口间距不宜超过 200m。

2. The distance between the pedestrian entrance and exit should not be more than 200m.

3. 最小纵坡不应小于 0.3%，最大纵坡应符合表 5.2.3 的规定；机动车与非机动车混行的道路，其纵坡宜按照或分段按照非机动车道要求进行设计。

3. The minimum longitudinal slope shall not be less than 0.3%, and the maximum longitudinal slope shall comply with the provisions of Table 5.2.3; the longitudinal slope of roads where motor vehicles and non-motor vehicles are mixed shall be designed according to the requirements of non-motor vehicle roads.

【居规划标】表 6.0.4：附属道路最大纵坡控制指标（%） 表 5.2.3

[Standard for urban residential area planning and design]Tab. 6.0.4: The Control Index (%) of the Maximum longitudinal Slope of Affiliated Roads　　Tab. 5.2.3

道路类别及其控制内容 Road category and the control content	一般地区 Common region	积雪或冰冻地区 Snow and frozen region
机动车道 Motor vehicle road	8.0	6.0
非机动车道 Non-motor vehicle road	3.0	2.0
步行道 Sidewalk	8.0	4.0

5.2.4 建筑基地地面排水

5.2.4 Ground Drainage of the Construction Base

5.2.4.1 排水设计原则

5.2.4.1 Drainage Design Principles

1.【城竖规】6.0.1 条：城市用地应结合地形、地质、水文条件及年均降雨量等因素合理选择地面排水方式，并与用地防洪、排涝规划相协调。

1. [Code for urban vertical planning] 6.0.1: For urban lands, the drainage modes of grounds shall be reasonably selected in combination with the factors of terrains, geological and hydrological conditions, average annual rainfall, etc., and coordinate the flood protection for the land and the drainage planning of stagnant water.

2. 低影响的排水系统见表 5.2.4.1。

2. Drainage system with low impact see Tab. 5.2.4.1.

低影响的排水系统　　　　　　　表 5.2.4.1
Drainage Systems with Low Impact　　　Tab. 5.2.4.1

道路横断面 Cross sections of roads	优化道路横坡坡向、路面与道路绿化带及周边绿化的竖向关系，便于径流雨水汇入绿地内的低影响开发设施 The vertical relationship between the directions of the transverse slope on the road, the pavement and the greening belt of the road and the surrounding greening shall be optimized to facilitate the flowing of the runoff rainwater into the low-impact development facilities in the green land
路面排水 Pavement drainage	生态排水方式，雨水 mode, that is, low-impact development facilities in which rainwater flows into the road greening belts and the s urrounding green spaces first
透水铺装 Permeable pavement	透水铺装的路面设计应满足路基路面强度和稳定性等要求 The permeable pavement design shall meet the requirements of roadbed and pavement strength, stability, etc.

5.2.4.2 室外运动场排水（图 5.2.4.2-1 ~ 图 5.2.4.2-3）

5.2.4.2 Drainage for outdoor venues（Fig. 5.2.4.2-1 ~ Fig. 5.2.4.2-3）

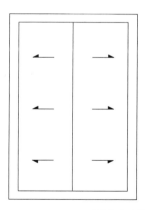 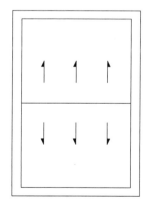 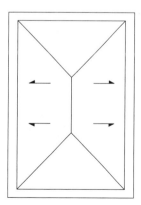

图 5.2.4.2-1　足球场地排水沟及排水坡度【09 措施】3.2.4-1
Fig. 5.2.4.2-1　Drainage ditch and drainage gradient of football field [Measures 2009] 3.2.4-1

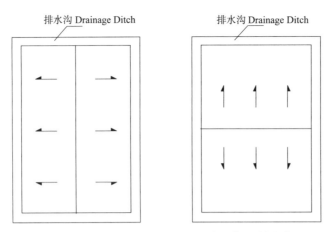

图 5.2.4.2-2　篮球场场地排水沟及排水坡度【09 措施】3.2.4-2

Fig. 5.2.4.2-2　Drainage ditches and drainage gradient of basketball court 3.2.4-2 of [Measures 2009]

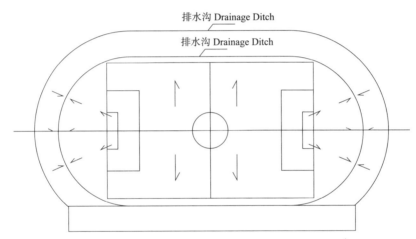

图 5.2.4.2-3　室外运动场场地排水沟及排水坡度【09 措施】3.2.4-4

Fig. 5.2.4.2-3　Drainage ditches and drainage slopes of outdoor fields [Measures 2009] 3.2.4-4

5.2.5　建筑物底层出入口处应采取措施防止室外地面雨水回流。

5.2.5　Proper measures shall be taken at the entrances and exits on the ground floors of buildings to prevent rainwater backflow from outdoor floors.

1.【地面规】3.1.5 条：建筑物的底层地面标高，应高出室外地面 150mm，当有生产、使用的特殊要求或建筑物预期较大沉降量等其他原因时，可适当增加室内外高差。

1. [Code for design of building ground] 3.1.5: The elevations of the ground floors of buildings shall be 150 mm more than that of the outdoor floor. If there exist special requirements of production and application, or the expected settlement of the building is greater, or other due reasons, the indoor and outdoor height difference can be increased appropriately.

2.【宿舍规】3.1.3 条：宿舍基地宜选择较平坦，且不利积水的地段。

2. [Code for design of dormitory building] 3.1.3: The bases of dormitories shall be selected

in the flat land sections unfavorable to water accumulation.

3.【城竖规】4.0.7 条：设置挡土墙高度大于 2m 的挡土墙和护坡，其上缘与建筑物的水平净距不应小于 3m，下缘与建筑物的水平净距不应小于 2m；高度大于 3m 的挡土墙与建筑物的水平净距还应满足日照标准要求。

3. [Code for urban vertical planning] 4.0.7: Set the retaining wall. Whose height is greater than 2m of the retaining wall and slope: the horizontal net distance between the reaining wall's upper edge and the building should not be less than 3m, and the one between its lower edge and the the buliding should not be less than 2m. The horizontal net distance between the retaning wall and the building with a height greater than 3m should also meet the requirements of the sunshine standard.

5.3 居住区道路及停车场

5.3 Roads and Parking Areas in Residential Areas

5.3.1【09 措施】4.1 条：一般规定

5.3.1 [Measures 2009] 4.1: General Provisions

1. 居住区道路及停车场要求见表 5.3.1。

1.Requirement for roads and parking areas in residential areas see Tab.5.3.1.

居住区道路及停车场要求　　　　　　　　　　　　　　表 5.3.1
Requirements for Roads and Parking Areas in Residential Areas　　　Tab. 5.3.1

类别 Category	系统要求 System requirements
道路系统 Road system	应有利于各类用地的功能分区和有机联系，以及建筑功能的合理布局，并有利于雨水排泄，便于管线敷设 To be favorable to the functional zoning and organic connection of various types of land, reasonable layout of building functions, rainwater drainage, and convenient pipeline laying
居住区道路 Roads in residential areas	应保障内外联系通畅、安全、避免迂回，便于消防车、救护车、货物、垃圾运输和居民小汽车通行 To ensure smooth and safe internal and external connection, avoid circuitousness, and facilitate the passage of fire trucks, ambulances, goods and garbage transport, and household vehicles
居住小区内道路出入口个数和距离 Number and distance of road entrances and exits in residential areas	应人车有序，主要道路至少有两个出入口（可以是两个方向也可是同一方向）。居住区规模较大时，应有两个方向与外界道路相连接。机动车道对外出入口间距不应小于 150m Walkways and driveways shall be arranged orderly, and the main road shall be provided with at least two entrances and two exits (in two directions or in the same direction) . When the scale of the residential area is large, it is connected with the outside roads in two directions. The distance between external entrances and exits of driveways shall be not less than 150m
居住小区内尽端式道路 Stub-end roads in residential areas	长度不宜大于 120m；应设置不小于 12m×12m 回车场 The length shall be appropriately not greater than 120m; The turnaround yards not smaller than 12 m × 12 m shall be arranged

2.【居规划标】8.0.5 四：回车场形式

2. [Standard for urban residential area planning and design] 8.0.5 IV：Shapes of turnaround yards

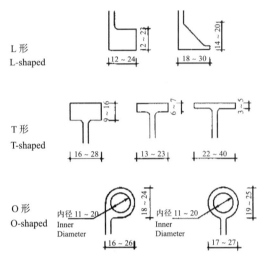

图 5.3.1　回车场形式【居规划标】8.0.5 四

Fig. 5.3.1　Shapes of Turnaround Yards [Standard for urban residential area planning and design] 8.0.5 IV

5.3.2　道路技术标准

5.3.2　Technical Standards for Roads

1. 居住区内道路纵坡控制指标应符合表 5.3.2-1 规定。

1. The longitudinal slope control indicators of the roads in residential areas shall conform to the provisions in Tab. 5.3.2-1.

【居规划标】表 8.0.3：居住区内道路纵坡控制指标　　　　表 5.3.2-1

[Code of urban residential areas planning and design] Tab. 8.0.3:
Longitudinal Slope Control Indicators of the Roads in Residential Areas　　Tab. 5.3.2-1

道路类型 Road type	最小纵坡 Minimum longitudinal slope	最大纵坡 Maximum longitudinal slope	多雪严寒地区最大纵坡 Maximum longitudinal slope in snowy and severely cold regions
机动车道 Driveways	≥ 0.2	≤ 8.0 $L ≤ 200m$	≤ 5 $L ≤ 600m$
非机动车道 Non-motorized vehicle lanes	≥ 0.2	≤ 3.0 $L ≤ 50m$	≤ 2 $L ≤ 100m$
步行道 Walkway	≥ 0.2	≤ 8.0	≤ 4

注：L 为坡长（m）。

Note: L is the slope length (m).

2. 消防车道路及扑救场地要求见表5.3.2-2。

2. Requirements for fire truck roads and fire fighting sites see Tab. 5.3.2-2.

【防火规】7.2条：消防车道路及扑救场地要求　　　　表 5.3.2-2
[Code for fire protection design of buildings] 7.2:
Requirements for Fire Truck Roads and Fire Fighting Sites　　Tab. 5.3.2-2

类别 Category			宽度（m） Width（m）	坡度（%） Gradient（%）		靠建筑外墙一侧边缘距离（m） Distance to the edge of one side of the exterior wall of the building（m）	转弯半径（m） Turning radius（m）
				纵坡 Longitudinal slope	横坡 Transverse slope		
消防车道 Fire lanes			≥ 4.0	≤ 3.0	≤ 3.0	≥ 5.0	9～12
登高操作场地 Ascending operation sites	建筑高度 Building height	≤ 50m	长度和宽度：15和10 Length and width: 15 and 10	≤ 3.0	≤ 3.0	不小于5m且不应大于10m Not less than 5m and not greater than 10m	—
		> 50m	长度和宽度：20和10 Length and width: 20 and 10	≤ 3.0	≤ 3.0	不小于5m且不应大于10m Not less than 5m and not greater than 10m	—

3. 居住区内道路边缘至建筑物、构筑物的最小距离见表5.3.2-3。

3. Minimum distance from the edge of the road in the residential area to the building or structure see Tab.5.3.2-3.

【居规划标】8.0.5：居住区内道路边缘至建筑物、构筑物的最小距离　　表 5.3.2-3
[Standard for urban residential areas planning and design] 8.0.5: Minimum Distance from the Edge of the Road in the Residential Area to the Building or Structure　　Tab. 5.3.2-3

与建、构筑物关系 Relationship with the structures and structures		道路级别 Road grade	居住区道路 Roads in residential areas	小区路 Roads in the community	组团及宅间小路 Roads in the residential cluster and paths between households
建筑物面向道路 Buildings facing roads	无出入口 Without entrances and exits	高层 High-rise buildings	5	3	2
		多层 Multi-storey buildings	3	3	2
	有出入口 With entrances and exits		—	5	2.5

续表

与建、构筑物关系 Relationship with the structures and structures	道路级别 Road grade	居住区道路 Roads in residential areas	小区路 Roads in the community	组团路及宅间小路 Roads in the residential cluster and paths between households
建筑物山墙面向道路 The gable wall of the building facing the road	高层 High-rise buildings	4	2	1.5
	多层 Multi-storey buildings	2	2	1.5
围墙面向道路 The enclosure wall facing the road		1.5	1.5	1.5

5.3.3 基地出入口与城市道路的连接的城市规划条件（表5.3.3-1、表5.3.3-2）

5.3.3 Urban Planning Conditions for the Connection between Entrances and Exits of the Base and Urban Roads（Tab. 5.3.3-1、Tab. 5.3.3-2）

【通则】4.1.2/6 条：基地出入口的数量要求　　　　表 5.3.3-1

[Code for design of civil buildings] 4.1.2/6: Requirements for the Quantities of Entrances and Exits of the Base　　Tab. 5.3.3-1

类别 Category	出入口数量 Quantities of entrances and exits		备注 Remarks
大型、特大型的文化娱乐、商业服务、体育、交通等人员密集建筑 Densely populated buildings for large and extra large cultural entertainment, business service, sports, transportation, etc.	2个以上不同方向 More than 2 different directions		基地应至少有一面直接邻接城市道路，该城市道路应有足够的宽度 At least one side of the base shall be directly adjacent to the urban road, and the urban road shall have sufficient width
小区内主要道路最少的出入口数量 Quantity of minimum entrances and exits of the main roads in the community	基地内建筑面积（m²） Area of buildings in bases（m²）		基地应至少有一面直接邻接城市道路，该城市道路应有足够的宽度 At least one side of the base shall be directly adjacent to the urban road, and the urban road shall have sufficient width
	< 3000 Less than 3000	> 3000 Greater than 3000	
	1个；基地道路的宽度不应小于4m 1; The width of the road in the base shall be not less than 7m	2个以上不同方向；基地道路的宽度不应小于4m More than 2 different directions; the width of the roads in the base shall be not less than 4m	

【通则】4.1.5 条：基地内机动车出入口位置的规定   表 5.3.3-2

[Code for design of civil buildings] 4.1.5: Regulations for Exits and Entrances for Motorized Vehicles in Bases   Tab. 5.3.3-2

类别 Category	城市带路设置 Arrangement of urban roads	距离（m） Distance（m）	备注 Remarks
基地内机动车出入口 Exits and entrances for motorized vehicles in bases	城市主干道 Trunk roads	70	
	人行横道线、人行地道（包括引道、引桥） Pedestrian crossings and pedestrian underpasses (including approach roads and approach bridges)	5	
	地铁出入口、公交站台边缘 Entrances and exits of railways and edges of platforms at bus station	15	
	公园、学校、儿童及残疾人使用建筑的出入口 Entrances and exits of parks, schools, and buildings for children and the disabled people	20	
	人行过街天桥、地道和桥梁、隧道引道 Pedestrian overpasses, underpasses and bridges, and approach roads of tunnels	50	【汽防规】第四条 [Code for fire protection design of garage, motor repair shop and parking area]
	交叉路口 Intersections	80	【汽防规】第四条 [Code for fire protection design of garage, motor repair shop and parking area]
	居住区内主要道路至少应有两个方向与外围道路相连且距离满足 The main roads in the community shall be connected with the peripheral roads in two directions at least and the requirements for distances shall be met	150	
	与城市道路交接时的交角 Intersection angle with the urban road	不宜小于 75° Not less than 75°	【住宅性能标准】5.2.4 条 [Technical standard for performance assessment of residential buildings] 5.2.4

5.3.4 【车库建规】3.1.6 条：建筑基地内地下车库的出入口设置

5.3.4 [Code for design of parking garage building] 3.1.6: Arrangement of Entrances and Exits of Underground Garages in Construction Bases

1. 基地出入口不应直接与城市快速路相连接，且不宜与城市主干路相连接。

1. The entrances and exits of the base shall not be directly connected with the urban expressway, and shall not be connected with the trunk roads.

2. 基地主要出入口的宽度不应小于 4m，并保证出入口与内部通道衔接的顺畅。

2. The width of main entrances and exits of bases shall be not less than 4m, and the entrances and exits shall be smoothly connected with interior passages.

3. 当需要在基地出入口办理车辆出入手续时，出入口处应设置候车道，且不应占用城市道路；机动车候车道宽度不应小于4m、长度不应小于10m，非机动城应留有等候空间。

3. When the access procedures of the vehicle are required to be handled at the entrances and exits of the base, the waiting lane shall be set at the entrances and exits and shall not occupy the urban roads; The width of the waiting lane for motorized vehicles shall be not less than 4m, the length not less than 10m, with the waiting space reserved for the non-motorized vehicles.

4. 机动车库基地出入口应具有通视条件，如图5.3.4所示，与城市道路连接的出入口地面坡度不宜大于5%。

4. The sighting conditions shall be provided at the entrances and exits of the bases of garages, as shown in Fig. 5.3.4 , and the gradient of the entrance and exit ground connected with the urban road shall be appropriately not greater than 5%.

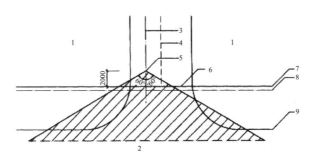

图5.3.4 【车库建规】3.1.6 图9：机动车库基地出入口通视要求示意图

Fig. 5.3.4 [Code for Design of Parking Garage Buildings] Fig. 9 in Clause 3.1.6：Schematic Diagram of Sighting Requirements for Entrances and Exits of Bases of Motor garages

1- 建筑基地；2- 城市道路；3- 车道中心线；4- 道路边线；5- 视点位置；
1-Construction bases; 2-Urban roads; 3-Centerlines of the driveway; 4-Road edge lines; 5-Viewpoint positions;
6- 基地机动车出入口；7- 基地边线；8- 道路红线；9- 道路缘石线
6- Exits and entrances for motorized vehicles in the base; 7 -Base edge lines; 8-Boundary lines of roads; 9-Curb lines

5. 机动车库基地出入口处的机动车道路转弯半径不宜小于6m，且应满足基地通行车辆最小转弯半径的要求。

5. The turning radius of driveways at entrances and exits of bases of garages shall be appropriately not less than 6m, and shall meet the requirements for minimum turning radius of the vehicle passage in the bases.

6. 相邻机动车库基地出入口之间的最小距离不应小于15m，且不应小于两出入口道路转弯半径之和。

6. The minimum distance between the entrances and exits of the bases of adjacent garages shall be not less than 15m, and shall be not less than the sum of the turning radius of the two

entrance and exit roads.

5.3.5 灭火救援设施（表 5.3.5-1、表 5.3.5-2）

5.3.5 Fire Fighting and Rescue Facilities（Tab.5.3.5-1、Tab.5.3.5-2）

【防火规】7.1 条：消防车道的设置　　　　　　　　　　　表 5.3.5-1
[Code for fire protection design of buildings] 7.1: Layout of Fire Lanes　　Tab. 5.3.5-1

序号 No	类别 Category	设置要求 Arrangement requirements
1	道路中心线间的距离 Distance between the road centerlines	街区内的道路应考虑消防车的通行，道路中心线间的距离不宜大于 160m For the roads in blocks, the passage of fire trucks shall be considered, and the distance between the road centerlines shall be appropriately not greater than 160m
2	当建筑物沿街道部分的长度大于 150m 或总长度大于 220m 时 When the length of buildings along urban roads is greater than 150m or the total length is greater than 220m	应设置穿过建筑物的消防车道。确有困难时，应设置环形消防车道 The fire lane penetrating through the building shall be arranged. If it is difficult to do so, the annular fire lane shall be arranged
3	高层民用建筑，超过 3000 个座位的体育馆，超过 2000 个座位的会堂，占地面积大于 3000m^2 的商店建筑、展览建筑等单、多层公共建筑 High-Rise civil buildings, gymnasiums with more than 3000 seats, assembly halls with more than 2000 seats, and single- or multi-storey public buildings such as shop buildings, exhibition buildings, etc., with the area greater than 3000m^2	应设置环形消防车道，确有困难时可沿建筑的两个长边设置消防车道；对于住宅建筑和山坡地或河道边临空建造的高层建筑，可沿建筑的一个长边设置消防车道，但该长边所在建筑立面应为消防车登高操作面 The annular fire lane shall be arranged. If it is difficult to do so, the fire lanes can be arranged along the two long sides of buildings; For the residential buildings and the high-rise buildings built on hillsides or riversides, the fire lane can be arranged along one long side of the building, but the facade of the building where the long side is located shall be the ascending operation surface for fire trucks
4	高层厂房，占地面积大于 3000m^2 的甲、乙、丙类厂房和占地面积大于 1500m^2 的乙、丙类仓库 High-rise plant buildings, and Classes A, B and C plant buildings with the floor area greater than 3000m^2 and Classes B and C warehouses with the floor area greater than 1500m^2	应设置环形消防车道，确有困难时，应沿建筑物的两个长边设置消防车道 The annular fire lane shall be arranged. If it is difficult to do so, the fire lanes shall be arranged along the two long sides of building
5	有封闭内院或天井的建筑物，当内院或天井的短边长度大于 24m 时 In buildings with closed inner courtyards or patios, when the length of the short side of the inner courtyard or patio is greater than 24m	宜设置进入内院或天井的消防车道 The fire lane into the inner courtyard or patio shall be appropriately arranged
6	当该建筑物沿街时 When the building is along the street	应设置连通街道和内院的人行通道（可利用楼梯间），其间距不宜大于 80m The pedestrian passage connecting with the street or the inner courtyard shall be arranged (the staircase can be used), and the spacing shall be appropriately not greater than 80m
7	在穿过建筑物或进入建筑物内院的消防车道两侧 Both sides of the fire lane penetrating through the building or entering the inner courtyard of the building	不应设置影响消防车通行或人员安全疏散的设施 The facilities affecting the passage of fire trucks or safe evacuation of personnel shall not be arranged

续表

序号 No	类别 Category	设置要求 Arrangement requirements
8	供消防车取水的天然水源和消防水池应设置消防车道 For the natural water source and fire pool for water intake of fire trucks, the fire lane shall be arranged	消防车道的边缘距离取水点不宜大于2m The distance from the edge of the fire lane to the water intake point shall be appropriately not greater than 2m
9	车道的净宽度和净空高度 Net width and clearance height of the lane	均不应小于4.0m Not less than 4.0m
10	转弯半径 Turning radius	8～12m 8～12m
11	消防车道与建筑之间 Between the fire lane and the building	不应设置妨碍消防车操作的树木、架空管线等障碍物 The obstacles such as trees, overhead pipelines, etc., hindering the operation of fire trucks shall not be arranged
12	消防车道靠建筑外墙一侧的边缘距离 Distance to the edge of one side of the fire lane near the exterior wall of the building	距离建筑外墙不宜小于5m The distance to the exterior wall of the building shall be appropriately not less than 5m
13	消防车道的坡度 Slope gradient of the fire lane	不宜大于8% Appropriately not greater than 8%
14	环形消防车道 Annular fire lane	至少应有两处与其他车道连通 Connected with other lanes at two places at least
15	尽头式消防车道应设置回车道或回车场 For the stub-end fire lane, the turnaround lane or turnaround yard shall be arranged	回车场的面积不应小于12m×12m；对于高层建筑，不宜小于15m×15m；供重型消防车使用时，不小于18m×18m The area of the turnaround yard shall be not less than 12m×12m; For high-rise buildings, the area shall be appropriately not less than 15m×15m; For heavy-duty fire trucks, the area shall be appropriately not less than 18m×18m
16	消防车道的路面、救援操作场地、消防车道和救援操作场地下面的管道和暗沟等承载力 Bearing capacity of the pipelines and gullies below the pavement of the fire lane, rescue operation site, fire lane and rescue operation site	登高平台车，举高喷射消防车，抢险救援车2.6～4.5t Ascending platform trucks, water tower fire trucks, emergency rescue vehicles 2.6～4.5t 消防通信指挥车2.16～3.23t Fire communication command vehicles 2.16～3.23t 火场供给消防车2.02～5.02t Supply fire trucks on the fire site 2.02～5.02t 供水车3.15t Water supply trucks 3.15t
17	消防车道不宜与铁路正线平交，确需平交时 The fire lane shall not be level-crossed with the railway, and if level-crossing is indeed necessary	应设置备用车道，且两车道的间距不应小于一列火车的长度 The spare lane shall be arranged, and the spacing of two lanes shall be not less than the length of a train

5 总平面设计 Site Planning

【防火规】7.2 条：消防车登高操作救援场地和入口

[Code for fire protection design of buildings] 7.2: Ascending Rescue Sites and Entrances for Fire Trucks

表 5.3.5-2 / Tab.5.3.5-2

	适用条件 Applicable conditions	高层建筑的设置要求 Arrangement requirements for high-rise buildings
消防车登高操作救援场地和入口 Ascending rescue sites and entrances for fire trucks	位置 Position	场地应与消防车道连通，场地靠建筑外墙一侧的边缘距离建筑外墙不宜小于 5m，且不应大于 10m，场地的坡度不宜大于 3% The site shall be connected with the fire lanes, the distance of the edge of one side near the exterior wall of the building to the exterior wall of the building in the site shall be appropriately not less than 5m, and not greater than 10m, and the gradient of sites shall be appropriately not greater than 3% 建筑物与消防车登高操作场地相对应的范围内，应设置直通室外的楼梯或直通楼梯间的入口 The staircases directly to the outdoor or entrances directly to the staircases shall be arranged in the corresponding range of the building and the ascending operation site of the fire trucks 场地与厂房、仓库、民用建筑之间不应设置妨碍消防车操作的树木、架空管线等障碍物和车库出入口 The obstacles such as trees, overhead pipelines, etc., hindering the operation of fire trucks and the entrances and exits of the garage shall not be arranged between plant buildings, warehouses and civil buildings
	窗口 Windows	厂房、仓库、公共建筑的外墙应在每层的适当位置设置可供消防救援人员进入的窗口 The window for firefighters to enter shall be arranged at appropriate positions on each floor of the exterior walls of plant buildings, warehouses and public buildings 供消防救援人员进入的窗口的净高度和净宽度均不应小于 1.0m，下沿距室内地面不宜大于 1.2m，间距不宜大于 20m 且每个防火分区不应少于 2 个，设置位置应与消防车登高操作场地相对应。窗口的玻璃应易于破碎，并应设置可在室外易于识别的明显标志 The net height and net width of the window for firefighters to enter shall be not less than 1.0m. The distance between the lower edge to the indoor ground shall be appropriately not greater than 1.2m. The space shall be appropriately not greater than 20m, and at least two windows shall be provided in each fire compartment. And the arrangement positions shall correspond to the ascending operation sites for fire trucks. The window glass shall be easy to be broken, and the obvious signs that can be easily identified outdoors shall be arranged
	长度和宽度 Length and width	场地的长度和宽度分别不应小于 15m 和 10m。对于建筑高度大于 50m 的建筑，场地的长度和宽度分别不应小于 20m 和 10m The length and width of the sites shall be not less than 15m and 10m respectively. For buildings higher than 50m, the length and width of sites shall be not less than 20m and 10m respectively

5.3.6 停车场
5.3.6 Parking Areas

1. 停车场设计要求见表 5.3.6-1。

1. Design requirements for parking areas see Tab.5.3.6-1.

停车场设计要求　　　表 5.3.6-1
Design Requirements for Parking Areas　　Tab.5.3.6-1

类别 Category	出入口设置 Arrangement of entrances and exits	要求 Requirements
停车场位置 Positions of parking areas	应符合行车视距的要求 The requirements for driving sight distance shall be met	应右转出入车道【城道规】8.1.8 条 The access to the lane shall be acquired by turning right [Code for transport planning on urban road] 8.1.8
	距离交叉口、桥隧坡道起止线 Distance to the starting and ending lines of the crossing and ramp of bridges and tunnels	50m【09 措施】4.5.3 条 50m [Measures 2009] 4.5.3
	城市主干道 Trunk roads	70m【09 措施】4.5.3 条 70m [Measures 2009] 4.5.3
	人行横道线、人行地道（包括一道、引桥） Pedestrian crossings and pedestrian underpasses (including approach roads and approach bridges)	5m【09 措施】4.5.3 条 5m [Measures 2009] 4.5.3
	地铁出入口、公交站台边缘 Entrances and exits of railways and edges of the platforms at bus station	15m【09 措施】4.5.3 条 15m [Measures 2009] 4.5.3
	公园、学校、儿童及残疾人使用建筑的出入口 Entrances and exits of parks, schools, and buildings for children and the disabled people	20m【09 措施】4.5.3 条 20m [Measures 2009] 4.5.3
	人行过街天桥、地道和桥梁、隧道引道 Pedestrian overpasses, underpasses, tunnels and bridges, and approach roads of tunnels	50m【停车场规】第四条 50m [Code for Design of Parking Areas] Clause IV
	交叉路口 Intersections	80m【停车场规】第四条 80m [Code for Design of Parking Areas] Clause IV
出入口宽度 Width of the entrances and exits	单向行驶 One-way driving	≥ 4m【车库建规】4.2.4 条 ≥ 4m [Code for design of parking garage buildings] 4.2.4
	双向行驶 Two-way driving	≥ 7m【车库建规】4.2.4 条 ≥ 7m [Code for design of parking garage buildings] 4.2.4
出入口间距 Entrance and exit spacing	机动车出入口 Exits and entrances for motorized vehicles	≥ 15m【车库建规】4.2.2 条 ≥ 15m [Code for design of parking garage buildings] 4.2.2

2. 机动车库出入口和车道数量见表5.3.6-2。

2.Quantities of entrances, exits and lanes of the motor garage see Tab.5.3.6-2.

【车库建规】4.2.6条：机动车库出入口和车道数量　　　　表 5.3.6-2

[Code for design of parking garage building] 4.2.6:
Quantities of Entrances, Exits and Lanes of the Motor Garage　　Tab.5.3.6-2

出入口和车道数量 Number of entrances, exits and lanes 停车当量 Parking equivalent 规模 Scale	特大型 Extra large	大型 Large		中型 Medium		小型 Small	
	>1000	501~1000	301~500	101~300	51~300	25~50	<25
机动车出入口数量 Number of entrances and exit of garages	≥3	≥2		≥2	≥1	≥1	
非居住建筑出入口车道数量 Number of lanes at entrances and exits of non-residential buildings	≥5	≥4	≥3	≥2		≥2	≥1
居住建筑出入口车道数量 Number of lanes at entrances and exits of residential buildings	≥3	≥2	≥2	≥2		≥2	≥1

3.【车库建规】3.2.13条：车库总平面内宜设置电动车辆的充电设施。

3. [Code for design of parking garage building] 3.2.13: In the general plane of the garage, the charging facilities for electrical vehicles shall be appropriately arranged.

5.3.7　机动车库停车场的要求

5.3.7　Requirements for Parking Areas in Motor Garages

5.3.7.1　【车库建规】4.1.1条：机动车库应根据停放车辆的设计车型外廓尺寸进行设计。机动车设计车型的外廓尺寸可按表5.3.7-1条取值。

5.3.7.1　[Code for design of parking garage building] 4.1.1: The garages shall be designed in accordance with the external dimensions of design vehicle types of vehicles parking. For the external dimensions of the design vehicle types of motorized vehicles, the values can be determined as per Tab. 5.3.7-1.

【车库建规】表 4.1.1 条：设计车型外廓尺寸　　　　表 5.3.7-1

[Code for design of parking garage building] Tab. 4.1.1: External Dimensions of Design Vehicle Types　　Tab. 5.3.7-1

设计车型尺寸 Dimensions of design vehicle type		外廓尺寸（m） External dimensions（m）		
		总长 Total length	总宽 Total width	总高 Total height
微型车 Minicars		3.80	1.60	1.80
小型车 Compact vehicles		4.80	1.80	2.00
轻型车 Light-duty vehicles		7.00	2.25	2.75
中型车 Medium-duty vehicles	客车 Passenger vehicles	9.00	2.50	3.20
	货车 Trucks	9.00	2.50	4.00
大型车 Heavy-duty vehicles	客车 Passenger vehicles	12.00	2.50	3.50
	货车 Trucks	11.50	3.50	4.00

5.3.7.2 机动车换算当量系数见表 5.3.7-2。

5.3.7.2 Conversion Equivalent Coefficients for Motorized Vehicles see Tab. 5.3.7-2.

【车库建规】表 4.1.2：机动车换算当量系数　　　　表 5.3.7-2

[Code for design of parking garage building] Tab. 4.1.2: Conversion Equivalent Coefficients for Motorized Vehicles　　Tab. 5.3.7-2

车型 Vehicle type	微型车 Minicars	小型车 Compact vehicles	轻型车 Light-duty vehicles	中型车 Medium-duty vehicles	大型车 Heavy-duty vehicles
换算系数 Conversion coefficient	0.7	1.0	1.5	2.0	2.5

5.3.7.3 非机动车：

5.3.7.3 Non-Motorized Vehicles：

1.【车库建规】6.1.1 条：非机动车设计车型的外廓尺寸可按表 5.3.7.3-1 的规定取值。

1. [Code for design of parking garage building] 6.1.1: The external dimensions of the design

vehicle types of non-motorized vehicles can be determined as per the provisions in Tab. 5.3.7.3-1.

【车库建规】表 6.1.1：非机动车设计车型外廓尺寸　　　表 5.3.7.3-1

[Code for design of parking garage building] Tab. 6.1.1: External Dimensions of Design Vehicle Types for Non-Motorized Vehicles　　Tab. 5.3.7.3-1

车型 Vehicle type	几何尺寸 Geometric dimensions	车辆几何尺寸（m） Geometric dimensions of vehicle (m)		
		长度 Length	宽度 Width	高度 Height
自行车 Bicycles		1.90	0.60	1.20
三轮车 Tricycles		2.50	1.20	1.20
电动自行车 Electric bicycles		2.00	0.80	1.20
机动轮椅车 Motorized wheelchairs		2.00	1.00	1.20

2.【车库建规】6.1.2 条：非机动车及二轮摩托车应以自行车为计算当量进行停车当量的换算，且车辆换算的当量系数应符合表 5.3.7.3-2 的规定。

2. [Code for design of parking garage buildings] 6.1.2: For non-motorized vehicles and two-wheeled motorcycles, the conversion of the parking equivalent shall be carried out as per the calculated equivalent for bicycles, and the equivalent coefficient for vehicle conversion shall conform to the provisions in Tab. 5.3.7.3-2.

【车库建规】表 6.1.2：非机动车及二轮摩托车车辆换算当量系数　　　表 5.3.7.3-2

[Code for design of parking garage building] Tab. 6.1.2: Conversion Equivalent Coefficient for Non-Motorized Vehicles and Two-Wheeled Motorcycles　　Tab. 5.3.7.3-2

车型 Vehicle type	非机动车 Non-motorized vehicle				二轮摩托车 Two-wheeled Motorcycle
	自行车 Bicycles	三轮车 Tricycles	电动自行车 Electric bicycles	机动轮椅车 Motorized wheelchair	
换算当量系数 Conversion equivalent coefficient	1.0	3.0	1.2	1.5	1.5

5.4 广场、商业步行区及室外活动运动场

5.4 Squares, Commercial Walking Areas and Outdoor Activity Venues

5.4.1 【通则】4.1.5 条：城市广场出入口位置应符合下列要求。

5.4.1 [Code for design of civil buildings] 4.1.5: City plaza access location should meet the following requirements.

1. 基地机动车出入口位置：

1. Base motor vehicle access location：

1）与大中城市主干道交叉口的距离，自道路红线交叉点量起不应小于 70m；

1）the distance from the intersection of the main road of large and medium-sized cities should not be less than 70m from the intersection of the road red line;

2）与人行横道线、人行过街天桥、人行地道（包括引道、引桥）的最边缘线不应小于 5m；

2）The most marginal line with the crosswalk line, pedestrian overpass, pedestrian tunnel (including lead road, lead bridge) should not be less than 5m;

3）距地铁出入口、公共交通站台边缘不应小于 15m；

3）It should not be less than 15m from the subway access and public transport platform side;

4）距公园、学校、儿童及残疾人使用建筑的出入口不应小于 20m；

4）The access of buildings used by parks, schools, children and persons with disabilities should not be less than 20m;

5）当基地道路坡度大于 8% 时，应设缓冲段与城市道路连接；

5）When the base road slope is greater than 8%, a buffer section should be arranged to connect with the urban roads;

6）与立体交叉口的距离或其他特殊情况，应符合当地城市规划行政主管部门的规定。

6）The distance from the stereo intersection or other special circumstances shall conform to the regulations of the local urban planning administrative department.

2. 大型、特大型的文化娱乐、商业服务、体育、交通等人员密集建筑的基地应符合下列规定：

2. Bases of large or oversized cultural entertainment, commercial services, sports, transportation and other densely populated buildings should meet the following requirements:

1）基地应至少有一面直接邻接城市道路，该城市道路应有足够的宽度，以减少人员

疏散时对城市正常交通的影响；

1）The base shall have at least one side directly connected to the urban road, and the urban road should be sufficiently wide to reduce the impact of evacuation on normal urban traffic;

2）基地沿城市道路的长度应按建筑规模或疏散人数确定，并至少不小于基地周长的 1/6；

2）The length of the base along the city road shall be determined by the size of the building or the number of evacuees, and at least not less than 1/6 of the circumference of the base;

3）基地应至少有两个或两个以上不同方向通向城市道路的（包括以基地道路连接的）出口；

3）The base shall have at least two or more than two exits in different directions to urban roads（including those connected by base roads）；

4）基地或建筑物的主要出入口，不得和快速道路直接连接，也不得直对城市主要干道的交叉口；

4）The main access of the base or building shall not be directly connected to the fast road, nor shall it direct to the intersection of the main trunk roads of the city;

5）建筑物主要出入口前应有供人员集散用的空地，其面积和长宽尺寸应根据使用性质和人数确定；

5）The main access of the building shall be open space for personnel distribution, the size and length and width of which shall be determined according to the nature and number of persons used;

6）绿化和停车场布置不应影响集散空地的使用，并不宜设置围墙、大门等障碍物。

6）Greening and parking lot layout should not affect the use of distributed open space, and it is not appropriate to set up walls, gates and other obstacles.

5.4.2 商业步行街应符合表 5.4.2 要求。

5.4.2 Commercial Pedestrian Streets should meet the following requirements（Tab.5.4.2）.

【防火规】5.3.6 条、【09 措施】5.2 条：商业步行街　　　　　表 5.4.2
[Code for fire protection design of building] 5.3.6, [Measures 2009] 5.2:
Commercial Pedestrian Streets　　　　　Tab.5.4.2

两侧建筑间距和总长度 Spacing and overall length of the buildings on both sides	不小于规范对相应高度建筑的防火间距要求且不应小于 9m。步行街的端部在各层均不宜封闭，确需封闭时，应在外墙上设置可开启的门窗，且可开启门窗的面积不应小于该部位外墙面积的一半。步行街的长度不宜大于 300m The fire separation of the buildings with the corresponding height shall be not less than 9m. The ends of the pedestrian street shall not be closed on all the floors. If they must be closed, openable doors and windows shall be arranged in the exterior walls, and the area of the openable doors and windows shall be not less than half of the area of the exterior wall at this position. The length of the pedestrian street shall be appropriately not greater than 300m

续表

商铺建筑面积 Building area of shops	不大于 300m² Not greater than 300m²
步行街两侧的建筑为多个楼层设置回廊或挑檐时 When corridors or overhanging eaves are arranged on multiple floors in the buildings on both sides of the pedestrian street	出挑宽度不应小于 1.2m The cantilever width shall be not less than 1.2m
步行街两侧的商铺在上部各层需设置回廊和连接天桥时 When corridors or connecting overpasses shall be arranged on all the upper floors of the stores on both sides of the pedestrian street	应保证步行街上部各层楼板的开口面积不应小于步行街地面面积的 37%，且开口宜均匀布置 The opening area of the slabs of all the upper floors on the pedestrian street shall be assured not less than 37% of the area of the pedestrian street, and the openings shall be evenly arranged
步行街两侧建筑内的疏散楼梯 Evacuation staircases in the buildings on both sides of the pedestrian street	宜直通室外，确有困难时，可在首层直接通至步行街；首层商铺的疏散门可直接通至步行街，步行街内任一点到达最近室外安全地点的步行距离不应大于 60m The staircase shall be directly connected to the outdoor. If it is difficult to do so, the staircase shall be directly connected to the pedestrian street on the first floor; The evacuation doors of the shops on the first floor can be directly connected to the pedestrian street, and the walking distance from any point in the pedestrian street to the nearest outdoor safe place shall be not greater than 60m
步行街两侧建筑二层及以上各层商铺的疏散门至该层最近疏散楼梯口或其他安全出口的直线距离 Straight-line distance from the evacuation doors of the stores on the second floor and the above floors of the buildings on both sides of the pedestrian street to the nearest evacuation stairway or other safety exit on the floor	不应大于 37.5m Not greater than 37.5m
步行街的顶棚下檐距地面的高度 Height of the lower edge of the ceiling of the pedestrian street to the ground	不应小于 6.0m Not less than 6.0m
道路要求 Road requirements	道路应满足送货车、清扫车和消防车通行的要求，道路宽度可采用 10～15m，每 500m 宜提供一处可供人们停留休憩的室外空间或配置小型广场 The roads shall meet the access requirements for delivery vehicles, sweepers and fire trucks. For the width of the roads, 10-15m can be selected. And one outdoor space for people to stay or rest or small squares arranged shall be provided every 500m
出口间距 Spacing between the exits	紧急安全疏散出口间隔距离不得大于 160m The separation distance of the emergency exits shall not be greater than 160m
与城市次干路的距离 Distance to urban secondary trunk roads	不宜大于 200m，步行出口距公共交通站的距离不宜大于 100m The distance shall be appropriately not greater than 200m and the distance from pedestrian exit to the public transport station shall be appropriately not greater than 100m

对周围出入口的要求 Requirements for the surrounding entrances and exits	附近应有相应规模的机动车、非机动车停车场、库,其距步行区进出口距离不宜大于100m The motorized and non-motorized vehicle parking areas and garages in corresponding scales shall be arranged in the vicinity, and the distance from it to the entrance and outlet of the walking area shall be appropriately not greater than 100m
卫生设施设置要求 Arrangement requirements of sanitary facilities	垃圾存放间、转运站及公厕宜设在建筑物内或邻次要道路 The garbage storage rooms, transfer stations and public toilets shall be appropriately arranged in the building or near the secondary road
无障碍设计 Accessibility design	应进行无障碍设计 The accessibility design shall be made

5.5 管线综合

5.5 Integrated Layout of Pipelines

5.5.1 【09措施】6.1.1条：场地内各种管线需与城市管线衔接，其中，雨水、污水管线标高要与城市相关管线标高协调。

5.5.1 [Measures 2009] 6.1.1: When various pipelines in the site shall be connected with the urban pipelines, coordination shall be made between the elevations of the rainwater and sewage pipelines and the elevations of the relevant urban rainwater and sewage pipelines.

5.5.2 【09措施】6.1.2条：管线布置应满足安全使用要求，并综合考虑其与建筑物、道路、环境相互关系和彼此间可能产生的影响。

5.5.2 [Measures 2009] 6.1.2: The pipeline layout shall meet the safety application requirements, and overall consideration shall be given to the interrelationship between buildings, roads, and the environment of mutual and the possible effects on each other.

5.5.3 【09措施】6.1.3条：管线走向宜与主体建筑、道路及相邻管线平行。地下管线应从建筑物向道路方向由浅至深敷设。

5.5.3 [Measures 2009] 6.1.3: The pipeline trend shall be appropriately parallel with the main buildings, roads and adjacent pipelines. The underground pipelines shall be laid from the shallow layer to the deep layer in the direction from the building to the road.

工程管线交叉时的最小垂直净距见表5.5。

Vertical distances between crossing pipelines see Tab.5.5.

【城管线规】4.1.14 条：工程管线交叉时的最小垂直净距（m） 表 5.5

[Code for design of urban pipelines] 4.1.14: Vertical Distances between Crossing Pipelines（m） Tab.5.5

序号 No	管线名称 Name of pipeline		给水管线 Water supply pipeline	污水、雨水管线 Sewage and rainwater drainage pipelines	热力管线 Heating pipeline	燃气管线 Gas pipeline	通信管线 Telecommunications pipeline		电力管线 Power pipeline		再生水管线 Reclaimed water pipeline
							直埋 Direct burial	保护管及通道 Protection tabes and passageway	直埋 No Protection (Direct-bury)	保护管 With Protection	
1	给水管线 Water supply pipeline		0.15								
2	污、雨水排水管线 Sewage and rainwater drainage pipelines		0.40	0.15							
3	热力管线 Heating pipeline		0.15	0.15	0.15						
4	燃气管线 Gas pipeline		0.15	0.15	0.15	0.15					
5	通信管线 Telecommunications pipeline	直埋 No Protection (Direct bury)	0.50	0.50	0.25	0.50	0.25	0.25			
		保护管、通道 Protection pipe trench	0.15	0.15	0.25	0.15	0.25	0.25			
6	电力管线 Power pipeline	直埋 Direct burial	0.50*	0.50*	0.50*	0.50*	0.50*	0.50*	0.50*	0.25	
		保护管 Protection pipe	0.25	0.25	0.25	0.15	0.25	0.25	0.25	0.25	
7	再生水管线 Reclaimed water pipeline		0.50	0.40	0.15	0.15	0.15	0.15	0.50*	0.25	0.15
7	管沟（foundation bottom） Ditch（foundation bottom）		0.15	0.15	0.15	0.15	0.25	0.25	0.50*	0.25	0.15
8	涵洞（基底） Culvert（foundation bottom）		0.15	0.15	0.15	0.15	0.25	0.25	0.50*	0.25	0.15
9	电车（轨底） Trolleys（track bottom）		1.00	1.00	1.00	1.00	1.00	1.00	1.00	1.00	1.00
10	铁路（轨底） Railways（track bottom）		1.00	1.20	1.20	1.20	1.50	1.50	1.00	1.00	1.00

注：1.* 用隔板分隔时不得小于 0.25m；
Notes: 1. *The partition by plates shall not be less than 0.25cm;
2. 燃气管线采用聚乙烯管材时，燃气管线与热力管线的最小垂直净距应按现行行业标准《聚乙烯燃气管道工程技术标准》CJJ 63 执行；
2.When used in PE materials, the minimum vertical net distance between gas pipelines and thermal pipelines shall be implemented in accordance with the current industry standard CJJ63 of *Technical Standard for Polyethylene Gas Pipeline Engineering*;
3. 铁路为时速大于等于 200km/h 客运专线时，铁路（轨底）与其他管线最小垂直净距为 1.50m。
3.As passenger dedicated lines which speed is greater than or equal to 200km/h, the minimum vertical net distance between railway (rail bottom) and other pipelines shall be 1.50m.

6

建筑防火一般规定
Fire Protection Design for Buildings

6.1 建筑分类和耐火等级

6.1 Building Classification and Refractory Grade

6.1.1 【防火规】5.1.1 条：民用建筑根据其建筑高度和层数可分为单、多层民用建筑和高层民用建筑。高层民用建筑根据其建筑高度、使用功能和楼层的建筑面积可分为一类和二类。民用建筑的分类应符合表 6.1.1 的规定。

6.1.1 [Code for fire protection design of buildings] 5.1.1: civil buildings can be divided into single, multi-storey civil buildings and high-rise civil buildings according to its building height and the number of floors. High-rise civil buildings can be divided into the Category I and Category II according to their building height, use function and floor building area. The classification of civil buildings shall conform to the provisions of Tab. 6.1.1.

6.1.2 【防火规】5.1.2 条：民用建筑的耐火等级可分为一、二、三、四级。

6.1.2 [Code for fire protection design of buildings] 5.1.2: civil construction of the refractory grade can be divided into I, II, III and IV.

6.1.3 【防火规】5.1.3 条：民用建筑的耐火等级应根据其建筑高度、使用功能、重要性和火灾扑救难度等确定，并应符合下列规定。

6.1.3 [Code for fire protection design of buildings] 5.1.3: The refractory grade of civil buildings shall be determined according to their building height, functions of use, importance and difficulty of fire fighting, and shall conform to the following provisions.

1. 地下或半地下建筑（室）和一类高层建筑的耐火等级不应低于一级；

1. The refractory grade of underground or semi-underground buildings (rooms) and high-rise buildings in Grade I should not be lower than Grade I;

2. 单、多层重要公共建筑和二类高层建筑的耐火等级不应低于二级。

2. The refractory grade of single-storey, multi-storey important public buildings and high-rise buildings in Grade II should not be less than Grade II.

6.1.4 【防火规】5.1.4 条：建筑高度大于 100m 的民用建筑，其楼板的耐火极限不应低于 2.00h。一、二级耐火等级建筑的上人平屋顶，其屋面板的耐火极限分别不应低于 1.50h 和 1.00h。

6.1.4 [Code for fire protection design of buildings] 5.1.4: For civil buildings with building height greater than 100m, the fire resistance limit of their floors should not be less than 2.00h.

6 建筑防火一般规定 Fire Protection Design for Buildings

【防火规】表 5.1.1：民用建筑的分类   表 6.1.1
[Code for fire protection design of buildings] Tab. 5.1.1: Classification of Civil Buildings   Tab. 6.1.1

名称 Name	高层民用建筑 High floor civil architecture		单、多层民用建筑 Single and multi-storey civil buildings
	一类 Category one	二类 Category two	
住宅建筑 Residential buildings	建筑高度大于 54m 的住宅建筑（包括设置商业服务网点的住宅建筑） Residential buildings with a height of more than 54m（including residential catch with commercial service outlets）	建筑高度大于 27m，但不大于 54m 的住宅建筑（包括设置商业服务网点的住宅建筑） Residential buildings with a height greater than 27m but not greater than 54m（including residential buildings with commercial service outlets）	建筑高度大于 27m 的住宅建筑（包括设置商业服务网点的住宅建筑） Residential buildings with less than 27m builders（including residential buildings with commercial service outlets）
公共建筑 Public buildings	1. 建筑高度大于 50m 的公共建筑； 1.Public buildings with a building height greater than 50m; 2. 建筑高度 24m 以上部分任一楼层建筑面积大于 1000m² 的商店、展览、电信、邮政、财贸金融建筑和其他多种功能组合的建筑； 2.Building height more than 24m parts of any floor building area larger than 1000m² stores, exhibitions, telecommunications, postal, vocational financial buildings and a variety of other functional combinations of buildings; 3. 医疗建筑、重要公共建筑； 3. Medical buildings, important public buildings; 4. 省级及以上的广播电视和防灾指挥调度建筑、网局级和省级电力调度建筑； 4. Provincial and higher radio and television and Disaster prevention Command and dispatch buildings, network Bureau and provincial power dispatching buildings; 5. 藏书超过 100 万册的图书馆、书库 5. Libraries, library books with a collection of more than 1 million	除一类高层公共建筑外的其他高层公共建筑 Other high-rise public buildings in addition to a class of high-rise public buildings	1. 建筑高度大于 24m 的单层公共建筑； 1. Single-storey public buildings with a building height greater than 24m; 2. 建筑高度不大于 24m 的其他公共建筑 2. Other public buildings with a building height not greater than 24m

注：1. 表中未列入的建筑，其类别应根据本表类比确定。
Notes: 1.The categories of buildings not included in the Table should be determined by analogy with this Table.
2. 除本规范另有规定外，宿舍、公寓等非住宅类居住建筑的防火要求，应符合本规范有关公共建筑的规定。
2.Except as otherwise provided in this code, the fire prevention requirements of residential buildings such as dormitories and apartments shall conform to the provisions of this code relating to public buildings.
3. 除本规范另有规定外，裙房的防火要求应符合本规范有关高层民用建筑的规定。
3.Except as otherwise provided in this code, the fire protection requirements of the skirt room shall conform to the provisions of this code relating to high-rise civil buildings.

The flat roof of buildings with refractory Grade I or II, the refractory limit of its roof panel should not be less than 1.50h and 1.00h respectively.

6.2 防火分区
6.2 Fire Compartments

6.2.1 【防火规】5.3.1 条：不同耐火等级建筑的允许建筑高度或层数、防火分区最大允许建筑面积应符合表 6.2.1 的规定。

6.2.1 [Code for fire protection design of buildings] Tab. 5.3.1: For buildings with different fire resistance ratings, the allowable building heights, floor numbers or maximum allowable building areas of the fire compartments shall meet the provisions in Tab. 6.2.1.

【防火规】表 5.3.1：不同耐火等级建筑的允许建筑高度或层数、防火分区最大允许建筑面积　表 6.2.1

[Code for fire protection design of buildings] Tab. 5.3.1: Allowable Building Heights, Floor Numbers or Maximum Allowable Building Areas of the Fire Compartments of the Buildings with Different Fire Resistance Ratings　Tab. 6.2.1

名称 Name	耐火等级 Fire resistance rating	允许建筑高度或层数 Allowable building heights or floor numbers	防火分区的最大允许建筑面积（m²） Maximum allowable building area of fire compartment（m²）	备注 Remarks
高层民用建筑 High-rise civil buildings	一、二级 Grade I and grade II	按本规范第 5.1.1 条确定 To be determined in accordance with clause 5.1.1 of the code	1500	对于体育馆、剧场的观众厅，防火分区的最大允许建筑面积可适当增加 As for the auditoriums of gymnasiums and theaters, the maximum allowable building area of fire compartments can be properly increased
单、多层民用建筑 Single-storey and multi-storey civil buildings	一、二级 Grade I and grade II	按本规范第 5.1.1 条确定 To be determined in accordance with clause 5.1.1 of the code	2500	
	三级 Grade III	5 层 5 storeys	1200	—
	四级 Grade IV	2 层 2 storeys	600	—

续表

名称 Name	耐火等级 Fire resistance rating	允许建筑高度或层数 Allowable building heights or floor numbers	防火分区的最大允许建筑面积（m²） Maximum allowable building area of fire compartment（m²）	备注 Remarks
地下或半地下建筑（室） Underground or semi-underground building（room）	一级 Grade I	—	500	设备用房的防火分区最大允许建筑面积不应大于1000m² The maximum allowable building area of the fire compartment of equipment rooms shall be not greater than 1000m²

注：1. 表中规定的防火分区最大允许建筑面积，当建筑内设置自动灭火系统时，可按本表的规定增加1.0倍；局部设置时，防火分区的增加面积可按该局部面积的1.0倍计算。
Notes: 1. When the automatic extinguishing system is arranged in the building, the maximum allowable building area of the fire compartment specified in the table can be increased by 1.0 time as per the provisions in the table; when the automatic extinguishing system is partially arranged, the increased area of the fire compartment can be calculated as per 1.0 time of the partial area.
2. 裙房与高层建筑主体之间设置防火墙时，裙房的防火分区可按单、多层建筑的要求确定。
2. When fire walls are arranged between the podium and the main body of the high-rise building, the fire compartments of the podium can be determined as per the requirements of single-storey and multi-storey buildings.

6.2.2 建筑内设置自动扶梯、敞开楼梯等上、下层相连通的开口及建筑内设置中庭时的设计要求见表6.2.2。

6.2.2 Design Requirements for the openings for escalators, open staircases connecting the upper and lower floors and atriums set in the building see Tab.6.2.2.

【防火规】5.3.2：建筑内设置自动扶梯、敞开楼梯等上、下层相连通的开口及建筑内设置中庭时的设计要求　　　　表6.2.2
[Code for fire protection design of buildings] 5.3.2:
Design Requirements for the Openings for Escalators, Open Staircases Connecting the Upper and Lower Floors and Atriums Set in the Building　　Tab. 6.2.2

建筑内设置自动扶梯、敞开楼梯等上、下层相连通的开口时 When an escalator is set up in the building, an open staircase is arranged, and the opening of the lower layer is connected	建筑内设置自动扶梯、敞开楼梯等上、下层相连通的开口时，其防火分区的建筑面积应按上、下层相连通的建筑面积叠加计算；当叠加计算后的建筑面积大于本规范第5.3.1条的规定时，应划分防火分区 For the openings for escalators and open staircases connecting the upper and lower floors, the building area of the fire compartment shall be calculated through the superposition method as per the building area of the connected upper and lower floors. When the building area calculated through the superposition method is greater than that specified in clause 5.3.1 of the code, the fire compartment shall be divided

续表

建筑内设置中庭时 When the atrium is set up in the building	建筑内设置中庭时，其防火分区的建筑面积应按上、下层相连通的建筑面积叠加计算；叠加计算后的建筑面积大于本规范第5.3.1条的规定时，应符合下列规定： When the atrium is arranged in the building, the building area of the fire compartment shall be calculated through the superposition method as per the building area of the connected upper and lower floors. When the building area calculated through the superposition method is greater than that specified in clause 5.3.1 of the code, the following provisions shall be met:
	1. 与周围连通空间应进行防火分隔：采用防火隔墙时，其耐火极限不应低于1.00h；采用防火玻璃墙时，其耐火隔热性和耐火完整性不应低于1.00h，采用耐火完整性不低于1.00h的非隔热性防火玻璃墙时，应设置自动喷水灭火系统进行保护；采用防火卷帘时，其耐火极限不应低于3.00h，并应符合本规范第6.5.3条的规定；与中庭相连通的门、窗，应采用火灾时能自行关闭的甲级防火门、窗 1.Fire separation shall be made for the space connected to the surroundings: When using fireproof partitions, their fire endurance shall not be less than 1.00 h . When using fireproof glass walls, the fire insulation and fire integrity shall not be less than 1.00h . When using non-heat-shielding fireproof glass walls with fire integrity not less than 1.00h , the automatic sprinkler system shall be arranged for protection. When using the fireproof roller shutters, their fire endurance shall not be less than 3.00h, and the provisions in clause 6.5.3 of the code shall be conformed to. Grade A fireproof doors and windows that can be automatically closed in case of fire shall be adopted as the doors and windows connected with the atrium
	2. 高层建筑内的中庭回廊应设置自动喷水灭火系统和火灾自动报警系统 2. The automatic sprinkler system and automatic fire alarm system shall be arranged in the atrium ambulatories in high-rise buildings
	3. 中庭应设置排烟设施 3. The smoke exhaust facilities shall be arranged in the atrium
	4. 中庭内不应布置可燃物 4. The combustible materials shall not be arranged in the atrium

6.2.3 【防火规】5.3.3条：防火分区之间应采用防火墙分隔，确有困难时，可采用防火卷帘等防火分隔设施分隔。采用防火卷帘分隔时，应符合本规范第6.5.3条的规定。

6.2.3 [Code for fire protection design of buildings] 5.3.3: Firewall partitions shall be separated with fire walls. Where difficulties exist, the fire separation facilities like fireproof roller shutter and the like can be adopted for separation. When the fireproof roller shutters are used for separation, the provisions in clause 6.5.3 of the code shall be conformed to.

6.2.4 一、二级耐火等级建筑内的商店营业厅、展览厅，当设置自动灭火系统和火灾自动报警系统并采用不燃或难燃装修材料时的防火分区面积见表6.2.4。

6.2.4 For the shops, business halls and exhibition halls in Grade I and Grade II fire-resistant buildings, automatic extinguishing system and automatic fire alarm system should be set, and fire compartment area of non-combustible or nonflammable decorative material shall be adopted in Tab. 6.2.4.

【防火规】5.3.4 条: 防火分区面积 [Code for fire protection design of buildings] 5.3.4: Fire Compartment Area	表 6.2.4 Tab. 6.2.4
1. 设置在高层建筑内时 1. When the fire compartments are arranged in high-rise buildings	≤ 4000m²
2. 设置在单层建筑或仅设置在多层建筑的首层内时 2. When the fire compartments are arranged in single-storey buildings or only on the first floors of the multi-storey buildings	≤ 10000m²
3. 设置在地下或半地下时 3. When the fire compartments are arranged in underground or semi-underground buildings	≤ 2000m²

6.3 安全疏散距离

6.3 Safety Evacuation Distance

6.3.1 【防火规】5.5.17.1 条：公共建筑直通疏散走道的房间疏散门至最近安全出口的直线距离不应大于表 6.3.1 的规定。

6.3.1 [Code for fire protection design of buildings] 5.5.17.1: The straight line distance between the straight-through evacuation door of a room and the nearest safe exit through the straight-through evacuation sidewalk of a public building shall not be greater than that specified in Tab. 6.3.1.

【防火规】表 5.5.17：直通疏散走道的房间疏散门至最近安全出口的直线距离（m） 表 6.3.1
[Code for fire protection design of buildings] Tab. 5.5.17: The Straight Distance from the Evacuation Doors of the Rooms of the Straight-through Evacuation Sidewalk to the Nearest Safe Exit (m) Tab. 6.3.1

名称 Name	位于两个安全出口之间的疏散门 Evacuation door between two safety exits			位于袋形走道两侧或尽端的 疏散门 Evacuation door at both sides or pocket-end corridor		
	一、二级 Grade I and Grade II	三级 Grade III	四级 Grade IV	一、二级 Grade I and Grade II	三级 Grade III	四级 Grade IV
托儿所、幼儿园、老年人建筑 Nursery and kindergarten buildings, and buildings for the aged	25	20	15	20	15	10

续表

名称 Name			位于两个安全出口之间的疏散门 Evacuation door between two safety exits			位于袋形走道两侧或尽端的 疏散门 Evacuation door at both sides or pocket-end corridor		
			一、二级 Grade I and Grade II	三级 Grade III	四级 Grade IV	一、二级 Grade I and Grade II	三级 Grade III	四级 Grade IV
歌舞娱乐放映游艺场所 Karaokes, dancing halls, entertainment rooms, video halls, amusement halls, etc.			25	20	15	9	—	—
医疗建筑 Medical buildings	单、多层 Single- or multiple-storey buildings		35	30	25	20	15	10
	高层 High-rise buildings	病房部分 Ward part	24	—	—	12	—	—
		其他部分 Other parts	30	—	—	15	—	—
教学建筑 Teaching buildings	单、多层 Single- or multiple-storey buildings		35	30	25	22	20	10
	高层 High-rise buildings		30	—	—	15	—	—
高层旅馆、展览建筑 High-rise hotels and exhibition buildings			30	—	—	15	—	—
其他建筑 Other buildings	单、多层 Single- or multiple-storey buildings		40	35	25	22	20	15
	高层 High-rise		40	—	—	20	—	—

注：1. 建筑内开向敞开式外廊的房间疏散门至最近安全出口的直线距离可按本表的规定增加 5m。
Notes：1.The straight-line distance from the room evacuation door opening to the open veranda to the nearest exit can be increased by 5m in accordance with the provisions in the table.
2. 直通疏散走道的房间疏散门至最近敞开楼梯间的直线距离，当房间位于两个楼梯间之间时，应按本表的规定减少5m；当房间位于袋形走道两侧或尽端时，应按本表的规定减少 2m。
2. When the room is between two staircases, the straight-line distance from the room evacuation door directly to the evacuation corridor to the nearest open staircase shall be decreased by 5m in accordance with the provisions in the table; and when the room is at both sides or the pocket-end corridor, the straight-line distance shall be decreased by 2m in accordance with the provisions in the table.
3. 建筑物内全部设置自动喷水灭火系统时，其安全疏散距离可按本表的规定增加 25%。
3.When the automatic sprinkler system is arranged in the entire building, the safety evacuation distance can be increased by 25% in accordance with the provisions in the table.

6.3.2 【防火规】5.5.17.3 条：房间内任一点至房间直通疏散走道的疏散门的直线距离，

不应大于表 6.3.1 规定的袋形走道两侧或尽端的疏散门至最近安全出口的直线距离。

6.3.2 [Code for fire protection design of buildings] 5.5.17.3: The straight line distance between any point in the room to the evacuation door of a straight-through evacuation sidewalk should not be greater than the straight line distance of the evacuation door to the nearest safe exit on either side of the bag-shaped sidewalk or the end as regulated in Tab. 6.3.1.

6.3.3 观众厅、展览厅、多功能厅、餐厅、营业厅室内任一点至最近疏散门或安全出口的直线距离见表 6.3.3。

6.3.3 Straight-line distance from any point in the room to the nearest evacuation door or safety exit see Tab. 6.3.3.

【防火规】5.5.17.4 条：观众厅、展览厅、多功能厅、餐厅、营业厅室内任一点至最近疏散门或安全出口的直线距离　　表 6.3.3

[Code for fire protection design of buildings] 5.5.17.4: Straight-line Distance from Any Point in the Room to the Nearest Evacuation Door or Safety Exit　　Tab. 6.3.3

名称 Name	疏散距离（m） Evacuation distance（m）	备注 Remarks
	耐火等级 一、二级 Fire resistance rating grade I and grade II	
观众厅、展览厅、多功能厅、餐厅、营业厅【防火规】5.5.17.4 条 Auditorium, exhibition hall, multi-function hall, dining hall and business hall [Code for fire protection design of buildings] 5.5.17.4	30	设置自动喷水灭火系统时增加 25% 25% is added when the automatic sprinkler system is arranged 当疏散门不能直通室外地面或疏散楼梯间时，应采用长度不大于 10m 的疏散走道通至最近的安全出口 When the evacuation door shall not be directly connected to the outdoor floor or evacuation staircase, the evacuation corridor with the length not greater than 10m shall be adopted to provide access to the nearest safety exit

6.3.4 【防火规】5.5.29 条：住宅建筑的安全疏散距离应符合下列规定。

6.3.4 [Code for fire protection design of buildings] 5.5.17.3: The safe evacuation distance of residential buildings shall comply with the following provisions.

1. 直通疏散走道的户门至最近安全出口的直线距离不应大于表 6.3.4 的规定。

1. The straight distance from the door of the a straight-through evacuation sidewalk to the nearest safe exit should not be greater than the ones given in Tab. 6.3.4.

【防火规】表 5.5.29：住宅建筑直通疏散走道的户门至最近安全出口的直线距离（m） 表 6.3.4

[Code for fire protection design of buildings] Tab. 5.5.29: The Straight Distance from the Door of the a Straight-through Evacuation Sidewalk to the Nearest Safe Exit of Residential Buildings (m) Tab. 6.3.4

住宅建筑类别 Residential Building Categories	位于两个安全出口之间的户门 At the door between two safe exits			位于袋形走道两侧或尽端的户门 Doors located on either side or end of the bag walkway		
	一、二级 Grade I、II	三级 Grade III	四级 Grade IV	一、二级 Grade I、II	三级 Grade III	四级 Grade IV
单、多层 Single, multi-layer	40	35	25	22	20	15
高层 High-rise	40	—	—	20	—	—

注：1. 开向敞开式外廊的户门至最近安全出口的最大直线距离可按本表的规定增加 5m。
Notes: 1.The maximum straight line distance from the door open to the open porch to the nearest safe exit can be increased by 5m in accordance with the provisions of this table.
2. 直通疏散走道的户门至最近敞开楼梯间的直线距离，当户门位于两个楼梯间之间时，应按本表的规定减少 5m；当户门位于袋形走道两侧或尽端时，应按本表的规定减少 2m。
2. If the door is located between the two stairwells, the straight line distance from the door of the straight-through evacuation sidewalk to the nearest straight distance between the staircase should be reduced by 5m according to the provisions of this Table; if the door is located on either side or end of the bag-shaped sidewalk, 2m shall be reduced in accordance with the provisions of this table.
3. 住宅建筑内全部设置自动喷水灭火系统时，其安全疏散距离可按本表的规定增加 25%。
3. If automatic sprinkler systems are arranged in all residential buildings, the safe evacuation distance can be increased by 25% in accordance with the provisions of this table.
4. 跃廊式住宅的户门至最近安全出口的距离，应从户门算起，小楼梯的一段距离可按其水平投影长度的 1.50 倍计算。
4. The distance from the door of the leaping house to the nearest safe exit shall be calculated from the door, and the distance of the small staircase may be 1.50 times of the length of its horizontal projection.

2. 楼梯间应在首层直通室外，或在首层采用扩大的封闭楼梯间或防烟楼梯间前室。层数不超过 4 层时，可将直通室外的门设置在离楼梯间不大于 15m 处。

2. The stairwell should be directly outside the first floor, or in the first floor with an expanded closed staircase or smoke-proof stairwell front room. When the number of floors does not exceeds 4, the door that goes straight outside can be arranged in a distance not less than 15m from the stairwell.

3. 户内任一点至直通疏散走道的户门的直线距离不应大于表 5.5.29 规定的袋形走道两侧或尽端的疏散门至最近安全出口的最大直线距离。

注：跃层式住宅，户内楼梯的距离可按其梯段水平投影长度的 1.50 倍计算。

3. The straight distance of any point in a room to the door of the straight-through evacuation sidewalk should not be greater than the maximum straight line distance between the evacuation

door on either side or end of the bag-shaped sidewalk to the nearest safe exit as Tab. 5. 5. 29.

Note: For leap-storey residential houses, the distance of the indoor staircase can be calculated by 1.50 times of the length of the horizontal projection of its staircases.

6.4 疏散走道

6.4 Evacuation Corridor

6.4.1 【防火规】5.5.18 条：公共建筑内疏散门和安全出口的净宽度不应小于0.90m，疏散走道和疏散楼梯的净宽度不应小于1.10m（表6.4.1）。

6.4.1 [Code for fire protection design of buildings] 5.5.18: The net width of evacuation doors and safety exits in public buildings should not be less than 0.90m, and the net width of evacuation sidewalks and stairs should not be less than 1.10m（Tab. 6.4.1）.

高层公共建筑内楼梯间的首层疏散门、首层疏散外门、疏散走道和楼梯的最小净宽度　　表6.4.1

The minimum net width of the first-floor evacuation door and first-floor exterior evacuation door of the staircase room, evacuation sidewalk and evacuation staircase in high-rise public buildings　　Tab.6.4.1

建筑类别 Building Category	楼梯间的首层疏散门、 首层疏散外门 First-floor evacuation door and first-floor exterior evacuation door of the staircase room	走道 Walkway		疏散楼梯 Evacuation staircase
		单面布房 With rooms arranged on one side	双面布房 With rooms arranged on both sides	
高层医疗建筑 High-Rise medical buildings	1.30	1.40	1.50	1.30
其他高层公共建筑 Other high-rise public buildings	1.20	1.30	1.40	1.20

6.4.2 剧场、电影院、礼堂等场所供观众疏散的所有内门、外门、楼梯和走道的各自总净宽度见表6.4.2。

6.4.2 Total net width of all the interior doors, exterior doors, stairs and walkways for audience evacuation in theater, cinema, auditorium, See Tab.6.4.2.

【防火规】表 5.5.20-1：剧场、电影院、礼堂等场所供观众疏散的所有内门、外门、楼梯和走道的各自总净宽度（m） 表 6.4.2

[Code for fire protection design of buildings] Tab. 5.5.20-1: Total Net Width of All the Interior Doors, Exterior Doors, Stairs and Walkways for Audience Evacuation in Theater, Cinema, Auditorium, and the Like (m)   Tab.6.4.2

观众厅座位数（座） Number of seats in audience halls (seats)			≤ 2500	≤ 1200
耐火等级 Fire resistance rating			一、二级 Grade I and Grade II	三级 Grade III
疏散部位 Evacuation positions	门和走道 Doors and walkways	平坡地面 Flat grounds	0.65	0.85
		阶梯地面 Stepped grounds	0.75	1.00
	楼梯 Staircase		0.75	1.00

6.4.3 体育馆每 100 人所需最小疏散净宽度见表 6.4.3。

6.4.3 Minimum net evacuation width required per 100 people in auditoriums See Tab.6.4.3.

【防火规】表 5.5.20-2：体育馆每 100 人所需最小疏散净宽度（m/100 人） 表 6.4.3

[Code for fire protection design of buildings] Tab. 5.5.20-2: Minimum Net Evacuation Width Required Per 100 People in Auditoriums   Tab.6.4.3

观众厅座位数范围（座） Scope of the number of the seats in audience halls (seats)			3000 ~ 5000	5001 ~ 10000	10001 ~ 20000
疏散部位 Evacuation positions	门和走道 Doors and walkways	平坡地面 Flat grounds	0.43	0.37	0.32
		阶梯地面 Stepped grounds	0.50	0.43	0.37
	楼梯 Staircase		0.50	0.43	0.37

6.4.4 【防火规】5.5.21 条：除剧场、电影院、礼堂、体育馆外的其他公共建筑每层的房间疏散门、安全出口、疏散走道和疏散楼梯的每 100 人最小疏散净宽度见表 6.4.4-1。

6.4.4 [Code for fire protection design of buildings] 5.5.21: Minimum net evacuation width (m/100 people) per 100 people of the evacuation doors, safety exits, evacuation corridors and evacuation staircases for the rooms on each floor of other public buildings except theaters, cinemas, auditoriums and gymnasiums see Tab. 6.4.4-1.

【防火规】表 5.5.21-1：每层的房间疏散门、安全出口、疏散走道和疏散楼梯的每 100 人最小疏散净宽度（m/100 人） 表 6.4.4-1

[Code for fire protection design of buildings] Tab. 5.5.21-1: Minimum Net Evacuation Width（m/100 People）Per 100 People of the Evacuation Doors, Safety Exits, Evacuation Corridors and Evacuation Staircases for the Rooms on Each Floor　　Tab. 6.4.4-1

建筑层数 Building storeys		建筑的耐火等级 Fire resistance rating of the building		
		一、二级 Grade I and Grade II	三级 Grade III	四级 Grade IV
地上楼层 Overground floors	1～2 层 1 to 2 storeys	0.65	0.75	1.00
	3 层 3 storeys	0.75	1.00	—
	≥ 4 层 ≥ 4 storeys	1.00	1.25	—
地下楼层 Underground floor	与地面出入口地面的高差 ΔH ≤ 10m Height difference ΔH between the floors of the entrance and exit ≤ 10m	0.75	—	—
	与地面出入口地面的高差 ΔH > 10m Height difference ΔH between the floors of the entrance and exit > 10m	1.00	—	—

1. 每层的房间疏散门、安全出口、疏散走道和疏散楼梯的各自总净宽度，应根据疏散人数按每 100 人的最小疏散净宽度不小于表 6.4.4-1 的规定计算确定。当每层疏散人数不等时，疏散楼梯的总净宽度可分层计算，地上建筑内下层楼梯的总净宽度应按该层及以上疏散人数最多一层的人数计算；地下建筑内上层楼梯的总净宽度应按该层及以下疏散人数最多一层的人数计算。

1. The respective total net width of the evacuation door, safety exit, evacuation corridor and evacuation staircase for the rooms on each floor shall be determined through calculation in accordance with the number of evacuees (the minimum net evacuation width per 100 people shall be not less than that in Tab. 6.4.4-1. When the number of evacuees of each floor is unequal, the total net width of the evacuation staircases shall be calculated respectively, and the total net width of the staircases of the lower floors in overground buildings shall be calculated as per the maximum number of the evacuees on and above the floor in the building; the total net width of the staircases of the upper floors in underground buildings shall be calculated as per the maximum number of the evacuees on and below the floor in the buildings.

2. 地下或半地下人员密集的厅、室和歌舞娱乐放映游艺场所,其房间疏散门、安全出口、疏散走道和疏散楼梯的各自总净宽度,应根据疏散人数按每100人不小于1.00m计算确定。

2. The respective total net width of the evacuation doors, safety exits, evacuation corridors and evacuation staircases for the rooms of the underground or semi-underground crowded halls, rooms and karaoke, dancing halls, entertainment rooms, video halls, amusement halls, etc., shall be calculated and determined in accordance with the number of evacuees not less than 1.00 m per 100 people.

3. 首层外门的总净宽度应按该建筑疏散人数最多一层的人数计算确定,不供其他楼层人员疏散的外门,可按本层的疏散人数计算确定。

3. The total net width of the first-floor exterior evacuation doors shall be calculated and determined as per the largest number of evacuees on the floor in the building, and the total net width of the exterior doors not for the personnel evacuation of other floors can be calculated and determined as per the number of evacuees on the floor.

4. 歌舞娱乐放映游艺场所中录像厅的疏散人数,应根据厅、室的建筑面积按不小于1.0人/m^2计算;其他歌舞娱乐放映游艺场所的疏散人数,应根据厅、室的建筑面积按不小于0.5人/m^2计算。

4. The number of evacuees in the halls in karaoke rooms, dancing halls, entertainment rooms, video halls, and amusement halls shall be calculated as at least 1.0 person/m^2; the number of the evacuees in other halls in karaoke rooms, dancing halls, entertainment rooms, video halls, and amusement halls shall be calculated as at least 0.5 person/m^2.

5. 有固定座位的场所,其疏散人数可按实际座位数的1.1倍计算。

5. The number of evacuees at places with fixed seats can be calculated as 1.1 times of the actual seats.

6. 展览厅的疏散人数应根据展览厅的建筑面积和人员密度计算,展览厅内的人员密度不宜小于0.75人/m^2。

6. The number of evacuees in exhibition halls shall be calculated in accordance with the building area and personnel density of the exhibition halls, and the personnel density of the exhibition halls shall be appropriately not less than 0.75 person/m^2.

7. 商店的疏散人数应按每层营业厅的建筑面积乘以表6.4.4-2规定的人员密度计算。对于建材商店、家具和灯饰展示建筑,其人员密度可按表6.4.4-2规定值的30%确定。

7. The number of evacuees in stores shall be calculated by multiplying the building area of the business hall in each floor by the personnel density specified in Tab. 6.4.4-2. As for building material stores and furniture and lighting display buildings, the personnel density can be

determined as 30% of the specified values in Tab. 6.4.4-2.

【防火规】表 5.5.21-2：商店营业厅内的人员密度（人 /m²）　　　表 6.4.4-2
[Code for fire protection design of buildings] Tab. 5.5.21-2:
Personnel Density of Shops and Business Halls (person/m²)　　　Tab. 6.4.4-2

楼层位置 Positions of the floors	地下第二层 Second floor underground	地下第一层 First floor underground	地上第一、二层 First and second floors overground	地上第三层 Third floor overground	地上第四层及以上各层 Fourth floor overground and above
人员密度 Personnel density	0.56	0.60	0.43 ~ 0.60	0.39 ~ 0.54	0.30 ~ 0.42

6.4.5　【防火规】5.5.30 条：住宅建筑的户门、安全出口、疏散走道和疏散楼梯的各自总净宽度应经计算确定，且户门和安全出口的净宽度不应小于 0.90m，疏散走道、疏散楼梯和首层疏散外门的净宽度不应小于 1.10m。建筑高度不大于 18m 的住宅中一边设置栏杆的疏散楼梯，其净宽度不应小于 1.0m。

6.4.5　[Code for fire protection design of buildings] 5.5.30: The total net width of the doors, safe exits, evacuation sidewalks and evacuation staircases of residential buildings in shall be determined by calculation, and the net width of the door and safe exit shall not be less than 0. 90m, the net width of evacuation sidewalk, evacuation staircase and the first-floor evacuation outer door should not be less than 1.10m. An evacuation staircase with railings on one side of a house with a height of not more than 18m should not have a net width of not less than 1.0m.

7

地下室和半地下室
Basements and Semi-basements

7.1 术语

7.1 Terminology

1.【防火规】2.1.7 条：地下室

1. [Code for fire protection design of buildings] 2.1.7: Basements

房间地面低于室外设计地面的平均高度大于该房间平均净高 1/2 者。

Rooms in which the floor height is less than the average height of the outdoor design floor and greater than 1/2 of the average net height of the rooms.

2.【防火规】2.1.6 条：半地下室

2. [Code for fire protection design of buildings] 2.1.6: Semi-basements

房间地面低于室外设计地面的平均高度大于该房间平均净高 1/3，且不大于 1/2 者。

Rooms in which the floor height is less than the average height of the outdoor design floor, and greater than 1/3 but not greater than 1/2 of the average net height of the rooms.

3.【汽防规】2.0.1 条：汽车库

3. [Code for fire protection design of garage, motor repair shop and parking area] 2.0.1: Garages

用于停放由内燃机驱动且无轨道的客车、货车、工程车等汽车的建筑物。

Buildings for the motor parking, such as passenger vehicles, trucks, and engineering vehicles without rails, driven by internal combustion engines.

4.【汽防规】2.0.2 条：修车库

4. [Code for fire protection design of garage, motor repair shop and parking area] 2.0.2: Motor repair shops

用于保养、修理由内燃机驱动且无轨道的客车、货车、工程车等汽车的建（构）筑物。

Buildings (structures) used for maintaining and repairing motors, such as passenger vehicles, trucks, and engineering vehicles without rails, driven by internal combustion engines.

5.【汽防规】2.0.4 条：地下汽车库

5. [Code for fire protection design of garage, motor repair shop and parking area] 2.0.4: Underground garages

地下室内地坪面与室外地坪面的高度之差大于该层车库净高 1/2 的汽车库。

Underground garages are the garages with the height differences between the underground indoor floor surfaces and outdoor floor surfaces greater than 1/2 of the net height of the garages on the floors.

6.【汽防规】2.0.5 条：半地下汽车库

6. [Code for fire protection design of garage, motor repair shop and parking area] 2.0.5: Semi-underground garages

地下室内地坪面与室外地坪面的高度之差大于该层车库净高 1/3 且不大于 1/2 的汽车库。

Semi-underground garages are garages with the height differences between the underground indoor floor surfaces and outdoor floor surfaces greater than 1/3 and not greater than 1/2 of the net height of the garages on the floors.

7.【汽防规】2.0.8 条：机械式汽车库

7. [Code for fire protection design of garage, motor repair shop and parking area] 2.0.8: Mechanical garages

采用机械设备进行垂直或水平移动等形式停放汽车的汽车库。

Mechanical garages are the garages for motor parking in the forms of vertical or horizontal movement with mechanical equipment, etc.

8.【汽防规】2.0.9 条：敞开式汽车库

8. [Code for fire protection sesign of garage, motor repair shop and parking area] 2.0.9: Open garages

任一层车库外墙敞开面积大于该层四周外墙体总面积的 25%，敞开区域均匀布置在外墙上且其长度不小于车库周长的 50% 的汽车库。

Open garages are the garages, the open areas of whose exterior wall of garages on any floors are 25% greater than the total areas of the surrounding exterior walls, and the open areas are evenly distributed on the exterior walls and the length of the open areas is not less than 50% of the garage perimeters.

9.【车库建规】2.0.7 条：独立式车库

9. [Code for design of parking garage building] 2.0.7: Stand-by garage

单独建造的，具有独立完整的建筑主体结构与设备系统的车库。

Separately-built garage with independent and complete main building structure and equipment system.

10.【车库建规】2.0.8 条：附建式车库

10. [Code for design of parking garage building] 2.0.8: Affiliated garages

与其他建筑物或构筑物结合建造，并共用或部分共用建筑主体结构与设备系统的车库。

Garages built in combination with other buildings and structures and sharing or partially sharing the main building structure and equipment system.

11.【车库建规】2.0.9 条：复式机动车库

11. [Code for design of parking garage building] 2.0.9: Duplex motor garages

室内有车道、有驾驶员进出的机械式机动车库。

Indoor mechanical garage with lanes and the access for drivers.

12.【车库建规】2.0.12 条：全自动机动车库

12. [Code for design of parking garage building] 2.0.12: Full-automatic motor garages

室内无车道，且无驾驶员进出的机械式机动车库。

Indoor mechanical garage without lanes and the access for drivers.

13.【车库建规】2.0.16 条：坡道式出入口

13. [Code for design of parking garage building] 2.0.16: Ramp-type exit and entrance

机动车库中通过坡道进行室内外车辆交通联系的部位。

Part of the motor garage where the traffic of indoor and outdoor vehicles are linked through the ramp.

14.【车库建规】2.0.17 条：升降梯式出入口

14. [Code for design of parking garage building] 2.0.17: Elevator-type entrance and exit

机动车库中通过升降梯进行室内外车辆交通联系的部位。

Part of the motor garage where the traffic of indoor and outdoor vehicles are linked through the elevator.

15.【车库建规】2.0.18 条：平入式出入口

15. [Code for design of parking garage building] 2.0.18: Straight-in entrance and exit

机动车库中由室外场地直接出入停车区域的部位。

Part of the garage through which the vehicles directly enter the parking area from the outdoor area.

16.【车库建规】2.0.13 条：车当量

16. [Code for design of parking garage building] 2.0.13: Vehicle equivalent

用于协调各种不同车型，便于统计与计算停车数量、停车位大小等数据而设定的标准参考车型单元。

Standard reference vehicle unit for the coordination of different vehicle types, facilitating the statistics and calculation of the number and size of the parking spaces and other data.

7 地下室和半地下室 Basements and Semi-basements

7.2 一般规定

7.2 General Provisions

7.2.1 【通则】6.3.1 条：地下室、半地下室应有综合解决其使用功能的措施，合理布置地下停车库、地下人防、各类设备用房等功能空间及各类出入口部；地下空间与城市地铁、地下人行道及地下空间之间应综合开发，相互连接，做到导向明确、流线简捷。

7.2.1 [Code for design of civil buildings] 6.3.1: The basements and semi-basements shall be provided with comprehensive measures of approaching the use functions. Reasonable arrangement shall be made for the functional spaces of underground parking garages, underground civil air defense, various equipment rooms, etc., and for their various entrances and exits. The underground spaces and the urban subways, underground walkways and underground spaces shall be developed comprehensively and connected to achieve clear guidance and simple streamline.

7.2.2 【通则】6.3.2 条：地下室、半地下室作为主要用房使用时，应符合安全、卫生的要求，并应符合下列要求。

7.2.2 [Code for design of civil buildings] 6.3.2: Where the basements and semi-basements are used as the main rooms, safety and hygiene requirements shall be conformed to, including the following requirements.

1. 严禁将幼儿、老年人生活用房设在地下室或半地下室。

1. Living rooms for children and the aged mustn't be set in basements or semi-basements.

2. 居住建筑中的居室不应布置在地下室内；当布置在半地下室时，必须对采光、通风、日照、防潮、排水及安全防护采取措施。

2. Habitable rooms in residential buildings shall not be arranged in basement. If they are arranged in semi-basements, relevant measures must be adopted for daylighting, ventilation, sun exposure, moisture proofing, drainage and safety protection.

3. 建筑物内的歌舞、娱乐、放映、游艺场所不应设置在地下二层及二层以下；当设置在地下一层时，地下一层地面与室外出入口地坪的高差不应大于 10m。

3. Karaoke rooms, dancing halls, entertainment rooms, video halls and amusement halls in a building shall not be arranged on Floor II underground and floors below Floor II. If they are arranged on the first floor underground, the height difference between the first floor underground and the ground floor at the outdoor entrance and exit shall not be greater than 10m.

7.2.3【通则】6.3.3条：地下室平面外围护结构应规整，其防水等级及技术要求除应符合现行国家标准《地下工程防水技术规范》GB 50108的规定外，尚应符合下列规定。

7.2.3 [Code for design of civil buildings] 6.3.3: The exterior enclosure structure of the basement plane shall be in regular shap. The waterproof grade and technical requirements shall not only conform to the provisions in the current national standard, *Technical Code for Waterproof of Underground Works* GB 50108, but also conform to the following provisions.

1. 地下室应在一处或若干处地面较低点设集水坑，并预留排水泵电源和排水管道；

1. For basements, the sump pit shall be arranged at one or several lower points on the ground, with power supply for the drainage pump and drainage pipeline;

2. 地下管道、地下管沟、地下坑井、地漏、窗井等处应有防止涌水、倒灌的措施。

2. At the positions of underground pipelines, underground pipe ditches, underground pits, floor drains, window wells, etc., preventive measures against water inrush and backflow shall be provided.

7.2.4【通则】6.3.4条：地下室、半地下室的耐火等级、防火分区、安全疏散、防排烟设施、房间内部装修等应符合防火规范的有关规定。

7.2.4 [Code for design of civil building] 6.3.4: For fire prevention of basement and semibasement, the relevant specifications in the code for fire resistance ratings, fire compartment, safty evacuation, smoke exhaust facilities, and interior decoration shall be met.

7.2.5【住设规】6.9.1条：卧室、起居室（厅）、厨房不应布置在地下室；当布置在半地下室时，必须对采光、通风、日照、防潮、排水及安全防护采取措施，并不得降低各项指标要求。

7.2.5 [Design code of residential buildings] 6.9.1: Bedrooms, living rooms（halls）and kitchens shall not be arranged in basement. When they are arranged in semibasement, measures must be adopted for daylighting, ventilation, sun exposure, moisture proofing, drainage and safety protection, and all the index requirements shall not be lowered.

7.2.6【住设规】6.9.2条：除卧室、起居室（厅）、厨房以外的其他功能房间可布置在地下室，当布置在地下室时，应对采光、通风、防潮、排水及安全防护采取措施。

7.2.6 [Design code of residential buildings] 6.9.2: Other functional rooms except bedrooms, living rooms（halls）and kitchens can be arranged in basement, When they are arranged in basement, measures for daylighting, ventilation, moisture proofing, drainage and safety protection shall be adopted.

7.2.7【车库建规】6.3.4条：非机动车库的停车区域净高不应小于2.0m。

7.2.7 [Code for design of parking garage building] 6.9.3: The net height of parking area in the non-motorized vehiclegarage shall not be less than 2.0m.

7.3 地下汽车库

7.3 Underground Garages

7.3.1 【车库建规】1.0.4 条:机动车车库建筑规模应按停车当量数划分为特大型、大型、中型、小型,非机动车车库应按停车当量数划分为大型、中型、小型(表 7.3.1)。

7.3.1 [Code for design of parking garage building] 1.0.4: The construction scale of the motor garage shall be divided into extra-large, large, medium and small types as per the quantity of parking equivalents. and that of non-motor garages shall be divided into large, medium and small types as per the quantity of parking equivalents(Tab.7.3.1).

车库建筑规模及停车当量数　　　　　　　表 7.3.1
The Parking Garage Building Scale　　　　Tab.7.3.1

规模 Scale / 当量数 Equivalent / 类型 Sort	特大型 Extra large	大型 Large	中型 Medium	小型 Small
机动车车库停车当量数 Equivalent parking spaces in the motor garage	>1000	301~1000	51~300	≤50
非机动车车库停车当量数 Equivalent parking spaces in the non-motor garage	—	>500	251~500	≤250

7.3.2 【汽防规】3.0.1 条:汽车库、修车库、停车场的分类见表 7.3.2。

7.3.2 [Code for fire protection design of garages, motor repair shop and parking area] 3.0.1: The classification of garages, motor repair shops and parking areas see Tab.7.3.2.

汽车库、修车库、停车场的分类　　　　　表 7.3.2
The Parking Garage Building Scale　　　　Tab.7.3.2

名称 Name		I	II	III	IV
汽车库 Garages	停车数量(辆) Parking quantity(set)	>300	151~300	51~150	≤50
	总建筑面积 S (m²) Total building area S (m²)	$S>10000$	$5000<S\leq10000$	$2000<S\leq5000$	$S\leq2000$

续表

名称 Name		I	II	III	IV
修车库 Motor repair shops	车位数（个） Number of the parking spaces	>15	6~15	3~5	≤2
	总建筑面积 S（m²） Total building area S (m²)	$S>3000$	$1000<S\leq3000$	$500<S\leq1000$	$S\leq500$
停车场 Parking area	停车数量（辆） Parking quantity (set)	>400	251~400	101~250	≤100

注：1. 当屋面露天停车场与下部汽车库共用汽车坡道时，其停车数量应计算在汽车库的车辆总数内。
Notes:1. When open parking areas on roofs share the same vehicle ramps with the lower garages, the number of parking spaces shall be calculated into the total number of vehicles in the garages.
2. 室外坡道、屋面露天停车场的建筑面积可不计入汽车库的建筑面积之内。
2. Building areas of outdoor ramps and open parking areas on roofs may not be calculated into the building areas of the garages.
3. 公交汽车库的建筑面积可按本表的规定值增加 2.0 倍。
3. Building areas of the bus garages can be increased by 2.0 times as per the specified values in the table.

7.3.3 【车库建规】4.1.2 条：机动车库应以小型车为计算当量进行停车当量的换算，各类车辆的换算当量系数见表 7.3.3。

7.3.3 [Code for design of parking garage building] 4.1.2: For motor garages, compact vehicles shall be used as the calculation equivalents for the conversion of parking equivalents, and the conversion coefficients for the equivalent number of various vehicles see Tab.7.3.3.

当量系数　　　　　　　　　　　　　　　　表 7.3.3
The Conversion Coefficients　　　　　　　　Tab.7.3.3

车型 Vehicle type	微型车 Minicars	小型车 Compact vehicles	轻型车 Light-duty vehicles	中型车 Medium-duty vehicles	大型车 Heavy-duty vehicles
换算系数 Conversion coefficient	0.7	1.0	1.5	2.0	2.5

7.3.4 【汽防规】5.1.1 条：汽车库防火分区的最大允许建筑面积应符合表 7.3.4 的规定。其中，敞开式、错层式、斜楼板式汽车库的上下连通层面积应叠加计算，每个防火分区的最大允许建筑面积不应大于表 7.3.4 规定的 2.0 倍；室内有车道且有人员停留的机械式汽车库，其防火分区最大允许建筑面积应按表 7.3.4 的规定减少 35%。

7.3.4 [Code for fire protection design of garage, motor repair shop and parking area] 5.1.1: The maximum allowable floor area of the fire protection zoning of the shall conform to the requirements of the Tab. 7.3.4. Among them, the upper and lower connected floor area of the open, staggered and oblique floor vehicle library should be superimposed and calculated. The maximum allowable floor area for each fireproof partition shall not be greater than 2.0 times of the one specified in Tab. 7.3.4; the maximum allowable floor area of the fireproof zoning shall be reduced by 35% in accordance with the provisions of Tab. 7.3.4 for mechanical vehicle depots with lanes in the room and with people staying.

汽车库防火分区的最大允许建筑面积（m²） 表 7.3.4
Maximum Allowable Building Area of Fire Compartments in Motor Garages (m²) Tab. 7.3.4

耐火等级 Fire resistance rating	单层汽车库 Single-storey garage	多层汽车库、半地下汽车库 Multi-storey garage and semi-underground garage	地下汽车库、高层汽车库 Underground garage and high-rise garage
一、二级 Grade I and Grade II	3000	2500	2000
三级 Grade III	1000	不允许 Not allowed	不允许 Not allowed

注：1. 除本规范另有规定外，防火分区之间应采用符合本规范规定的防火墙、防火卷帘等分隔。
Notes: 1. Unless otherwise specified in the code, fire walls and fireproof roller shutters and other partitions, which conform to the provisions of the Code, shall be adopted for separation between fire compartments.
2. 设置自动灭火系统的汽车库，其每个防火分区的最大允许建筑面积不应大于本表规定的 2.0 倍。
2. Set up the automatic fire extinguishing system of the vehicle depot, the maximum allowable floor area of each fireproof section should not be greater than the 2.0 times of area specified in this Table.

7.3.5 【汽防规】5.1.3 条：室内无车道且无人员停留的机械式汽车库，应符合下列规定。

7.3.5 [Code for fire protection design of garages, motor repair shop and parking area] 5.1.3: The indoor mechanical garages without lanes and without resident people shall conform to the following provisions.

1. 当停车数量超过 100 辆时，应采用无门、窗、洞口的防火墙分隔为多个停车数量不大于 100 辆的区域，但当采用防火隔墙和耐火极限不低于 1.00h 的不燃性楼板分隔成多个停车单元，且停车单元内的停车数量不大于 3 辆时，应分隔为停车数量不大于 300 辆的区域；

1. The garage with more than 100 parking spaces shall be divided into multiple areas with no more than 100 parking spaces whose fire walls are without doors, windows and openings, but when the garage is divided into multiple parking units with fire partition and incombustible floor slabs of the fire endurance not less than 1.00 h, and when the number of parking spaces in the parking unit is not more than 3, the garage shall be divided into areas with no more than 300 parking spaces;

2. 汽车库内应设置火灾自动报警系统和自动喷水灭火系统，自动喷水灭火系统应选用快速响应喷头；

2. The automatic fire alarm system and automatic sprinkler system shall be set up in garages, and quick-response spray nozzles shall be equipped for the automatic sprinkler system;

3. 楼梯间及停车区的检修通道上应设置室内消火栓；

3. Indoor fire hydrants shall be set on the maintenance passages in staircases and parking areas;

4. 汽车库内应设置排烟设施，排烟口应设置在运输车辆的通道顶部。

4. Smoke exhaust facilities shall be set up in garages, and the smoke exhaust ports shall be set up on the tops of the accesses for transport vehicles.

7.3.6 【汽防规】6.0.1条：汽车库、修车库的人员安全出口和汽车疏散出口应分开设置。设置在工业与民用建筑内的汽车库，其车辆疏散出口应与其他场所的人员安全出口分开设置。

7.3.6 [Code for fire protection design of garage, motor repair shop and parking area] 6.0.1: The safety exits of the garage and motor repair shop and vehicle evacuation exit shall be separately arranged. For the garages arranged in industrial and civil buildings, the vehicle evacuation exits shall be separately arranged from the safety exits for the personnel in other places.

7.3.7 【车库建规】4.2.8条：机动车库的人员出入口与车辆出入口应分开设置，机动车升降梯不得替代乘客电梯作为人员出入口，并应设置标识。

7.3.7 [Code for design of parking garage building] 4.2.8: The accesses for personnel of the motor garages shall be arranged separately with those for vehicles . The elevators for motor vehicles shall not be used as the accesses for personnel in the place of passenger elevators, and relevant signs shall be provided.

7.3.8 【汽防规】6.0.2条：除室内无车道且无人员停留的机械式汽车库外，汽车库、修车库内每个防火分区的人员安全出口不应少于2个，Ⅳ类汽车库和Ⅲ、Ⅳ类修车库可设置1个。

7.3.8 [Code for fire protection design of garage, motor repair shop and parking area] 6.0.2: Except for the indoor mechanical garage without lanes and resident people, at least 2 safety exits for personnel in the fire compartments of the garage and motor repair shop shall be arranged, and one can be arranged for Class IV garage and Classes III and IV motor repair shops.

7.3.9 【汽防规】6.0.3条：汽车库、修车库的疏散楼梯应符合下列规定。

7.3.9 [Code for fire protection design of garage, motor repair shop and parking area] 6.0.3: The evacuation staircases of the garage and motor repair shop shall conform to the following provisions.

1. 建筑高度大于32m的高层汽车库、室内地面与室外出入口地坪的高差大于10m的地下汽车库应采用防烟楼梯间，其他汽车库、修车库应采用封闭楼梯间；

1. The smoke-proof staircases shall be adopted in the high-rise garage with the building higher than 32 m and the underground garage with the height difference between the indoor floor and the floor at the outdoor access greater than 10 m. Enclosed staircases shall be adopted in other garages and motor repair shops;

2. 楼梯间和前室的门应采用乙级防火门，并应向疏散方向开启；

2. Grade B fire doors shall be used as doors of the staircase and the anteroom, and they shall be opened in the evacuation direction;

3. 疏散楼梯的宽度不应小于1.1m。

3. The width of the evacuation staircases shall not be less than 1.1 m.

7.3.10 【汽防规】6.0.6条：汽车库室内任一点至最近人员安全出口的疏散距离不应大于45m，当设置自动灭火系统时，其距离不应大于60m。对于单层或设置在建筑首层的汽车库，室内任一点至室外最近出口的疏散距离不应大于60m。

7.3.10 [Code for fire protection design of garage, motor repair shop and parking area] 6.0.6: The evacuation distances from any points in the garages to the nearest safety exits for personnel shall not be greater than 45m. When the automatic extinguishing systems are set, its distances shall not be greater than 60m. For single-storey garages or garages set on the first floor of buildings, the evacuation distances from any indoor point to the nearest outdoor exit shall not be greater than 60m.

7.3.11 【汽防规】6.0.7条：与住宅地下室相连通的地下汽车库、半地下汽车库，人员疏散可借用住宅部分的疏散楼梯；当不能直接进入住宅部分的疏散楼梯间时，应在汽车库与住宅部分的疏散楼梯之间设置连通走道，走道应采用防火隔墙分隔，汽车库开向该走道的门均应采用甲级防火门。

7.3.11 [Code for fire protection design of garage, motor repair shop and parking area] 6.0.7: For underground garages and semi-underground garages connected with basements of residences, the evacuation staircases of the residence parts can be used for personnel evacuation. When it is not possible to directly enter the evacuation staircases of the residence parts, corridors shall be arranged to connect the garages with the evacuation staircases of the residence parts, and the corridors shall be partitioned with fire partitions. For all the doors of the garages opening to the corridors, Grade A fire doors shall be used.

7.3.12 【汽防规】6.0.8条：室内无车道且无人员停留的机械式汽车库可不设置人员安全出口，但应按下列规定设置供灭火救援用的楼梯间。

7.3.12 [Code for fire protection design of garage, motor repair shop and parking area] 6.0.8: The safety exit for personnel may not be arranged in the indoor mechanical garages without lanes and without resident people, but the staircase for firefighting and rescue shall be arranged as per the following provisions.

1. 每个停车区域当停车数量大于 100 辆时，应至少设置 1 个楼梯间；

1. When more than 100 vehicles are parked in each parking area, at least 1 staircase shall be arranged;

2. 楼梯间与停车区域之间应采用防火隔墙进行分隔，楼梯间的门应采用乙级防火门；

2. The staircase and the parking area shall be separated with a fireproof partition, and the Grade B fire doors shall be used as the doors of the staircases;

3. 楼梯的净宽不应小于 0.9m。

3. The net width of the staircase shall not be less than 0.9 m.

7.3.13 【车库建规】4.2.6 条：机动车库出入口和车道数量应符合表 7.3.13 的规定，且当车道数量大于等于 5 且停车当量大于 3000 辆时，机动车出入口数量应经过交通模拟计算确定。

7.3.13 [Code for design of parking garage building] 4.2.6: The number of entrances, exits and lanes of the garage shall conform to the provisions in Tab. 7.3.13. When the number of lanes is greater than or equal to 5, and when the parking equivalent is more than 3000, the number of entrances and exits of the garage shall be determined through traffic simulation calculation.

【车库建规】4.2.6 条：机动车库出入口和车道数量　　表 7.3.13
[Code for design of parking garage building] 4.2.6:
Quantities of Entrances, Exits and Lanes of Motor Garage　　Tab. 7.3.13

规模 Scale 停车当量 Parking equivalent 出入口和车道数量 Number of entrances, exits and lanes	特大型 Extrza large	大型 Large		中型 Medium		小型 Small	
	>1000	501~1000	301~500	101~300	51~100	25~50	<25
机动车出入口数量 Number of entrances and exit of motor garages	≥3	≥2		≥2	≥1	≥1	
非居住建筑出入口车道数量 Number of lanes at entrances and exits of non-residential buildings	≥5	≥4	≥3	≥2		≥2	≥1
居住建筑出入口车道数量 Number of lanes at entrances and exits of residential buildings	≥3	≥2	≥2	≥2		≥2	≥1

7.3.14 【车库建规】4.2.7 条：对于停车当量小于 25 辆的小型车库，出入口可设一个单车道，并应采取进出车辆的避让措施。

7.3.14 [Code for design of parking garage building] 4.2.7: For small garages with the quantities of parking equivalents less than 25, one single lane can be set at the entrance and exit, and avoidance measures for vehicles entering and leaving shall be taken.

7.3.15 【汽防规】6.0.9 条和 6.0.10 条：汽车库、修车库的汽车疏散出口总数不应少于 2 个，且应分散布置。当符合下列条件之一时，汽车库、修车库的汽车疏散出口可设置 1 个。

7.3.15 [Code for fire protection design of garage, motor repair shop and parking area] 6.0.9 and 6.0.10: The total number of vehicle evacuation exits of garages and motor repair shops shall not be less than 2, and the exits shall be arranged in a dispersive way. When one of the following conditions is met, one vehicle evacuation exit can be set for garages and motor repair shops.

1. IV 类汽车库；

1. Class IV garages;

2. 设置双车道汽车疏散出口的 III 类地上汽车库；

2. Class III overground garages with two-lane vehicle evacuation exits;

3. 设置双车道汽车疏散出口、停车数量小于或等于 100 辆且建筑面积小于 4000 m^2 的地下或半地下汽车库；

3. Underground or semi-underground garages with two-lane vehicle evacuation exits, vehicle parking number less than or equal to 100 and building areas less than 4000 m^2;

4. II、III、IV 类修车库。

4. Classes II, III and IV motor repair shops.

7.3.16 【汽防规】6.0.11 条：I、II 类地上汽车库和停车数量大于 100 辆的地下、半地下汽车库，当采用错层或斜楼板式，坡道为双车道且设置自动喷水灭火系统时，其首层或地下一层至室外的汽车疏散出口不应少于 2 个,汽车库内其他楼层的汽车疏散坡道可设置 1 个。

7.3.16 [Code for Fire Protection Design of garage, motor repair shop and parking area] 6.0.11: For Class I and II overground motor garages and underground or semi-underground garages with the number of parking spaces more than 100, when the split-level or inclined-floor type is adopted for the parking spaces, double-lane passage is adopted for the ramps. When automatic sprinkler systems are set for the ramps, the number of the vehicle evacuation exits on the first floors or from the first floors underground to outdoor places shall be not less than 2, and the number of the vehicle evacuation ramps on other floors of the garages can be 1.

7.3.17 【汽防规】6.0.12 条：IV 类汽车库设置汽车坡道有困难时，可采用汽车专用升降机作汽车疏散出口，升降机的数量不应少于 2 台，停车数量少于 25 辆时，可设置 1 台。

7.3.17 [Code for Fire Protection Design of garage, motor repair shop and parking area] 6.0.12: When it is difficult to arrange vehicle ramps for Class IV garages, the special vehicle lifters can be adopted as evacuation exits for vehicles. The number of the lifters shall not be less than 2, but when the parking quantity is less than 25, one set can be arranged.

7.3.18 【车库建规】4.2.4 条：车辆出入口宽度，双向行驶时不应小于 7m，单向行驶时不应小于 4m。

7.3.18 [Code for design of parking garge building] 4.2.4: The width of vehicle entrances and exits shall be not less than 7m in the case of two-way driving, and shall be not less than 4m in the case of one-way driving.

7.3.19 【车库建规】4.2.2 条：车辆出入口的最小间距不应小于 15m，并宜与基地内部道路相接通。

7.3.19 [Code for design of parking garge building] 4.2.2: The minimum spacing distance between vehicle entrances and exits shall be not less than 15m, and they shall be appropriately connected with the internal roads in the base.

7.3.20 【汽防规】6.0.14 条：除室内无车道且无人员停留的机械式汽车库外，相邻两个汽车疏散出口之间的水平距离不应小于 10m；毗邻设置的两个汽车坡道应采用防火隔墙分隔。

7.3.20 [Code for fire protection design of garage, motor repair shop and parking area] 6.0.14: Except for the indoor mechanical garage without lanes and without resident people, the horizontal distance between two adjacent vehicle evacuation exits shall be not less than 10m; for two adjacent vehicle ramps, fire partitions shall be adopted for separation.

7.3.21 【车库建规】4.2.1 条：按出入方式，机动车库出入口可分为平入式、坡道式、升降梯式三种类型。

7.3.21 [Code for design of parking garage building] 4.2.1: The entrances and exits of motor garages can be divided into three types: straight-in type, ramp type and elevator type, as per the access modes.

7.3.22 【车库建规】4.2.5 条：车辆出入口及坡道的最小净高应符合表 7.3.22 的规定。

7.3.22 [Code for design of parking garage building] 4.2.5: The minimum net height of the vehicle entrances and exits and ramps shall conform to the provisions in Tab. 7.3.22.

【车库建规】表 4.2.5：车辆出入口及坡道的最小净高　　　　表 7.3.22
[Code for design of parking garage building] Tab. 4.2.5:
Minimum Net height of Vehicle Entrances and Exits and Ramps　　Tab. 7.3.22

车 型 Vehicle type	最小净高（m） Minimum net height（m）
微型车、小型车 Minicars and small-sized vehicles	2.20
轻型车 Light-duty vehicles	2.95

续表

车型 Vehicle type	最小净高（m） Minimum net height（m）
中型车、大型客车 Medium-duty vehicles and heavy-duty passenger vehicles	3.70
中型、大型货车 Medium and heavy-duty trucks	4.20

注：净高指从楼地面面层（完成面）至吊顶、设备管道、梁或其他构件底面之间的有效使用空间的垂直高度。
Note: the net height refers to the vertical height of the effective use space between the floor ground (finished surface) to the bottom surface of the suspended ceiling, equipment pipeline, beam or other components.

7.3.23 【车库建规】4.2.9 条：平入式出入口应符合下列规定。

7.3.23 [Code for design of parking garage building] 4.2.9: The straight-in entrances and exits shall conform to the following provisions.

1. 平入式出入口室内外地坪高差不应小于 150mm，且不宜大于 300mm；

1. The height difference between indoor and outdoor floors at straight-in entrances and exits shall be not less than 150 mm, and shall be appropriately not greater than 300mm;

2. 出入口室外坡道起坡点与相连的室外车行道路的最小距离不宜小于 5.0m；

2. The minimum distance between the starting points of the outdoor ramps at the entrances or exits to the outdoor roadways connected shall be appropriately not less than 5.0m;

3. 出入口的上部宜设有防雨设施；

3. The upper parts of the entrances and exits shall be provided with rainproof facilities;

4. 出入口处宜设置遥控启闭的大门。

4. Doors for remote opening and closing shall be arranged at the entrances and exits.

7.3.24 【车库建规】4.2.10 条：坡道式出入口应符合下列规定。

7.3.24 [Code for design of parking garage building] 4.2.10: The ramp-up exit and entrance shall conform to the following provisions.

1. 出入口可采用直线坡道、曲线坡道和直线与曲线组合坡道，其中直线坡道可选用内直坡道式、外直坡道式。

1. The linear ramp, curved ramp and linear-curve-combined ramp can be adopted for the entrance and exit, and therein, internal linear-ramp type and external linear-ramp type can be adopted for the linear ramp.

2. 出入口可采用单车道或双车道，坡道最小净宽应符合表 7.3.24-1 的规定。

2. Single-lane or double-lane can be adopted for the entrance and exit, and the minimum net width of the ramp shall conform to the provisions in Tab. 7.3.24-1.

【车库建规】表 4.2.10-1：坡道最小净宽　　　　　　　　　　表 7.3.24-1

[Code for design of parking garage building] Tab. 4.2.10-1:
Minimum Net width of Ramps　　　　　　　　　　　　　　Tab. 7.3.24-1

形式 Forms	最小净宽（m） Minimum net width（m）	
	微型、小型车 Micro and compact vehicles	轻型、中型、大型 Light-duty, medium-duty and heavy-duty
直线单行 Linear one-way driving	3.0	3.5
直线双行 Linear two-way driving	5.5	7.0
曲线单行 Curved one-way driving	3.8	5.0
曲线双行 Curved two-way driving	7.0	10.0

注：此宽度不包括道牙及其他分隔带宽度。当曲线比较缓时，可以按直线宽度进行设计。
Note: The width of curbs and other separation belts shall not be included in this width. When the curve is gentle, the net width can be designed as per the width of the linear ramp.

3. 坡道的最大纵向坡度应符合表 7.3.24-2 的规定。

3. The maximum longitudinal gradient of ramps shall conform to the provisions in Tab. 7.3.24-2.

【车库建规】表 4.2.10-2：坡道的最大纵向坡度　　　　　　表 7.3.24-2

[Code for design of parking garage building] Tab. 4.2.10-2:
Maximum Longitudinal Gradient of Ramps　　　　　　　Tab. 7.3.24-2

车型 Vehicle type	直线坡道 Linear ramp		曲线坡道 Curved ramp	
	百分比（%） Percentage（%）	比值（高:长） Ratio（height：length）	百分比（%） Percentage（%）	比值（高:长） Ratio（height：length）
微型、小型车 Mini and small-sized vehicles	15.0	1：6.67	12	1：8.3
轻型车 Light-duty vehicles	13.2	1：7.5	10	1：10.0
中型车 Medium-duty vehicles	12.0	1：8.3		
大型客车、大型货车 Heavy-duty passenger vehicles and heavy-duty trucks	10.0	1：10.0	8	1：12.5

4. 当坡道纵向坡度大于10%时，坡道上、下端均应设缓坡坡段，其直线缓坡段的水平长度不应小于3.6m，缓坡坡度应为坡道坡度的1/2；曲线缓坡段的水平长度不应小于2.4m，曲率半径不应小于20m，缓坡段的中心为坡道原起点或止点（图7.3.24）；大型车的坡道应根据车型确定缓坡的坡度和长度。

4. When longitudinal gradients of ramps are greater than 10%, the upper and lower ends of the ramps shall be provided with gentle slope sections. The horizontal length of the gentle linear slopes shall be not less than 3.6m, and the gradients of the gentle slopes shall be 1/2 of the ramp gradients. The horizontal length of the gentle curved slopes shall be not less than 2.4m, and curvature radiuses shall be not less than 20m, with the center of the gentle slope section as the original starting points or ending points of the ramps (Fig. 7.3.24) ; For ramps for heavy-duty vehicles, the gradients and lengths of the gentle slopes shall be determined in accordance with the vehicle types.

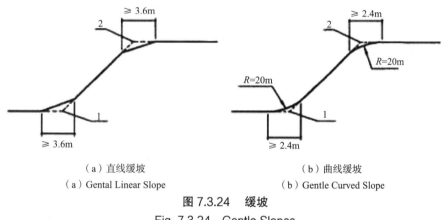

（a）直线缓坡　　　　　　　　　　（b）曲线缓坡
(a) Gental Linear Slope　　　　　　(b) Gentle Curved Slope

图 7.3.24　缓坡
Fig. 7.3.24　Gentle Slopes
1- 坡道起点；2- 坡道止点
1-Starting point of the ramp; 2-Ending point of the ramp

5. 微型车和小型车的坡道转弯处的最小环形车道内半径（r_0）不宜小于表7.3.24-3的规定；其他车型的坡道转弯处的最小环形车道内半径应按计算确定。

5. The minimum inner radiuses (r_0) of the circular lanes at turnings of ramps for mini-vehicles and small-sized vehicles are not appropriately less than the provisions in Tab. 7.3.24-3; the minimum inner radiuses of circular lanes at turnings of ramps for other vehicle types shall be determined by calculation.

【车库建规】表 4.2.10-3：坡道转弯处的最小环形车道内半径（r_0） 表 7.3.24-3

[Code for design of parking garage building] Tab. 4.2.10-3:
Minimum Inner Radiuses of Circular Lanes at Turnings of Ramps (r_0) Tab. 7.3.24-3

角度 Angle 半径 Radius	坡道转向角度 α Turning angles of ramps α		
	$α \leq 90°$	$90° < α < 180°$	$α \geq 180°$
最小环形车道内半径 r_0 Minimum inner radius of the circular lane r_0	4m	5m	6m

注：坡道转向角度为机动车转弯时的连续转向角度。
Note: The turning angle of the ramp is the continuous steering angle when a motor vehicle turns.

6. 环形坡道处弯道超高宜为 2% ~ 6%。

6. For the circular ramp, the supper-elevation of the ramp turning shall be 2% ~ 6%.

7.3.25 【车库建规】4.2.11 条：升降梯式出入口应符合下列规定。

7.3.25 [Code for design of parking garage building] 4.2.11: The elevator-type entrance and exit shall conform to the following provisions.

1. 当小型机动车库设置机动车坡道有困难时，可采用升降梯作为机动车库出入口，升降梯可采用汽车专用升降机等提升设备，且升降梯的数量不应少于两台，停车当量少于 25 辆的可设一台；

1. When it is difficult to arrange a motor vehicle ramp in small garage, elevators can be adopted as entrances and exits of the garage. The lifting equipment of special vehicle lifter and the like can be adopted as elevators. At least two elevators shall be arranged, and one elevator can be set when the parking equivalent is less than 25;

2. 机动车出口和入口宜分开设置；

2. The entrance and exit for motor vehicles shall be arranged separately;

3. 升降梯宜采用通过式双向门，当只能为单侧门时，应在进（出）口处设置车辆等候空间；

3. The pass-type two-way doors shall be adopted for elevators. When only the one-side doors are allowed, the vehicle waiting space shall be reserved at the position of the entrance (exit);

4. 升降梯出入口处应设有防雨设施，且升降梯底坑应设有机械排水系统；

4. The rainproof facilities shall be provided at the entrance and exit of the elevator, and the mechanical drainage system shall be arranged in the pits of elevators;

5. 机动车库应在每层出入口处的明显部位设置楼层和行驶方向的标志，并宜在驾驶员方便触及的部位，设置升降梯的操纵按钮；

5. The signs for floor sequence and driving directions shall be provided at obvious positions at the entrances and exits on each floor in the garage, and the control buttons of the elevators shall be arranged at the positions with easy access for drivers;

6. 当采用升降平台时，应在每层周边设置安全护栏和防坠落等措施；

6. When the lifting platform is adopted, safety guardrails and anti-dropping facilities and the like shall be arranged;

7. 升降梯出入口处应设限高和限载标志。

7. The signs of height limit and load limit shall be arranged at the entrances and exits of the elevators.

7.3.26 【车库建规】4.3.1 条：停车区域应由停车位和通车道组成。

7.3.26 [Code for design of parking garage building] 4.3.1: The parking area shall be composed of the parking spaces and the traffic lanes.

7.3.27 【车库建规】4.3.2 条：停车区域的停车方式应排列紧凑、通道短捷、出入迅速、保证安全和与柱网相协调，并应满足一次进出停车位要求。

7.3.27 [Code for design of parking garage building] 4.3.2: For the parking modes in the parking areas, the arrangement shall be compact, with short passage, easy access, safety guaranty and coordination with the column network. The requirements for entering or getting out of the parking space in one time shall be met.

7.3.28 【车库建规】4.3.3 条：停车方式可采用平行式、斜列式（倾角30°、45°、60°）和垂直式（图 7.3.28），或混合式。

7.3.28 [Code for design of parking garage building] 4.3.3: The parking modes can be the parallel type, diagonal type (with the inclination angle of 30°, 45° and 60°) and vertical type (Fig. 7.3.28) or mixed type.

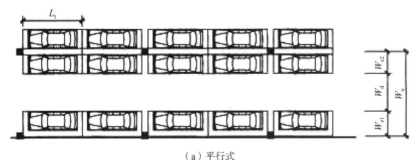

（a）平行式

(a) Parallel Type

图 7.3.28 停车方式（一）

Fig. 7.3.28 Parking Mode 1

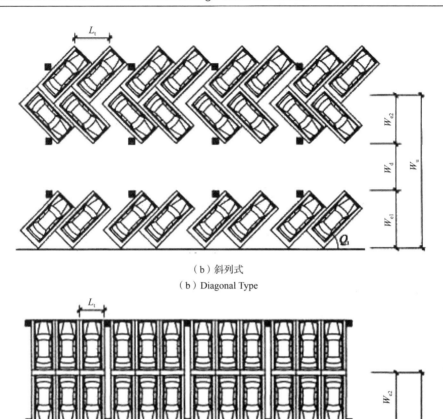

（b）斜列式
(b) Diagonal Type

（c）垂直式
(c) Vertical Type

图 7.3.28 停车方式（二）
Fig. 7.3.28 Parking Mode 2

注：W_u 为停车带宽度；W_{e1} 为停车位毗邻墙体或连续分隔物时，垂直于通（停）车道的停车位尺寸；W_{e2} 为停车位毗邻时，垂直于通（停）车道的停车位尺寸；W_d 为通车道宽度；L_t 为平行于通车道的停车位尺寸；Q_t 为机动车倾斜角度

Note: W_u means the width of the parking belt; W_{e1} means the size of the parking space adjacent to the wall or continuous partition perpendicular to the driving (parking) lane; W_{e2} means the size of the adjacent parking space perpendicular to the driving (parking) lane; W_d means the width of the traffic lane; L_t means the size of the parking space parallel to the traffic lane; Q_t means the inclination angle of the motor vehicle

7.3.29 【车库建规】4.3.4 条：机动车最小停车位、通（停）车道宽度可通过计算或作图法求得，且库内通车道宽度应大于或等于 3.0m。小型车的最小停车位、通（停）车道宽度宜符合表 7.3.29 的规定。

7.3.29 [Code for design of parking garage building] 4.3.4: The minimum widths of parking spaces and driving (parking) lanes for motor vehicles can be determined through calculation or the plotting method, and the widths of the traffic lanes in the garages shall be greater than or equal to 3.0m. The minimum widths of parking spaces and driving (parking) lanes for small-sized vehicles shall appropriately meet the provisions in Tab. 7.3.29.

【车库建规】表 4.3.4：小型车的最小停车位、通（停）车道宽度　　　表 7.3.29

[Code for fire design of parking garage building] Tab. 4.3.4: Minimum Widths of Parking Spaces and Driving (Parking) Lanes for Compact vehicles　　Tab. 7.3.29

停车方式 Parking mode			垂直通车方向的最小停车位宽度（m） Minimum widths (m) of parking spaces perpendicular to the direction of traffic lanes		平行通车方向的最小停车位宽度（m） Minimum widths (m) of parking spaces parallel to the direction of traffic lanes	通（停）车道最小宽度 W_d（m） Minimum widths W_d (m) of driving (parking) lanes
			W_{e1}	W_{e2}		
平行式 Parallel type		后退停车 Back-up parking	2.4	2.1	6.0	3.8
斜列式 Diagonal type	30°	前进（后退）停车 Forward (backward) parking	4.8	3.6	4.8	3.8
	45°	前进（后退）停车 Forward (backward) parking	5.5	4.6	3.4	3.8
	60°	前进停车 Forward parking	5.8	5.0	2.8	4.5
	60°	后退停车 Back-up parking	5.8	5.0	2.8	4.2
垂直式 Vertical		前进停车 Forward parking	5.3	5.1	2.4	9.0
		后退停车 Back-up parking	5.3	5.1	2.4	5.5

7.3.30【车库建规】4.3.5 条：微型车和小型车的环形通车道最小内半径不得小于 3.0m。

7.3.30 [Code for design of parking garage building] 4.3.5: The minimum inner radius of the circular driving lanes for mini-vehicles and compact vehicles shall not be less than 3.0 m.

7.3.31【车库建规】4.3.6 条：停车区域净高不应小于本书第 7.3.22 条规定的出入口及坡道处净高要求。

7.3.31 [Code for design of parking garage building] 4.3.6: The net height of the parking area shall be not less than the requirements for net heights of entrances, exits and ramps specified in clause 7.3.22 of this Guide.

7.3.32【车库建规】4.3.7 条：根据停车楼板的形式，停车区域可分为平楼板式、错层式和斜楼板式。错层式可分为二段式错层和三段式错层；斜楼板式可分为直坡形斜楼板式和螺旋形斜楼板式。

7.3.32 [Code for design of parking garage building] 4.3.7: In accordance with the form of the parking floor, the parking area can be divided into flat-floor type, split-level type and inclined-floor type. The split-level type can be divided into two-section type and three-section type; the inclined-floor type can be divided into straight-ramp type and spiral type.

7.3.33 【车库建规】4.3.8 条：错层式停车区域应符合下列规定。

7.3.33 [Code for design of parking garage building] 4.3.8: The split-level parking area shall conform to the following provisions.

1. 两直坡道之间的水平距离应使车辆在停车层作 180° 转向，两段坡道中心线之间的距离不应小于 14.0m；

1. The horizontal distance between two straight ramps shall be appropriate to ensure that the vehicle can make a 180° turn on the parking floor, and the distance between the centerlines of the two sections of the ramp shall be not less than 14.0m;

2. 三段错层式停车区域的通车道应限定车辆行驶路线；

2. The driving route of the vehicle shall be limited in the traffic lane in the three-section split-level parking area;

3. 错层式停车区域内楼面空间可以叠交，叠交水平尺寸不应大于 1.5m。

3. The floor space in the split-level parking areas can be overlapped, and the horizontal overlapping size shall be not greater than 1.5m.

7.3.34 【车库建规】4.3.9 条：斜楼板式停车区域的楼板坡度、停车位应符合下列规定。

7.3.34 [Code for design of parking garage building] 4.3.9: The gradient of the floor slab and the parking space in the inclined-floor parking area shall conform to the following provisions.

1. 楼板坡度不应大于 5%；

1. The gradient of the floor slab shall be not greater than 5%;

2. 当停车位采用斜列式停车时，其停车位的长向中线与斜楼板的纵向中线之间的夹角不应小于 60°。

2. When diagonal type parking is adopted for the parking space, the angle between the longitudinal centerline of the parking space and the longitudinal centerline of the inclined floor shall be not less than 60°.

7.3.35 【车库建规】4.3.10 条：对于斜楼板式停车区域，必要时可设转向的中间通车道，为防止行车高峰堵车，可增设螺旋坡道。

7.3.35 [Code for design of parking garage building] 4.3.10: If necessary, for the inclined-floor parking areas, the middle traffic lane for steering can be arranged. In order to prevent traffic jam, the spiral ramps can be additionally arranged.

7.3.36 【车库建规】4.3.11 条：当机动车停车库内设有修理车位时，应集中布置，且应符合本标准的规定。

7.3.36 [Code for design of parking garage building] 4.3.11: When the parking spaces for repair are arranged in the motor parking garage, they shall be arranged in a concentrated way, and shall conform to the provisions in this code.

7.3.37 【车库建规】4.4.1 条：对于有防雨要求的出入口和坡道处，应设置不小于出入口和坡道宽度的截水沟和耐轮压沟盖板以及闭合的挡水槛。出入口地面的坡道外端应设置防水反坡。

7.3.37 [Code for design of parking garage building] 4.4.1: At the entrances and exits and ramps with the rainproof requirements, the intercepting ditch, wheel pressure resistant groove cover and closed water retaining sill with the width not less than that of the entrance, exit and ramp shall be arranged. At the outer end of the ramp at the ground floors of the entrance and exit, the adverse water preventive slope shall be arranged.

7.3.38 【车库建规】4.4.2 条：通往地下的坡道低端宜设置截水沟；当地下坡道的敞开段无遮雨设施时，在坡道敞开段的较低处应增设截水沟。

7.3.38 [Code for design of parking garage building] 4.4.2: The underground low ends of the ramps shall be provided with intercepting ditches; if there are no rain-shelter facilities on the open sections of underground ramps, the lower parts of the open sections of the ramps shall be provided with intercepting ditches.

7.3.39 【车库建规】4.4.3 条：机动车库的楼地面应采用强度高、具有耐磨防滑性能的不燃材料，并应在各楼层设置地漏或排水沟等排水设施。地漏（或集水坑）的中距不宜大于 40m。敞开式车库和有排水要求的停车区域应设不小于 0.5% 的排水坡度和相应的排水系统。

7.3.39 [Code for design of parking garage building] 4.4.3: For the floors of garages, the incombustible materials with the properties of high strength, abrasion resistant and skidproof function shall be adopted, and the drainage facilities such as floor drains or drainage ditches shall be arranged on all the floors. The distance between the centers of floor drains (or sumps) shall be appropriately not greater than 40m. The open garages and parking areas with drainage requirements shall be provided with the drainage ditches with the gradients not less than 0.5% and the corresponding drainage systems.

7.3.40 【车库建规】4.4.4 条：机动车库内通车道和坡道的楼地面宜采取限制车速的措施。

7.3.40 [Code for design of parking garage building] 4.4.4: Measures to limit the vehicle speed shall be adopted for the floors of the traffic lanes and ramps in the garages.

7.3.41 【车库建规】4.4.5条：机动车库内通车道和坡道面层应采取防滑措施，并宜在柱子、墙阳角凸出结构等部位采取防撞措施。

7.3.41 [Code for design of parking garage building] 4.4.5: Skidproof measures shall be adopted for the surfaces of the traffic lanes and ramps in the garages, and collision preventive measures shall be adopted at the positions of columns, protruding structures of the outer corners of walls, etc.

7.3.42 【车库建规】4.4.6条：机动车库内停车位应设车轮挡，车轮挡宜设于距停车位端线为机动车前悬或后悬的尺寸减0.2m处，其高度宜为0.15m，且车轮挡不得阻碍楼地面排水。

7.3.42 [Code for design of parking garage building] 4.4.6: For parking spaces in the garages, vehicle wheel stoppers shall be arranged at the positions where the distances to the endlines of the parking spaces are the sizes of the front overhangs or rear overhangs of the motor vehicles minus 0.2m, with the height of 0.15m. The vehicle wheel stoppers shall not block the floor drainage.

7.3.43 【车库建规】4.4.7条：通往地下的机动车坡道应设置防雨和防止雨水倒灌至地下车库的设施。敞开式车库及有排水要求的停车区域楼地面应采取防水措施。

7.3.43 [Code for design of parking garage building] 4.4.7: Rain-proof facilities and facilities for preventing rainwater from pouring back into the underground garages shall be set on motor vehicle ramps to underground floors. Waterproof measures shall be taken for floors of open garages and parking areas with drainage requirements.

7.3.44 【车库建规】4.4.8条：通往车库的出入口和坡道的上方应有防坠落物设施。

7.3.44 [Code for design of parking garage building] 4.4.8: Preventive facilities against falling objects shall be set above the entrances and exits to garages and ramps.

7.3.45 【车库建规】4.4.9条：严寒和寒冷地区机动车库室外坡道应采取防雪和防滑措施。

7.3.45 [Code for design of parking garage building] 4.4.9: Snow preventive and skidproof measures shall be taken for outdoor ramps of garages in severely cold and cold regions.

7.3.46 【车库建规】4.4.10条：当机动车库坡道横向内（或外）侧无实体墙体时，应在无实体墙处设护栏和道牙。道牙宽度不应小于0.30m，高度不应小于0.15m。

7.3.46 [Code for design of parking garage building] 4.4.10: When there are no solid walls on the transverse inner (or outer) sides of garage ramps, guardrails and curbs shall be set on the places without solid walls. The widths of the curbs shall be not less than 0.30m, and the heights shall be not less than 0.15m.

7.3.47 【防火规】5.5.6 条：直通建筑内附设汽车库的电梯，应在汽车库部分设置电梯候梯厅，并应采用耐火极限不低于 2.00h 的防火隔墙和乙级防火门与汽车库分隔。

7.3.47 [Code for fire protection design of buildings] 5.5.6: For elevators directly leading to the garages affiliated to buildings, elevator waiting halls shall be set up in the garage sections, and the halls shall be partitioned from the garages with fire partitions and Class B fire doors with the fire endurance not less than 2.00 hours.

7.3.48 【汽防规】5.2.5 及 8.1.6 条：汽车库内，可燃气体和甲、乙类液体管道严禁穿过防火墙，防火墙内不应设置排气道。防火墙或防火隔墙上不应设置通风孔道，也不宜穿过其他管道（线）；当管道（线）穿过防火墙或防火隔墙时，应采用防火封堵材料将孔洞周围的空隙紧密填塞。风管应采用不燃材料制作，且不应穿过防火墙、防火隔墙，当必须穿过时，尚应符合下列规定。

7.3.48 [Code for fire protection design of garage, motor repair shop and parking area] 5.2.5 and 8.1.6: In garages, combustible gas and Classes A and B liquid pipelines are strictly prohibited from penetrating fire walls, and no exhaust passages shall be arranged in the fire walls. No ventilation hole shall be arranged in the fire walls or fire partitions, and other pipes (pipelines) shall not be penetrative through the fire walls or fire partitions. When the pipes (pipelines) penetrate the fire walls or fire partitions, the gaps around the holes shall be filled with fireproof blocking materials. Air ducts shall be made of incombustible materials and shall not penetrate the fire walls or fire partitions. When the air ducts are required to pass through the fire walls and fire partitions, the following provisions shall be met.

1. 应在穿过处设置防火阀，防火阀的动作温度宜为 70℃；

1. Fire dampers with the operating temperature of 70℃ shall be arranged at the penetrating positions;

2. 位于防火墙、防火隔墙两侧各 2m 范围内的风管绝热材料应为不燃材料。

2. The duct insulation materials within 2m of both sides of the fire wall and fire partition shall be incombustible.

7.3.49 【汽防规】5.3.3 条：除敞开式汽车库、斜楼板式汽车库外，其他汽车库内的汽车坡道两侧应采用防火墙与停车区隔开，坡道的出入口应采用水幕、防火卷帘或甲级防火门等与停车区隔开；但当汽车库和汽车坡道上均设置自动灭火系统时，坡道的出入口可不设置水幕、防火卷帘或甲级防火门。

7.3.49 [Code for fire protection design of garage, motor repair shop and parking area] 5.3.3: Except for the open garages and inclined-floor type garages, both sides of the vehicle ramp in other garages shall be separated from the parking area by fire walls. The exit and entrance of the ramp shall be separated from the parking area by the water curtain, fireproof roller shutter or Grade A fire door, etc.; But when the automatic extinguishing systems are arranged in the garage and on the vehicle ramp, the water curtain, fireproof roller shutters or Grade A fire doors are not required at the exit and entrance of the ramp.

7.4 地下商业

7.4 Underground Commercial Buildings

7.4.1 【商店建规】1.0.4 条：商店建筑的规模应按单项建筑内的商店总建筑面积进行划分，并应符合表 7.4.1 的规定。

7.4.1 [Code for design of store buildings] 1.0.4: The scale of commercial buildings shall be divided in accordance with the total building area of the store in the single building, and conform to the provisions in Tab. 7.4.1.

【商店建规】表 1.0.4：商店建筑的规模划分　　　表 7.4.1

[Code for design of store buildings] Tab. 1.0.4: Scale Division of Commercial Buildings　　Tab. 7.4.1

规模 Scale	小型 Small	中型 Medium	大型 Large
总建筑面积 Total building area	< 5000m²	5000 ~ 20000m²	> 20000m²

7.4.2 【商店建规】4.1.10 及 4.1.11 条：商店建筑宜利用天然采光和自然通风。商店建筑采用自然通风时，其通风开口的有效面积不应小于该房间（楼）地板面积的 1/20。

7.4.2 [Code for design of store buildings] 4.1.10 and 4.1.11: Natural daylighting and ventilation shall be adopted in the commercial buildings. When natural ventilation is adopted in commercial buildings, the effective area of the ventilation opening shall be not less than 1/20 of the floor area of the room.

7.4.3 【商店建规】4.2.10 条：大型和中型商店建筑内连续排列的商铺之间的公共通道最小净宽度应符合表 7.4.3 的规定。

7.4.3 [Code for design of store buildings] 4.2.10: The minimum net width of the public sideway between the continuously arranged shops in large and medium-sized store buildings shall be in accordance with the provisions of Tab. 7.4.3.

【商店建规】表 4.2.10：连续排列的商铺之间的公共通道最小净宽度　　　　表 7.4.3

[Code for design of store buildings] Tab. 4.2.10: The Minimum Net Width of the Public Sideway between the Continuously Arranged Shops　　　Tab. 7.4.3

通道名称 Channel name	最小净宽度（m） Minimum net width（m）	
	通道两侧设置商铺 Set up shops on both sides of the aisle	通道一侧设置商铺 Set up shops on one side of the channel
主要通道 Main channels	4.00，且不小于通道长度的 1／10 4.00, And not less than 1/10 of the channel length	3.00，且不小于通道长度的 1/15 3.00, And not less than 1/15 of the channel length
次要通道 Secondary channels	3.00	2.00
内部作业通道 Internal job Channels	1.80	—

注：主要通道长度按其两端安全出口间距计算。

Note: The length of main sideways is calculated by the distance between the safe exits at both ends.

7.4.4 【商店建规】4.3.8 条：当商店建筑的地下室、半地下室用作商品临时储存、验收、整理和加工场地时，应采取防潮、通风措施。

7.4.4　[Code for design of store buildings] 4.3.8: When basements and semi-basements of the commercial buildings are used as temporary storage, acceptance, finishing and processing sites of goods, moisture-proof and ventilation measures shall be adopted.

7.4.5 【商店建规】5.1.5 条：商店营业厅的吊顶和所有装修饰面，应采用不燃材料或难燃材料，并应符合建筑物耐火等级要求和现行国家标准《建筑内部装修设计防火规范》GB 50222 的规定。

7.4.5　[Code for design of store buildings] 5.1.5: For the suspended ceilings and all decorative surfaces of the business hall of the store, the non-combustible materials or non-flammable materials shall be adopted. They shall conform to the requirements for the fire resistance rating of the building and the provisions in the current national standard, *Code for Fire Prevention in Design of Interior Decoration of Buildings* GB 50222.

7.4.6 【防火规】5.3.4.3 条：一、二级耐火等级建筑内的商店营业厅、展览厅，当设置自动灭火系统和火灾自动报警系统并采用不燃或难燃装修材料时，设置在地下或半地下时，其每个防火分区的最大允许建筑面积不应大于 2000m^2。

7.4.6　[Code for fire protection design of buildings] 5.3.4.3: Business halls and exhibition halls in buildings in fire-resistant Grade I and II, when setting automatic fire extinguishing system and automatic fire alarm system and using non-flammable or refractory decoration materials in the ground or semi-underground, the maximum allowable floor area of each fireproof partition should not be greater than 2000m^2.

7.4.7 【防火规】5.3.5条：总建筑面积大于20000m² 的地下或半地下商店，应采用无门、窗、洞口的防火墙、耐火极限不低于2.00h的楼板分隔为多个建筑面积不大于20000m² 的区域。相邻区域确需局部连通时，应采用下沉式广场等室外开敞空间、防火隔间、避难走道、防烟楼梯间等方式进行连通。

7.4.7 [Code for fire protection design of buildings] 5.3.5: The underground or semi-underground shop with a total building area more than 20000 m² shall be divided into multiple areas with the building area within 20000m² via the fire walls without doors, windows and openings, and the floor slabs with fire resistance not lower than 2.00h. If partial connection is required, the adjacent areas shall be connected in the modes of outdoor open spaces such as sunken squares, or fire compartments, refuge sidewalks, smoke-proof staircases, etc.

7.4.8 【防火规】5.4.3条：营业厅、展览厅不应设置在地下三层及以下楼层。地下或半地下营业厅、展览厅不应经营、储存和展示甲、乙类火灾危险性物品。

7.4.8 [Code for fire protection design of buildings] 5.4.3: The business halls and exhibition halls shall not be arranged on Floor III underground and floors below. The business halls and exhibition halls underground or semi-underground shall not be used to sell, store or display Classes A and B fire hazard materials.

7.4.9 【防火规】5.5.9条：一、二级耐火等级公共建筑内的安全出口全部直通室外确有困难的防火分区，可利用通向相邻防火分区的甲级防火门作为安全出口，但应符合下列要求。

7.4.9 [Code for fire protection design of buildings] 5.5.9: In all the public buildings with Grade I and Grade II fire-resistant, if actual difficulties exist for safety exits to be arranged to directly connect to the outdoor fire compartments, the Grade A fire door to the adjacent fire compartment may be adopted as a safety exit, but the requirements in the following shall be met.

1. 利用通向相邻防火分区的甲级防火门作为安全出口时，应采用防火墙与相邻防火分区进行分隔。

1. When being used as a safety exit, the Grade A fire door to the adjacent fire compartment shall be separated from the adjacent fire compartment by the means of fire walls.

2. 建筑面积大于1000m² 的防火分区，直通室外的安全出口不应少于2个；建筑面积不大于1000m² 的防火分区，直通室外的安全出口不应少于1个。

2. For the fire compartments greater than 1000 m², the number of the safety exits directly to the outdoor shall be not less than 2; for the fire compartments with the building area not greater than 1000 m², the number of the safety exits directly to the outdoor shall be not less than 1.

3. 该防火分区通向相邻防火分区的疏散净宽度不应大于其按【防火规】5.5.21条（见本书6.4.4条）规定计算所需疏散总净宽度的30%，建筑各层直通室外的安全出口总净宽度不应小于按照【防火规】5.5.21条（见本书6.4.4条）规定计算所需疏散总净宽度。

3. The net width of the evacuation corridors from the fire compartments to the adjacent fire compartments shall be not greater than 30% of the required total net evacuation width calculated as per the specifications in Clause 5.5.21 of the Code. The total net width of the safety exits directly to the outdoor on all the floors of the buildings shall be not less than the required total net evacuation width calculated as per the specifications in Clause 5.5.21 of the Code.

7.4.10 【防火规】5.5.14 条：公共建筑内的客、货电梯宜设置电梯候梯厅，不宜直接设置在营业厅、展览厅、多功能厅等场所内。

7.4.10 [Code for fire protection design of buildings] 5.5.14: The passenger and freight elevators in public buildings shall be provided with elevator waiting halls, and shall not be directly arranged in the places of business halls, exhibition halls, multi-function halls, etc.

7.4.11 【防火规】5.5.17.4 条：一、二级耐火等级建筑内疏散门或安全出口不少于 2 个的观众厅、展览厅、多功能厅、餐厅、营业厅等。其室内任一点至最近疏散门或安全出口的直线距离不应大于 30m；当疏散门不能直通室外地面或疏散楼梯间时，应采用长度不大于 10m 的疏散走道通至最近的安全出口。当该场所设置自动喷水灭火系统时，室内任一点至最近安全出口的安全疏散距离可分别增加 25%。

7.4.11 [Code for fire protection design of buildings] 5.5.17.4: For auditoriums, exhibition halls, multi-function halls, dining halls, business halls and others, with the number of the evacuation doors or safety exits not less than 2 in Grade I and Grade II fire-resistant buildings, the straight-line distance from any point in the room to the nearest evacuation door or safety exit shall not be greater than 30 m. When the evacuation door shall not be directly connected to the outdoor floor or evacuation staircase, the evacuation sidewalk with the length not greater than 10 m shall be adopted, to be connected to the nearest safety exit. When the automatic sprinkler system is arranged in this place, the safety evacuation distance from any indoor point the nearest safety exit can be increased by 25% respectively.

7.4.12 【防火规】5.5.21.2 条：地下或半地下人员密集的厅、室和歌舞娱乐放映游艺场所，其房间疏散门、安全出口、疏散走道和疏散楼梯的各自总净宽度，应根据疏散人数按每 100 人不小于 1.00m 计算确定。

7.4.12 [Code for fire protection design of buildings] 5.5.21.2: The respective total net width of evacuation doors, safety exits, evacuation sidewalks and evacuation staircases for rooms of underground or semi-underground crowded halls, rooms and karaoke bars, dancing halls, entertainment rooms, video halls, amusement halls, and others shall be determined in accordance with the number of evacuees as not less than 1.00 m per 100 people.

7.4.13 【防火规】5.5.21.3 条：首层外门的总净宽度应按该建筑疏散人数最多一层的人数计算确定，不供其他楼层人员疏散的外门，可按本层的疏散人数计算确定。

7.4.13 [Code for fire protection design of buildings] 5.5.21.3: The total net width of the first-floor exterior evacuation doors shall be calculated and determined as per the largest number of evacuees at the floor in the building. The total net width of the interior doors not for the personnel evacuation of other floors can be calculated and determined as per the number of evacuees of the floor.

7.4.14 【防火规】5.5.21.7 条：商店的疏散人数应按每层营业厅的建筑面积乘以表7.4.14规定的人员密度计算。对于建材商店、家具和灯饰展示建筑，其人员密度可按表7.4.14规定值的30%确定。

7.4.14 [Code for fire protection design of buildings] 5.5.21.7: The number of stores evacuees shall be calculated by multiplying the building area of the business hall on each floor by the personnel density specified in table 7.4.14. for building material stores, furniture and lighting display buildings, the personnel density can be determined by 30% of the specified values in table7.4.14.

【防火规】表 5.5.21-2：商店营业厅内的人员密度（人/m²）　　　　表 7.4.14

[Code for fire protection design of buildings] Tab. 5.5.21-2: Personnel Density of the Shops and Business Halls (person/m²)　　Tab. 7.4.14

楼层位置 Position of floor	地下第二层 Second floor underground	地下第一层 First floor underground	地上第一、二层 First and second floors above the ground	地上第三层 Third floor above the ground	地上第四层 Fourth floor above the ground 及以上各层 and all the floors above
人员密度 Personnel density	0.56	0.60	0.43 ~ 0.60	0.39 ~ 0.54	0.30 ~ 0.42

7.4.15 【防火规】6.4.12 条：用于防火分隔的下沉式广场等室外开敞空间，应符合下列规定。

7.4.15 [Code for fire protection design of buildings] 6.4.12: The outdoor open spaces such as sunken squares, etc., for the fire separation shall conform to the following provisions.

1. 分隔后的不同区域通向下沉式广场等室外开敞空间的开口最近边缘之间的水平距离不应小于13m。室外开敞空间除用于人员疏散外不得用于其他商业或可能导致火灾蔓延的用途，其中用于疏散的净面积不应小于169m²。

1. The horizontal distance between the different partitioned areas and the nearest edge of the openings of the outdoor open spaces such as sunken plaza, etc., shall be not less than 13m. The outdoor open spaces shall not be used for other commercial purposes or purposes that may cause the fire to spread except for personnel evacuation, and therein, the net area for evacuation shall be not less than 169m².

2. 下沉式广场等室外开敞空间内应设置不少于 1 部直通地面的疏散楼梯。当连接下沉广场的防火分区需利用下沉广场进行疏散时，疏散楼梯的总净宽度不应小于任一防火分区通向室外开敞空间的设计疏散总净宽度。

2. Not less than one evacuation staircase directly to the ground shall be arranged in outdoor open spaces such as sunken squares, etc., If the sunken square is required for evacuation in the fire compartment connecting the sunken square, the total net width of the evacuation staircase shall be not less than the total net width of the designed evacuation corridor from any fire compartment to the outdoor open space.

3. 确需设置防风雨篷时，防风雨篷不应完全封闭，四周开口部位应均匀布置，开口的面积不应小于该空间地面面积的 25%，开口高度不应小于 1.0m；开口设置百叶时，百叶的有效排烟面积可按百叶通风口面积的 60% 计算。

3. If the wind prevention canopy is required, the wind prevention canopy shall not be enclosed. The opening positions in the surroundings shall be evenly arranged, and the area of the openings shall be not less than 25% of the floor area of the space and the height of the openings shall be not less than 1.0m; When the louver is arranged at the opening, the effective smoke exhaust area of the louver can be calculated as per 60% of the area of the vent louver.

7.4.16 【防火规】6.4.13 条：防火隔间的设置应符合下列规定。

7.4.16 [Code for fire protection design of buildings] 6.4.13: The arrangement of the fire compartments shall conform to the following provisions.

1. 防火隔间的建筑面积不应小于 $6.0m^2$；

1. The building area of fire compartments shall be not less than $6.0m^2$;

2. 防火隔间的门应采用甲级防火门；

2. For the doors of fire compartments, Grade A fire doors shall be adopted;

3. 不同防火分区通向防火隔间的门不应计入安全出口，门的最小间距不应小于 4m；

3. The doors of different fire compartments to fire compartments shall be not included into the safety exits, and the minimum spacing of the door shall be not less than 4m;

4. 防火隔间内部装修材料的燃烧性能应为 A 级；

4. The combustion performance of decorative materials in the fire compartments shall be of Grade A;

5. 不应用于除人员通行外的其他用途。

5. The fire compartments shall not be used for other purposes except for personnel passage.

7.4.17 【防火规】6.4.14 条：避难走道的设置应符合下列规定。

7.4.17 [Code for fire protection design of buildings] 6.4.14: The arrangement of refuge corridors shall conform to the following provisions.

1. 避难走道防火隔墙的耐火极限不应低于 3.00h，楼板的耐火极限不应低于 1.50h。

1. The fire endurance of the fire partitions in the refuge corridors shall be not less than 3.00 hours, and the fire endurance of the floor slabs shall be not less than 1.50 hours.

2. 避难走道直通地面的出口不应少于2个，并应设置在不同方向；当避难走道仅与一个防火分区相通且该防火分区至少有1个直通室外的安全出口时，可设置1个直通地面的出口。任一防火分区通向避难走道的门至该避难走道最近直通地面的出口的距离不应大于60m。

2. The number of the exits of the refuge corridors directly to the ground shall be not less than 2, and the exits shall be arranged in different directions; When the refuge corridors are only connected to one fire compartment and at least 1 safety exit directly to the outdoor floor is arranged for the fire compartment, and 1 exit directly to the ground can be arranged. The distance from the door of any fire compartment to the refuge corridor to the nearest exit of the refuge corridor directly to the ground shall be not greater than 60m.

3. 避难走道的净宽度不应小于任一防火分区通向该避难走道的设计疏散总净宽度。

3. The net width of refuge corridors shall be not less than the total net width of the designed evacuation of any fire compartments to the refuge corridors.

4. 避难走道内部装修材料的燃烧性能应为A级。

4. The combustion performance of the interior decorative materials for refuge corridors shall be of Grade A.

5. 防火分区至避难走道入口处应设置防烟前室，前室的使用面积不应小于$6.0m^2$，开向前室的门应采用甲级防火门，前室开向避难走道的门应采用乙级防火门。

5. The smoke-free ante-room shall be arranged at the entrance of the fire compartment to the refuge corridor, the usable floor area of the ante-room shall be not less than $6.0m^2$. Grade A fire door shall be used as the door opening to the ante-room. The Grade B fire door shall be used as the door of the ante-room opening to the refuge corridor.

6. 避难走道内应设置消火栓、消防应急照明、应急广播和消防专线电话。

6. Fire hydrants, fire emergency lighting, emergent broadcast and special line for fire protection shall be arranged in refuge corridors.

7.4.18 【防火规】7.3.1.3条：设置消防电梯的建筑的地下或半地下室、埋深大于10m且总建筑面积大于$3000m^2$的其他地下或半地下建筑（室）应设置消防电梯。

7.4.18 [Code for fire protection design of buildings] 7.3.1.3: Underground or semi-underground rooms in buildings provided with fire elevators and other underground or semi-underground（building）rooms with the burial depth greater than 10 m and the total building area greater than $3000m^2$.

7.4.19 【防火规】7.3.2条：消防电梯应分别设置在不同防火分区内，且每个防火分区不应少于1台。

7.4.19 [Code for fire protection design of buildings] 7.7.2: Fire elevators shall be separately set in different fire compartments with at least 1 set for each fire compartment.

7.4.20 【防火规】7.3.4 条：符合消防电梯要求的客梯或货梯可兼作消防电梯。

7.4.20 [Code for fire protection design of buildings] 7.3.4: Passenger elevators or freight elevators meeting the requirements of fire elevators can also be used as fire elevators.

7.5 地下室防水

7.5 Waterproof of Basements

7.5.1 【地下防水规】1.0.3 条：地下工程防水的设计和施工应遵循"防、排、截、堵相结合，刚柔相济，因地制宜，综合治理"的原则。

7.5.1 [Technical code for waterproofing of underground works] 1.0.3: For the design and construction for the waterproof of underground works, the principles of combination of waterproof, drainage, intercepting and blocking, adoption of rigid and flexible waterproof materials, adjustment as per specific conditions and comprehensive treatment shall be followed.

7.5.2 【地下防水规】1.0.4 条：地下工程防水的设计和施工应符合环境保护的要求，并应采取相应措施。

7.5.2 [Technical code for waterproofing of underground works] 1.0.4: The design and construction for waterproof of underground works shall conform to the environmental protection requirements, and the appropriate measures shall be adopted.

7.5.3 【地下防水规】1.0.5 条：地下工程的防水，应积极采用经过试验、检测和鉴定并经实践检验质量可靠的新材料、新技术、新工艺。

7.5.3 [Technical code for waterproofing of underground works] 1.0.5: For the waterproof of underground works, new materials, new technologies and new processes tested and identified and proven to be reliable in quality shall be adopted.

7.5.4 【地下防水规】3.1.1 条：地下工程应进行防水设计，并应做到定级准确、方案可靠、施工简便、耐久适用、经济合理。

7.5.4 [Technical code for waterproofing of underground works] 3.1.1: Waterproof design shall be carried out for the underground works, and the key points of exact gradation, reliable planning, convenient construction, high applicability and lasting durability, and economic rationality shall be followed.

7.5.5 【地下防水规】3.1.2条：地下工程防水方案应根据工程规划、结构设计、材料选择、结构耐久性和施工工艺等确定。

7.5.5 [Technical code for waterproofing of underground works] 3.1.2: The waterproof program for underground works shall be determined in accordance with engineering planning, structural design, material selection, structural durability, construction techniques, etc.

7.5.6 【地下防水规】3.1.3条：地下工程的防水设计，应根据地表水、地下水、毛细管水等的作用，以及由于人为因素引起的附近水文地质改变的影响确定。单建式的地下工程，宜采用全封闭、部分封闭的防排水设计；附建式的全地下或半地下工程的防水设防高度，应高出室外地坪高程500mm以上。

7.5.6 [Technical code for waterproofing of underground works] 3.1.3: The waterproof design of the underground works shall be determined in accordance with the action of surface water, underground water, capillary pipe water, etc., and the influence of the changes in hydrogeological characteristics of nearby underground water caused by human factors. The enclosed or partially enclosed water preventive and drainage design shall be adopted for separately-built underground works; the waterproof fortification height of affiliated underground or semi-underground works shall be 500mm more than the elevation of outdoor floor.

7.5.7 【地下防水规】3.1.4条：地下工程迎水面主体结构应采用防水混凝土，并应根据防水等级的要求采取其他防水措施。

7.5.7 [Technical code for waterproofing of underground works] 3.1.4: The main structures of the water-facing surfaces of the underground works shall be made of waterproof concrete, and other waterproof measures shall be taken as per the requirements of waterproof grades.

7.5.8 【地下防水规】3.1.6条：地下工程的排水管沟、地漏、出入口、窗井、风井等，应采取防倒灌措施；寒冷及严寒地区的排水沟应采取防冻措施。

7.5.8 [Technical code for waterproofing of underground works] 3.1.6: Anti-backflow measures shall be taken for the drainage trenches, floor drains, entrances and exits, window wells, air wells, etc., of the underground works; and the protective measures shall be adopted for the drainage ditches in cold and severely cold regions.

7.5.9 【地下防水规】3.1.7条：地下工程的防水设计，应根据工程的特点和需要搜集下列资料。

7.5.9 [Technical code for waterproofing of underground works] 3.1.7: The following information shall be collected in accordance with characteristics and requirements of the project for waterproof design of the underground works.

1. 最高地下水位的高程、出现的年代，近几年的实际水位高程和随季节变化情况；

1. Elevation and occurrence age of the highest groundwater level and the elevations and seasonal changes of the actual water levels in recent years;

2. 地下水类型、补给来源、水质、流量、流向、压力；

2. Type, supply source, water quality, flow, flow direction and pressure of groundwater;

3. 工程地质构造，包括岩层走向、倾角、节理及裂隙，含水地层的特性、分布情况和渗透系数，溶洞及陷穴、填土区、湿陷性土和膨胀土层等情况；

3. Geological structures of the construction, including the conditions of the strike direction, inclination, joint and fracture of the stratum, the characteristics, distribution, permeability coefficient of the water-bearing ground, karst cave and sink hole, landfill, collapsible soil and expansive soil layer, etc.;

4. 历年气温变化情况、降水量、地层冻结深度；

4. Temperature changes, precipitation and freezing depth of the strata over the years;

5. 区域地形、地貌、天然水流、水库、废弃坑井以及地表水、洪水和给水排水系统资料；

5. Data of regional terrain, landform, natural water flow, reservoir, abandoned pit and surface water, flood and water supply and drainage systems;

6. 工程所在区域的地震烈度、地热，含瓦斯等有害物质的资料；

6. The information on the seismic intensity, geotherm, harmful substance content of gas, etc., of the area of the project;

7. 施工技术水平和材料来源。

7. Technical level and material source for construction.

7.5.10 【地下防水规】3.1.8 条：地下工程防水设计，应包括下列内容。

7.5.10 [Technical code for waterproofing of underground works] 3.1.8: The waterproof design of the underground works shall include the following contents.

1. 防水等级和设防要求；

1. Waterproof grade and fortification requirements;

2. 防水混凝土的抗渗等级和其他技术指标、质量保证措施；

2. Impermeability grade of the waterproof concrete and other technical indexes and quality assurance measures;

3. 其他防水层选用的材料及其技术指标、质量保证措施；

3. Materials selected for the other waterproof layers and the technical indexes and quality assurance measures;

4. 工程细部构造的防水措施，选用的材料及其技术指标、质量保证措施；

4. Waterproof measures and materials selected of the detail structure of the project and the technical indexes and quality assurance measures;

5. 工程的防排水系统、地面挡水、截水系统及工程各种洞口的防倒灌措施。

5. Water preventing and draining system and the ground water-blocking and water-intercepting

systems for the project and the anti-backflow measures for various openings of the project.

7.5.11 【地下防水规】3.2.1 条和 3.2.2 条：地下工程的防水等级应分为四级，其各等级防水标准和适用范围见表 7.5.11。

7.5.11 [Technical code for waterproofing of underground works] 3.2.1 and 3.2.2: The waterproof grades of the underground works shall be divided into four grades, and for the criteria and application scope of the waterproof of all the grades, see Tab. 7.5.11.

【地下防水规】表 3.2.1 及表 3.2.2：地下工程防水等级和适用范围　　　　表 7.5.11
[Technical code for waterproofing of underground works] Tab. 3.2.1 and Tab. 3.2.2:
Waterproof grade and Application Scope for Underground Works　　Tab. 7.5.11

防水等级 Waterproof grade	防水标准 Waterproof criteria	适用范围 Application scope
一级 Grade I	不允许渗水，结构表面无湿渍 Water seepage is not allowed, and no wet stain shall be found on the structure surface	人员长期停留的场所；因有少量湿渍会使物品变质、失效的贮物场所及严重影响设备正常运转和危及工程安全运营的部位；极重要的战备工程、地铁车站 Places with people staying for a long term; Storage places with a small amount of wet stains that will lead to deterioration and ineffectiveness of materials and positions that seriously affect the normal operation of equipment and endanger the safe operation of projects; Extremely important war readiness engineering and subway stations
二级 Grade II	不允许漏水，结构表面可有少量湿渍。 Water leakage is not allowed, and a small amount of wet stains can be allowed on the structure surface. 工业与民用建筑：总湿渍面积不应大于总防水面积（包括顶板、墙面、地面）的 1/1000；任意 100m² 防水面积上的湿渍不超过 2 处，单个湿渍的最大面积不大于 0.1m²。 Industrial and civil buildings: the total area of wet stains shall be not greater than 1/1000 of the total waterproof area (including top board, wall surface and floor); the number of wet stain places on any 100 m² waterproof area shall not exceed 2, and the maximum area of a single wet stain shall not be greater than 0.1 m². 其他地下工程：总湿渍面积不应大于总防水面积的 2/1000；任意 100m² 防水面积上的湿渍不超过 3 处，单个湿渍的最大面积不大于 0.2m²；其中，隧道工程还要求平均渗水量不大于 0.05L/（m²·d），任意 100m² 防水面积上的渗水量不大于 0.15L/（m²·d） Other underground works: the total area of wet stains shall be not greater than 2/1000 of the total waterproof area; the wet stains on any 100m² waterproof area shall not exceed 3 places, and the maximum area of a single wet stain shall be not greater than 0.2 m²; therein, the average water seepage in the tunnel works shall be not greater than 0.05 L/（m²·d）, and the water seepage in any 100m² waterproof area shall be not greater than 0.15 L/（m²·d）	人员经常活动的场所；在有少量湿渍的情况下不会使物品变质、失效的贮存物场所及基本不影响设备正常运转和工程安全运营的部位；重要的战备工程 Places with frequent people activities; storage places where materials will not be deteriorated and invalid in case of a small amount of wet stain and positions basically not affecting the normal operation of the equipment and safe operation of the project; important war readiness engineering

续表

防水等级 Waterproof grade	防水标准 Waterproof criteria	适用范围 Application scope
三级 Grade III	有少量漏水点，不得有线流和漏泥砂； There are few water leakage points, and line flow and muddy sand leakage are not allowed； 任意100m² 防水面积上的漏水点数不超过7处，单个漏水点的最大漏水量不大于2.5L/d，单个湿渍的最大面积不大于0.3m² The number of water leakage points on any 100m² waterproof area shall not exceed 7, the maximum water leakage amount at a single water leakage point shall be not greater than 2.5L/d, and the maximum area of a single wet stain shall be not greater than 0.3m²	人员临时活动的场所；一般战备工程 Places with temporary people activities; general war readiness engineering
四级 Grade IV	有漏水点，不得由线流和漏泥砂； There are water leakage points, and line flow and muddy sand leakage are not allowed； 整个工程平均漏水量不大于2L/（m²·d），任意100m² 防水面积上的平均漏水量不大于4L/（m²·d） The average water leakage amount of the whole project shall be not greater than 2L/（m²·d）; and the average water leakage amount on any 100m² waterproof area shall be not greater than 4L/（m²·d）	对渗漏水无严格要求的工程 Projects without strict requirements for water leakage

7.5.12【地下防水规】3.3.1 条：地下工程的防水设防要求，应根据使用功能、使用年限、水文地质、结构形式、环境条件、施工方法及材料性能等因素确定。

7.5.12 [Technical code for waterproofing of underground works] 3.3.1: The waterproof fortification requirements for underground works shall be determined in accordance with the factors of using function, service life, hydrological geology, structural form, environmental conditions, construction methods, material properties, etc.

1. 明挖法地下工程的防水设防要求应按表 7.5.12-1 选用；

1. The fortification requirements for waterproof for underground works in the open-cut method shall be determined as per Tab. 7.5.12-1;

2. 暗挖法地下工程的防水设防要求应按表 7.5.12-2 选用。

2. The fortification requirements for waterproof for underground works in the subsurface excavation method shall be determined as per Tab. 7.5.12-2.

7.5.13【地下防水规】4.1.1 条：防水混凝土可通过调整配合比，或掺加外加剂、掺合料等措施配制而成，其抗渗等级不得小于 P6。

7.5.13 [Technical code for waterproofing of underground works] 4.1.1: The waterproof concrete can be mixed through measures of adjusting mixing ratio, adding additive and admixture, etc., and the impermeability grade shall not be less than P6.

7.5.14【地下防水规】4.1.2 条：防水混凝土的施工配合比应通过试验确定，试配混凝土的抗渗等级应比设计要求提高 0.2MPa。

124 建筑设计技术指南 Architecture Design Technical Guide

[地下防水规] 表 3.3.1-1: 明挖法地下工程的防水设防要求　　　　表 7.5.12-1

[Technical code for waterproofing of underground works] Tab. 3.3.1-1:
Fortification Requirements for Waterproof for Underground Works in the Open-cut Method　　Tab. 7.5.12-1

工程部位 Positions of engineering	防水措施 Waterproof measures																							
		主体结构 Main structure							施工缝 Construction joint								后浇带 Post-pouring belt					变形缝（诱导缝） Deformation joint (induction joint)			
防水等级 Waterproof level		防水混凝土 Waterproof concrete	防水卷材 Waterproof coiled material	防水涂料 Waterproof coating	塑料防水板 Plastic waterproof board	膨润土防水材料 Waterproof bentonite materials	防水砂浆 Waterproof mortar	金属防水板 Metal waterproof board	遇水膨胀止水条（胶） Water-swelling strip (rubber)	外贴式止水带 Externally bonded waterstop	中埋式止水带 Buried waterstop	外抹防水砂浆 Waterproof mortar plastered outside	外涂防水涂料 Overcoated waterproof coating	水泥基渗透结晶型防水涂料 Cement-based permeable crystalline waterproof coating	预埋注浆管 Pre-embedded grouting pipe	补偿收缩混凝土 Shrinkage compensation concrete	外贴式止水带 Externally bonded waterstop	预埋注浆管 Pre-embedded grouting pipe	遇水膨胀止水条（胶） Water-swelling strip (rubber)	防水密封材料 Waterproof sealing material	中埋式止水带 Buried waterstop	外贴式止水带 Externally bonded waterstop	可卸式止水带 Detachable waterstop	防水密封材料 Waterproof sealing material	外贴式防水卷材 Externally bonded waterproof coiled material	外涂防水涂料 Overcoated waterproof coating
一级 Grade I	应选 To be selected	应选一至二种 One or two types shall be selected							应选二种 Two types shall be selected							应选 To be selected	应选二种 Two types shall be selected				应选 To be selected	应选二种 Two types shall be selected			
二级 Grade II	应选 To be selected	应选一种 One type shall be selected							应选一至二种 One or two types shall be selected							应选 To be selected	应选一至二种 One or two types shall be selected				应选 To be selected	应选一至二种 One or two types shall be selected			
三级 Grade III	应选 To be selected	宜选一种 One type shall be selected							宜选一至二种 One or two types shall be selected							应选 To be selected	宜选一至二种 One or two types shall be selected				应选 To be selected	宜选一至二种 One or two types shall be selected			
四级 Grade IV	应选 shall be selected	—							宜选一种 One type shall be selected							应选 To be selected	宜选一种 One type shall be selected				应选 To be selected	宜选一种 One type shall be selected			

7 地下室和半地下室 Basements and Semi-basements

[地下防水规] 表 3.3.1-2：暗挖法地下工程的防水设防要求
[Technical code for waterproofing of underground works] Tab. 3.3.1-2
Fortification Requirements for Waterproof for Underground Works in the Subsurface Excavation Method

表 7.5.12-2 / Tab. 7.5.12-2

工程部位 Positions of engineering	防水混凝土 Waterproof concrete	衬砌结构 Lining structure					内衬砌施工缝 Construction joint of inner lining						内衬砌变形缝（诱导缝）Deformation joint (induction joint) of inner lining				
防水措施 Waterproof measures		塑料防水板 Plastic waterproof board	防水砂浆 Waterproof mortar	防水涂料 Waterproof coating	防水卷材 Waterproof coiled material	金属防水板 Metal waterproof board	外贴式止水带 Externally bonded waterstop belt	预埋注浆管 Pre-embedded grouting pipe	遇水膨胀止水条（胶）Water-swelling strip (rubber)	防水密封材料 Waterproof sealing material	中埋式止水带 Buried waterstop	水泥基渗透结晶型防水涂料 Cement-based permeable crystalline waterproof coating	中埋式止水带 Buried waterstop	外贴式止水带 Externally bonded waterstop	可卸式止水带 Detachable waterstop	防水密封材料 Waterproof sealing material	遇水膨胀止水条（胶）Water-swelling strip (rubber)
一级 Grade I	必选 Required	应选一至二种 One or two types shall be selected					应选一至二种 One or two types shall be selected				应选 To be selected		应选 To be selected	应选一至二种 One or two types shall be selected			
二级 Grade II	应选 To be selected	应选一种 One type shall be selected					应选一种 One type shall be selected				应选 To be selected		应选 To be selected	应选一种 One type shall be selected			
三级 Grade III	宜选 Suitable for selection	宜选一种 advisable to choose one					宜选一种 advisable to choose one				宜选 Should choose		宜选 Should choose	宜选一种 advisable to choose one			
四级 Grade IV	宜选 Suitable for selection	宜选一种 advisable to choose one					宜选一种 advisable to choose one				宜选 Should choose		宜选 Should choose	宜选一种 advisable to choose one			

防水等级 Waterproof grade

7.5.14 [Technical code for waterproofing of underground works] 4.1.2: The construction mixing ratio of the waterproof concrete shall be determined through testing, and the impermeability grade of tested concrete shall be improved by 0.2MPa more than the grade specified in the design requirements.

7.5.15【地下防水规】4.1.3 条：防水混凝土应满足抗渗等级要求，并应根据地下工程所处的环境和工作条件，满足抗压、抗冻和抗侵蚀性等耐久性要求。

7.5.15 [Technical code for waterproofing of underground works] 4.1.3: The waterproof concrete shall meet the impermeability grade requirements, and meet the durability requirements for compression resistance, frost resistance, erosion resistance, etc., as per the environment and working conditions of underground works.

7.5.16【地下防水规】4.1.4 条：防水混凝土的设计抗渗等级，应符合表 7.5.16 的规定。

7.5.16 [Technical code for waterproofing of underground works] 4.1.4: The designed impermeability grade of the waterproof concrete shall conform to the provisions in Tab. 7.5.19.

【地下防水规】表 4.1.4：防水混凝土的设计抗渗等级　　　　表 7.5.16

[Technical code for waterproofing of underground works] Tab. 4.1.4: Designed Impermeability Grade of Waterproof Concrete　　Tab. 7.5.16

工程埋置深度 H（m） Engineering embedment depth H (m)	设计抗渗等级 Designed impermeability grade
$H < 10$	P6
$10 \leq H < 20$	P8
$20 \leq H < 30$	P10
$H \geq 30$	P12

注：1. 本表适用于 Ⅰ、Ⅱ、Ⅲ类围岩（土层及软弱围岩）。
Notes: 1. This Table is applicable to Categories Ⅰ, Ⅱ and Ⅲ surrounding rock (soil layer and weak surrounding rock).
2. 山岭隧道防水混凝土的抗渗等级可按国家现行有关标准执行。
2. The anti-seepage grade of waterproof concrete in mountain tunnel can be implemented according to the relevant standards in the country.

7.5.17【地下防水规】4.1.5 条：防水混凝土的环境温度不得高于 80℃；处于侵蚀性介质中防水混凝土的耐侵蚀要求应根据介质的性质按有关标准执行。

7.5.17 [Technical code for waterproofing of underground works] 4.1.5: The ambient temperature of waterproof concrete shall be not higher than 80℃; the corrosion resistance requirements for waterproof concrete in erosive mediums shall be performed as per the properties of the medium in accordance with the relevant standards.

7.5.18【地下防水规】4.1.6 条：防水混凝土结构底板的混凝土垫层，强度等级不应小于 C15，厚度不应小于 100mm，在软弱土层中不应小于 150mm。

7.5.18 [Technical code for waterproofing of underground works] 4.1.6: The strength grade of the concrete cushion layer of the waterproof concrete structure soleplate shall be not less than C15, and the thickness shall be not less than 100mm and shall be not less than 150mm in the soft soil layer.

7.5.19 【地下防水规】4.1.7 条：防水混凝土结构，应符合下列规定。

7.5.19 [Technical code for waterproofing of underground works] 4.1.7: The waterproof concrete structure shall conform to the following provisions.

1. 结构厚度不应小于 250mm；

1. The thickness of the structure shall be not less than 250mm;

2. 裂缝宽度不得大于 0.2mm，并不得贯通；

2. The crack width shall be not greater than 0.2mm, and the cracks shall not be penetrative;

3. 钢筋保护层厚度应根据结构的耐久性和工程环境选用，迎水面钢筋保护层厚度不应小于 50mm。

3. The thickness of the reinforcement protection layer shall be selected as per the durability of the structure and the engineering environment, and the thickness of the reinforcement protection layer on the upstream face shall be not less than 50mm.

7.5.20 【地下防水规】4.3.23 及 4.4.5 条：地下室防水典型构造如图 7.5.20-1、图 7.5.20-2A、图 7.5.20-2B 所示。

7.5.20 [Technical code for waterproofing of underground works] 4.3.23 and 4.4.5: Typical structure diagram for waterproof of basement see Tab.7.5.20-1、Tab.7.5.20-2A、Tab.7.5.20-2B.

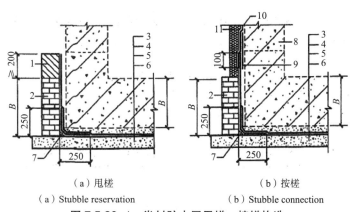

（a）甩槎　　　　　　　　（b）按槎
(a) Stubble reservation　　(b) Stubble connection

图 7.5.20-1 卷材防水层甩槎、接槎构造
Fig.7.5.20-1 Stubble Reservation and Connection Structures on Coil Waterproof Layer

1- 临时保护墙；2- 永久保护墙；3- 细石混凝土保护层；4- 卷材防水层；5- 水泥砂浆找平层；6- 混凝土垫层；7- 卷材加强层；8- 结构墙体；9- 卷材加强层；10- 卷材防水层；11- 卷材保护层

1–Temporary protection wall; 2–Permanent protection wall; 3–Fine-aggregated concrete protective layer; 4–Coiled material waterproof layer; 5–Cement mortar leveling layer; 6–Concrete cushion layer; 7–Coil reinforcement layer; 8–Structural wall; 9–Coil reinforcement layer; 10–Coiled material waterproof layer; 11–Coil protection layer

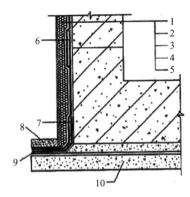

图7.5.20-2A 【地下防水规】图4.4.5-1：防水涂料外防外涂构造
Fig.7.5.20-2A [Technical code for waterproofing of underground works] Fig. 4.4.5-1: Exterior Prevention and Exterior Coating Structure of Waterproof Coating

1- 保护墙；2- 砂浆保护层；3- 涂料防水层；4- 砂浆找平层；5- 结构墙体；6- 涂料防水层加强层；7- 涂料防水加强层；
8- 涂料防水层搭接部位保护层；9- 涂料防水层搭接部位；10- 混凝土垫层
1-Protective Wall; 2-Protective Mortar Layer; 3-Waterproof Coating Layer; 4-Mortar Leveling Layer; 5-Structural Wall; 6-Reinforcement Layer of Waterproof Coating Layer; 7- Waterproof Reinforcement Coating Layer; 8-Protective Layer of the Overlapping Position of the Waterproof Coating Layer; 9-Overlapping Position of the Waterproof Coating Layer; 10-Concrete Cushion Layer

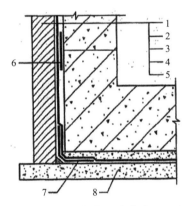

图7.5.20-2B 防水涂料外防内涂构造【地下防水规】图4.4.5-2
Fig. 7.5.20-2B Exterior Prevention and Interior Coating Structure of Waterproof Coating [Technical code for waterproofing of underground works] Fig. 4.4.5-2

1- 保护墙；2- 涂料保护层；3- 涂料防水层；4- 找平层；5- 结构墙体；6- 涂料防水层加强层；7- 涂料防水加强层；8- 混凝土垫层
1- Protective Wall; 2-Protective Coating Layer; 3-Waterproof Coating Layer; 4-Leveling Layer; 5-Structural Wall; 6-Reinforcement Layer of Waterproof Coating Layer; 7-Waterproof Reinforcement Coating Layer; 8-Concrete Cushion Layer

7.5.21 【地下防水规】4.8.1条：地下工程种植顶板的防水等级应为一级。

7.5.21 [Technical code for waterproofing of underground works] 4.8.1: The waterproof grade of planting top boards in underground works shall be Grade I.

7.5.22 【地下防水规】4.8.2条：种植土与周边自然土体不相连，且高于周边地坪时，应按种植屋面要求设计。【地下防水规】4.8.3条：地下工程种植顶板结构应符合下列规定。

7.5.22 [Technical code for waterproofing of underground works] 4.8.2: When the planting soil is not connected with the ambient natural soil and is higher than the peripheral floor, the boards shall be designed as per the requirements of planting roofs. [Technical code for waterproofing of underground works] 4.8.3: Planting top board structures in underground works shall meet the following provisions.

1. 种植顶板应为现浇防水混凝土，结构找坡，坡度宜为1%~2%；

1. The planting top boards shall be cast-in-place waterproof concrete boards with the sloped structure, and the gradient shall be 1%~2%;

2. 种植顶板厚度不应小于250mm，最大裂缝宽度不应大于0.2mm，并不得贯通。

2. The thickness of planting top board shall be not less than 250mm, and the maximum crack width shall not be greater than 0.2mm, and the cracks shall not be penetrative.

7.5.23 【地下防水规】4.8.9条：地下工程种植顶板的防排水构造应符合下列要求。

7.5.23 [Technical code for waterproofing of underground works] 4.8.9: Waterproof and drainage structures of planting top boards in underground works shall meet the requirements in the following.

1. 耐根穿刺防水层应铺设在普通防水层上面。

1. The root-penetration-resistant waterproof layer shall be laid on the ordinary waterproof layer.

2. 耐根穿刺防水层表面应设置保护层，保护层与防水层之间应设置隔离层。

2. A protective layer shall be set on the surface of the root-penetration-resistant waterproof layer, and an isolation layer shall be set between the protective layer and waterproof layer.

3. 排（蓄）水层应根据渗水性、储水量、稳定性、抗生物性和碳酸盐含量等因素进行设计；排（蓄）水层应设置在保护层上面，并应结合排水沟分区设置。

3. The water drainage (storage) layer shall be designed in accordance with the factors of water permeability, water storage capacity, stability, biological resistance, carbonate content, etc.; the water drainage (storage) layer of the drainage ditch shall be arranged on the protective layer and arranged in areas in combination with the drainage ditch.

4. 排（蓄）水层上应设置过滤层，过滤层材料的搭接宽度不应小于200mm。

4. A filter layer shall be set on the water drainage (storage) layer, and the overlapping width of the material of the filter layer shall be not less than 200mm.

5. 种植土层与植被层应符合国家现行标准《种植屋面工程技术规程》JGJ 155的有关规定。

5. The planting soil layer and vegetation layer shall conform to the relevant provisions of the current national standard, *Technical Specification for Planting Roof Engineering* JGJ 155.

7.5.24 【地下防水规】4.8.10条：地下工程种植顶板防水材料应符合下列要求。

7.5.24 [Technical code for waterproofing of underground works] 4.8.10: The waterproof materials for planting top boards in the underground works shall conform to the requirements in the following.

1. 绝热（保温）层应选用密度小、压缩强度大、吸水率低的绝热材料，不得选用散状绝热材料；

1. For thermal insulation layers, the insulation materials with low density, high compressive strength and low water absorption rate shall be selected, rather than bulk insulation materials;

2. 耐根穿刺层防水材料的选用应符合国家相关标准的规定或具有相关权威检测机构出具的材料性能检测报告；

2. In the selection of waterproof materials for root-penetration-resistant layers, the provisions in relevant national standards shall be met or the material performance test reports issued by relevant authoritative testing institutions shall be provided;

3. 排（蓄）水层应选用抗压强度大且耐久性好的塑料排水板、网状交织排水板或轻质陶粒等轻质材料。

3. For water drainage (storage) layers, the lightweight materials with high compressive strength and durability, such as plastic drainage boards, mesh interwoven drainage boards, lightweight porcelain granule boards, etc., shall be selected.

8

墙体
Walls

8.1 墙体类型与材料

8.1 Types and Materials of Walls

8.1.1 【09措施】4.1.1条：墙体的类型。墙体按其所处部位和性能分为：

8.1.1 [Measures 2009] 4.1.1: Wall types. The walls shall be divided into the following as per the wall positions and performance:

1. 外墙：包括承重墙、非承重墙（如框架结构填充墙）及幕墙。

1. Exterior walls: including bearing walls, non-bearing walls (such as infill walls with frame structures) and curtain walls.

2. 内墙：包括承重墙、非承重墙（包括固定式和灵活隔断式）。

2. Interior walls: including bearing walls and non-bearing walls (including the fixed type and flexible partition type).

8.1.2 【09措施】4.1.2条：墙体的常用材料。

8.1.2 [Measures 2009] 4.1.2: Common materials for the walls.

1. 常用于承重墙的材料有：

1. The materials commonly used for the bearing walls are shown as follows:

1）钢筋混凝土。

1) Reinforced concrete.

2）蒸压类：主要有蒸压加气混凝土砌块、蒸压灰砂砖、蒸压粉煤灰砖等。

2) Autoclaved types: mainly include autoclaved aerated concrete blocks, autoclaved sand-lime bricks, autoclaved fly ash brick, etc.

3）混凝土空心砌块类：主要有普通混凝土小型空心砌块。

3) Hollow concrete blocks: mainly include small common concrete hollow masonry blocks.

4）多孔砖类：主要有烧结多孔砖（孔洞率应不小于25%）、混凝土多孔砖（孔洞率应不小于30%）；烧结多孔砖主要有：黏土、页岩、粉煤灰及煤矸石等品种。

4) Perforated bricks: mainly include sintered perforated bricks (the porosity shall be not less than 25%) and concrete perforated bricks (the porosity shall be not less than 30%); the sintered perforated bricks mainly include the varieties of clay, shale, fly ash, coal gangue, etc.

5）实心砖类：主要有黏土、页岩、粉煤灰及煤矸石等品种（孔洞率不大于25%）。

5) Solid bricks: mainly include the varieties of clay, shale, fly ash, coal gangue, etc. (the porosity shall be not greater than 25%).

2. 常用于非承重墙的砌块材料有：蒸压加气混凝土砌块（包括砂加气混凝土和粉煤灰加气混凝土）、复合保温砌块、装饰混凝土小型空心砌块、轻集料混凝土小型空心砌块（轻集料主要包括：黏土陶粒、页岩陶粒、粉煤灰陶粒、浮石、火山渣、炉渣、自然煤矸石、膨胀矿渣珠、膨胀珍珠岩等材料,轻集料的粒径不宜大于10mm）、石膏砌块(包括实心、空心)、多孔砖（包括烧结多孔砖和混凝土多孔砖）、实心砖（包括烧结实心砖和蒸压实心砖）等。

2. The masonry materials commonly used for non-bearing walls include autoclaved aerated concrete blocks (including sand aerated concrete and coal ash aerated concrete), compound insulation blocks, decorative small hollow concrete blocks, small lightweight-aggregate concrete hollow masonry blocks (the light aggregates mainly include the materials of clay ceramsite, shale ceramsite, fly ash ceramsite, pumice, volcanic slag, cinder, natural coal gangue, expanded slag beads, expanded perlite, etc., and the grain sizes of light aggregates shall be appropriately not greater than 10mm), gypsum blocks (including solid and hollow gypsum blocks), perforated bricks (including sintered perforated bricks and concrete perforated bricks), solid bricks (including sintered solid bricks and autoclaved solid bricks), etc.

3. 常用于非承重墙的板材有：预制钢筋混凝土或GRC墙板、钢丝网抹水泥砂浆墙板、彩色钢板或铝板墙板、轻集料混凝土墙板、加气混凝土墙板、石膏圆孔墙板、轻钢龙骨石膏板或硅钙板等板材类、玻璃隔断等。

3. The plates commonly used for the non-bearing walls include precast reinforced concrete or GRC wall panel, steel wire cement mortar plastered wallboard, color steel plate or aluminum plate wall panel, light aggregate concrete wall panel, aerated concrete wall panel, gypsum round hole wall panel, light steel keel gypsum board, calcium silicate board, glass partition, etc., and glass partitions.

8.1.3 【环控规】4.3.1条:民用建筑工程室内不得使用国家禁止使用、限制使用的建筑材料。

8.1.3 [Code for indoor environmental pollution control of civil building engineering] 4.3.1: Civil Construction engineering indoor shall not use the national prohibited use, restricted use of building materials.

8.1.4 【砌体结构规范】3.1条：砌体结构房屋墙体的一般构造要求

8.1.4 [Code for design of masonry structures] 3.1: General Structural Requirements for Walls of Buildings with Masonry Structures

1. 砌体结构墙体砌块和砂浆的强度等级见表8.1.4-1～表8.1.4-4.

1. For strength grades of blocks and mortars for walls with masonry structures, see Tab. 8.1.4-1 ~ Tab. 8.1.4-4.

【砌体结构规范】3.1.1～3.1.3条：烧结普通砖、烧结多孔砖和砂浆的强度等级　　表 8.1.4-1

[Code for design of masonry structures] 3.1.1 ~ 3.1.3: Strength Grades of the Ordinary Sintered Brick, Sintered Perforated Brick and Mortar　　Tab. 8.1.4-1

材料名称 Material name	强度等级划分 Classification of strength grades					
烧结普通砖、烧结多孔砖 Ordinary sintered brick and sintered perforated brick	MU30	MU25	MU20	MU15	MU10	
砂浆 Mortar	按结构计算 Calculated as per structure			M15	M10	

【砌体结构规范】3.1.1～3.1.3条：蒸压灰砂砖、蒸压粉煤灰砖和砂浆的强度等级　　表 8.1.4-2

[Code for design of masonry structures] 3.1.1 ~ 3.1.3: Strength Grade of Autoclaved Sand-lime Brick, Autoclaved Fly Ash Brick and Mortar　　Tab. 8.1.4-2

材料名称 Material name	强度等级划分 Classification of strength grades					
蒸压灰砂砖、蒸压粉煤灰砖 Autoclaved sand-lime brick and autoclaved fly ash brick	MU25	MU20	MU15	MU10	MU7.5	MU5
砂浆 Mortar	按结构计算 Calculated as per structure		Mb15	Mb10	Mb7.5	Mb5

【砌体结构规范】3.1.1～3.1.3条：混凝土砌块、轻集料混凝土砌块和砂浆的强度等级　　表 8.1.4-3

[Code for design of masonry structures] 3.1.1 ~ 3.1.3: Strength Grades of Concrete Block, Lightweight-aggregate Concrete Block and Mortar　　Tab. 8.1.4-3

材料名称 Material name	强度等级划分 Classification of strength grades				
混凝土砌块、轻集料混凝土砌块 Concrete blocks and lightweight-aggregate concrete blocks	MU20	MU15	MU10	MU7.5	MU5
砂浆 Mortar	按结构计算 Calculated as per structure	Mb15	Mb10	Mb7.5	Mb5

【砌体结构规范】3.1.1～3.1.3条：石材和砂浆的强度等级　　表 8.1.4-4

[Code for design of masonry structures] 3.1.1 ~ 3.1.3: Strength Grades of Stone and Mortar　　Tab. 8.1.4-4

材料名称 Material name	强度等级划分 Classification of strength grades						
石材 Stone	MU100	MU80	MU60	MU50	MU40	MU30	MU20
砂浆 Mortar		M7.5	M7.5	M7.5	M7.5	M7.5	M7.5

注：【砌体结构规范】3.1.3 注：确定砂浆等级时应采用同类块体为砂浆强度试块底膜。

Note: [Code for design of masonry structures] Note of 3.1.3: The blocks of the same type shall be used as the bottom moulds for mortar strength testing blocks.

8.2 墙体防潮、防水、隔汽

8.2 Moisture Proofing, Waterproof and Vapor Proofing of Walls

8.2.1 【通则】6.9.3条：墙身防潮应符合下列要求。

8.2.1 [Code for design of civil buildings] 6.9.3: Wall base Moisture-proof wall moisture should meet the following requirements.

1. 砌体墙应在室外地面以上，位于室内地面垫层处设置连续的水平防潮层；室内相邻地面有高差时，应在高差处墙身侧面加设防潮层。

1. Masonry wall should be above the outdoor ground, located in the indoor floor cushion to set a continuous horizontal moisture-proof layer, indoor adjacent ground has a high difference, should be in the high difference wall side to add moisture-proof layer.

2. 湿度大的房间的外墙或内墙内侧应设防潮层。

2. The room outside wall or inner side of inside wall with large humidity should be set moisture-proof layer.

3. 室内墙面有防水、防潮、防污、防碰等要求时，应按使用要求设置墙裙。

3. If indoor wall has waterproof, moisture-proof, anti-fouling, anti-collision and other requirements, wainscot should be set according to the use requirements.

注：地震区防潮层应满足墙体抗震整体连接的要求。
Note: The moisture-proof layer of the seismic zone should meet the requirements of the overall seismic connection of the wall.

8.2.2 【09措施】4.2.2条：墙面防水。

8.2.2 [Measures 2009] 4.2.2: Waterproof of wall surfaces.

1. 内隔墙：石膏板隔墙用于卫浴间、厨房时，应做墙面防水处理，根部应做C20混凝土条形基础，条形基础高度距完成面不低于100mm。

1. Interior partitions: when gypsum board partition are used in bathrooms and kitchens, the wall surface waterproof treatment shall be made; At the base, the C20 concrete strip foundations shall be made, and the distance from the height of the strip foundation to the finished surface shall be not less than 100mm.

2. 外墙：建筑物外墙应根据工程性质、当地气候条件、所采用的墙体材料及饰面材料等因素确定防水做法。一般外墙防水做法采用防水砂浆，设计时应注意细节的构造处理，如：

2. Exterior walls: for exterior walls of buildings, the waterproof method shall be determined as per the factors of engineering properties, local climatic conditions, wall materials and finish materials adopted, etc. In the waterproof method of the general exterior walls, waterproof mortar will be adopted, and during the design, attention shall be paid to the treatment of the structural details, such as:

1）不同墙体材料交接处应在饰面找平层中铺设钢丝网或玻纤网格布。

1）At the junctions of various wall materials, wire meshes or fiberglass meshes shall be paved in the finish leveling layers.

2）对于墙体采用空心砌块或轻质砖的建筑，基本风压值大于 0.6kPa 或雨量充沛地区，以及对防水有较高要求的建筑等，外墙或迎风面外墙宜采用 20mm 防水砂浆或 7mm 厚聚合物水泥砂浆抹面后，再做外饰面层。

2）For the exterior walls or exterior walls on the windward side in the buildings with hollow blocks or light-weight bricks for walls in the areas with a basic wind pressure greater than 0.6kPa or abundant rainfall and the buildings with higher requirements for waterproof, etc., after the 20mm waterproof mortar is adopted, and the 7 mm thick polymer cement mortar is plastered, the exterior finish layer shall be made.

3）加气混凝土外墙应采用配套砂浆砌筑，配套砂浆抹面或加钢丝网抹面。

3）For aerated concrete exterior walls, supporting mortar shall be adopted in building, and supporting mortar shall be adopted in plastering, or steel wire meshes shall be added in plastering.

4）填充墙与框架梁柱间加 200mm 宽 20mm×20mm 网格 $\phi 1$ 的钢丝网或玻纤网格布抹灰。

4）Between the infill wall and the frame beam column, the 200mm wide wire meshes or fiberglass meshes with the specifications of 20mm×20mm shall be adopted in plastering.

5）突出外墙面的横向线脚、窗台、挑板等出挑构件上部与墙交接处应做成小圆角并向外找坡不小于 3%，以利于排水，且下部应做滴水槽。

5）At the junction between the upper parts of the overhanging components of transverse architrave, windowsill, cantilever board, etc., protruding from the exterior wall and the walls, small round corners shall be made, and external sloping with the gradient not less than 3% shall be made, to facilitate drainage, and the water drips shall be made at the lower part.

6）外门窗洞口四周的墙体与门窗框之间应采用发泡聚氨酯等柔性材料填塞严密，且最外表的饰面层与门窗框之间应留约 7mm×7mm 的凹槽，并满嵌耐候防水密封膏。

6）The gaps between the wall surfaces around the exterior door and window openings and the door and window frames shall be tightly filled with flexible materials such as foamed polyurethane, etc., And the 7mm×7mm grooves shall be reserved between the outermost finish layer, and the door and window frames, and in the grooves, the weatherproof and waterproof sealing paste shall be fully filled.

7）安装在外墙上的构件、管道等均宜采用预埋方式连接，也可用螺栓固定，但螺栓需用树脂粘结严密。

7) The components and pipelines, etc., installed on exterior walls shall be appropriately connected through the embedment mode, or fixed with bolts, and at this time, the bolts shall be tightly bonded with resin.

8.2.3 【09措施】4.2.3条：墙面防潮和隔汽。

8.2.3 [Measures 2009] 4.2.3: Moisture proofing and vapor proofing of wall surface.

1. 处于高湿度环境的墙体应采用混凝土或混凝土砌块等耐水性好的材料。不宜采用吸湿性强的材料，更不应采用因吸水变形、腐烂导致强度降低的材料。墙面应有防潮措施。高湿度房间（如卫浴间、厨房）的墙或有直接被淋水的墙（如淋浴间、小便槽处），应做墙面防水隔离层。受水冲淋的部位应尽量避免靠外墙设置。

1. The materials with high water resistance such as concrete, concrete block, etc., shall be adopted for walls in the high humidity environment. The high-hygroscopicity materials or materials with strength reduced due to water absorption deformation and decay shall not be adopted . The moisture prevention measures shall be adopted on walls. The waterproof isolation layer of wall surfaces shall be made for walls of rooms with high humidity（such as bathrooms and kitchens）or walls with direct contact to water. It shall be avoided to arrange the water-washed parts against the exterior walls.

2. 室内温度低的房间（如冷藏间）的墙，应在其内侧做隔汽层再做绝热层。两个室内温差很大的房间之间，有可能在墙体内部和另一墙体引起结露时，应根据实际情况采用单面隔汽层或双面隔汽层。

2. For walls of a room（such as a cold room）with low indoor temperature, the insulation layer shall be constructed on the inner side after a vapor barrier layer is made in the same side. For two rooms with a large indoor temperature difference, when there is a possibility that condensation may occur in the wall and on the other wall, the single-side or double-side vapor barrier layer shall be adopted in accordance with the actual situation.

8.2.4 【通则】6.9.3条：墙身防潮应符合下列要求。

8.2.4 [Code for design of civil buildings] 6.9.3: For moisture prevention for walls, the requirements in the following shall be conformed to.

1. 砌体墙应在室外地面上，位于室内地面垫层处设置连续的水平防潮层；室内相邻地面有高差时，应在高差处墙身侧面加设防潮层。

1. The masonry walls shall be arranged on the outdoor floor, and the continuous horizontal

moisture prevention layer shall be arranged on the cushion layer of the indoor floor; When there is a height difference between the adjacent floors in the room, a moisture prevention layer shall be added on the side of the wall with height difference.

2. 湿度大的房间的外墙或内墙内侧应设防潮层。

2. The exterior walls or the inside of the interior walls of the rooms with high humidity shall be set with a moisture prevention layer.

3. 室内墙面有防水、防潮、防污、防碰等要求时，应按使用要求设置墙裙。

3. When there are requirements for waterproof, moisture prevention, fouling prevention, collision, etc., to interior walls, the wall skirting shall be arranged as per the application requirements.

注：地震区防潮层应满足墙体抗震整体连接的要求。
Note: The moisture prevention layers in the seismic (active) areas shall meet the integral connection requirements of the wall for earthquake resistance.

8.3 墙体防火

8.3 Fireproofing of Walls

8.3.1 【防火规】6.1 条：防火墙。

8.3.1 [Code for fire protection design of buildings] 6.1: Fire walls.

1.【防火规】6.1.1 条：防火墙应直接设置在建筑的基础或框架、梁等承重结构上，框架、梁等承重结构的耐火极限不应低于防火墙的耐火极限。

1. [Code for fire protection design of buildings] 6.1.1: The fire walls shall be directly arranged on the bearing structures of foundations or frames, beams, etc., of the building, and the fire endurance of the bearing structures of frames, beams, etc., shall be not less than the fire endurance of the fire walls.

防火墙应从楼地面基层隔断至梁、楼板或屋面板的底面基层。当高层厂房（仓库）屋顶承重结构和屋面板的耐火极限低于 1.00h，其他建筑屋顶承重结构和屋面板的耐火极限低于 0.50h 时，防火墙应高出屋面 0.5m 以上。

The fire walls shall be separated from the base layers of the floor grounds to the base layers on the bottom surfaces of the beams, floor slabs or roof boards. If the fire endurance of the roof load-bearing structure and roof slab of the high-rise plant building is less than 1.00h and the fire endurance of bearing structures for roofs and roof boards of other buildings is less than 0.50h, the fire wall shall be 0.5m higher than the roof.

2.【防火规】6.1.2 条：防火墙横截面中心线水平距离天窗端面小于 4.0m，且天窗端面为可燃性墙体时，应采取防止火势蔓延的措施。

2. [Code for fire protection design of buildings] 6.1.2: The horizontal distance from the cross section centerline of the fire wall to the end surface of the skylight shall be less than 4.0m. If the end surface of the skylight is flammable wall, the measures to prevent fire from spreading shall be taken.

3.【防火规】6.1.3 条：建筑外墙为难燃性或可燃性墙体时，防火墙应凸出墙的外表面 0.4m 以上，且防火墙两侧的外墙均应为宽度均不小于 2.0m 的不燃性墙体，其耐火极限不应低于外墙的耐火极限。

3. [Code for fire protection design of buildings] 6.1.3: If the exterior walls of the building are non-flammable or flammable, the fire walls shall be higher than the outer surface of the walls by more than 0.4m. The exterior walls on both sides of the fire walls shall be the non-combustible walls with the width not less than 2.0m, and the fire endurance shall be not less than the fire endurance of the exterior walls.

建筑外墙为不燃性墙体时，防火墙可不凸出墙的外表面，紧靠防火墙两侧的门、窗、洞口之间最近边缘的水平距离不应小于 2.0m；采取设置乙级防火窗等防止火灾水平蔓延的措施时，该距离不限。

If the exterior walls of the building are non-combustible, the fire walls can be not higher than the outer surface of the walls. The horizontal distance from the nearest edge to the doors, windows and openings on both sides of the fire wall shall be not less than 2.0m; If the measures are taken to prevent fire from spreading horizontally, such as to set up Class B fire windows, etc., the distance shall be not limited.

4.【防火规】6.1.4 条：建筑内的防火墙不宜设置在转角处，确需设置时，内转角两侧墙上的门、窗、洞口之间最近边缘的水平距离不应小于 4.0m；采取设置乙级防火窗等防止火灾水平蔓延的措施时，该距离不限。

4. [Code for fire protection design of buildings] 6.1.4: The fire walls inside the building shall not be arranged at the corners, and if the arrangement is required, the horizontal distance from the nearest edge to the doors, windows and openings on both sides of the inner corners shall be not less than 4.0m; If the measures are taken to prevent fire from spreading horizontally, such as to set up Class B fire windows, etc., the distance shall be not limited.

5.【防火规】6.1.5 条：防火墙上不应开设门、窗、洞口，确需开设时，应设置不可开启或火灾时能自动关闭的甲级防火门、窗。

5. [Code for fire protection design of buildings] 6.1.5: Doors, windows and openings shall not be

arranged in fire walls. .If they are required to be arranged indeed, the Grade A fire doors and windows unopenable or capable of automatic closing in case of a fire shall be arranged.

可燃气体和甲、乙、丙类液体的管道严禁穿过防火墙。防火墙内不应设置排气道。

It is prohibited that the pipelines for combustible gases and Classes A, B and C liquids penetrate the fire walls. Exhaust passages shall not be arranged in the fire walls.

6. 除本指南 6.4.4.1 第 5 条外的其他管道不宜穿过防火墙，确需穿过时，应采用防火封堵材料将墙与管道之间的空隙紧密填实，穿过防火墙处的管道的保温材料，应采用不燃烧材料；当管道为难燃及可燃材料时，应在防火墙两侧的管道上采取防火措施。

6. The pipelines other than those specified in Clause 5 of 6.4.4.1 shall not penetrate the fire walls. If the pipelines are required to penetrate the fire walls, the gaps between walls and pipelines shall be tightly filled with fireproof blocking materials. The insulation materials of the pipelines penetrating the fire walls shall be noncombustible. When the materials for the pipelines are flame-retardant and combustible, the fire prevention measures shall be adopted for the pipelines on both sides of the fire walls.

7.【防火规】6.1.7 条：防火墙的构造应能在防火墙任意一侧的屋架、梁、楼板等受到火灾的影响而破坏时，不会导致防火墙倒塌。

7. [Code for fire protection design of buildings] 6.1.7: For the structures of fire walls, the collapse of the fire walls will not be caused when the roof trusses, beams, floor slabs, etc., on either side of the fire walls are affected and damaged in a fire.

8.3.2 【防火规】5.1.2 条：民用建筑的耐火等级可分为一、二、三、四级。除本规范另有规定外，不应低于表 8.3.2 的规定。

8.3.2 [Code for fire protection design of buildings] 5.1.2: The fire resistance ratings of civil buildings can be divided into Grades I, II, III and IV. Unless otherwise specified in the Code, the ratings shall be not less than that of Tab. 8.3.2.

【防火规】表 5.1.2: 不同耐火等级建筑相应构件的燃烧性能和耐火极限（h） 表 8.3.2

[Code for fire protection design of buildings] Tab. 5.1.2: Combustion Performance and Fire Endurance (h) of Corresponding Components of Buildings with Different Fire Resistance Ratings　Tab. 8.3.2

构件名称 Component name		耐火等级 Fire resistance rating			
		一级 Grade I	二级 Grade II	三级 Grade III	四级 Grade IV
墙 Walls	防火墙 Fire walls	不燃性 Incombustibility 3.00	不燃性 Incombustibility 3.00	不燃性 Incombustibility 3.00	不燃性 Incombustibility 3.00

续表

构件名称 Component name		耐火等级 Fire resistance rating			
		一级 Grade I	二级 Grade II	三级 Grade III	四级 Grade IV
墙 Walls	承重墙 Bearing walls	不燃性 Incombustibility 3.00	不燃性 Incombustibility 2.50	不燃性 Incombustibility 2.00	难燃性 Flame retardancy 0.50
	非承重外墙 Non-bearing exterior walls	不燃性 Incombustibility 1.00	不燃性 Incombustibility 1.00	不燃性 Incombustibility 0.50	可燃性 Combustibility
	楼梯间和前室的墙 Walls of the staircases and anterooms 电梯井的墙 Walls in elevator shafts 住宅建筑单元之间的墙和分户墙 Walls and household dividing walls between residential building units	不燃性 Incombustibility 2.00	不燃性 Incombustibility 2.00	不燃性 Incombustibility 1.50	难燃性 Flame retardancy 0.50
	疏散走道两侧的隔墙 Partition walls on both sides of the evacuation walkway	不燃性 Incombustibility 1.00	不燃性 Incombustibility 1.00	不燃性 Incombustibility 0.50	难燃性 Flame retardancy 0.25
	房间隔墙 Room partitions	不燃性 Incombustibility 0.75	不燃性 Incombustibility 0.50	难燃性 Flame retardancy 0.50	难燃性 Flame retardancy 0.25
柱 Columns		不燃性 Incombustibility 3.00	不燃性 Incombustibility 2.50	不燃性 Incombustibility 2.00	难燃性 Flame retardancy 0.50
梁 Beams		不燃性 Incombustibility 2.00	不燃性 Incombustibility 1.50	不燃性 Incombustibility 1.00	难燃性 Flame retardancy 0.50
楼板 Floor slabs		不燃性 Incombustibility 1.50	不燃性 Incombustibility 1.00	不燃性 Incombustibility 0.50	可燃性 Combustibility
屋顶承重构件 Load-bearing components of the roof		不燃性 Incombustibility 1.50	不燃性 Incombustibility 1.00	可燃性 Combustibility 0.50	可燃性 Combustibility
疏散楼梯 Evacuation staircases		不燃性 Incombustibility 1.50	不燃性 Incombustibility 1.00	不燃性 Incombustibility 0.50	可燃性 Combustibility
吊顶（包括吊顶搁栅） Suspended ceilings (including ceiling grilles)		不燃性 Incombustibility 0.25	难燃性 Flame retardancy 0.25	难燃性 Flame retardancy 0.15	可燃性 Combustibility

注：1. 除本规范另有规定外，以木柱承重且墙体采用不燃材料的建筑，其耐火等级应按四级确定。
Notes:1. Unless otherwise specified in the code, the fire resistance ratings of the buildings with bearing wood columns and incombustible materials for walls shall be determined as per Level IV.
2. 住宅建筑构件的耐火极限和燃烧性能可按现行国家标准《住宅建筑规范》GB 50368 的规定执行。
2. The fire endurance and combustion performance of components for residential buildings shall conform to the provisions in the current national standard *Code for Residential Buildings* GB 50368.

8.3.3 建筑保温和外墙装饰。

8.3.3 building insulation and exterior wall decoration.

1.【防火规】6.7.1条：建筑的内、外保温系统，宜采用燃烧性能为A级的保温材料，不宜采用B2级保温材料，严禁采用B3级保温材料；设置保温系统的基层墙体或屋面板的耐火极限应符合本规范的有关规定。

1. [Code for fire protection design of buildings] 6.7.1: building internal and external insulation system, it is advisable to use combustion performance of A grade insulation materials, should not use B2 grade insulation materials, strictly prohibit the use of B3-grade insulation materials, set up insulation system of the base wall or House panel fire resistance limit should comply with the relevant provisions of this specification.

2.【防火规】6.7.2条：建筑外墙采用内保温系统时，保温系统应符合下列规定。

2. [Code for fire protection design of buildings] 6.7.2: when using an internal insulation system for building exterior walls, the insulation system shall comply with the following requirements.

1）对于人员密集场所，用火、燃油、燃气等具有火灾危险性的场所以及各类建筑内的疏散楼梯间、避难走道、避难间、避难层等场所或部位，应采用燃烧性能为A级的保温材料。

1）For personnel intensive places, with fire, fuel, gas and other places with fire hazards, as well as all kinds of buildings in the evacuation stairwell, refuge walkway, refuge room, refuge layer and other places or parts, should use combustion performance of A grade insulation materials.

2）对于其他场所，应采用低烟、低毒且燃烧性能不低于B1级的保温材料。

2）For other places, should be low smoke, low toxicity and combustion performance of no less than B1 level insulation materials.

3）保温系统应采用不燃材料做防护层。采用燃烧性能为B1级的保温材料时，防护层的厚度不应小于10mm。

3）The insulation system should use non-combustible materials to do protective layer. The thickness of the protective layer should not be less than 10mm when the thermal insulation material with combustion performance of Grade B1 is used.

3.【防火规】6.7.3条：建筑外墙采用保温材料与两侧墙体构成无空腔复合保温结构体时，该结构体的耐火极限应符合本规范的有关规定；当保温材料的燃烧性能为B1、B2级时，保温材料两侧的墙体应采用不燃材料且厚度均不应小于50mm。

3. [Code for fire protection design of buildings] 6.7.3: Building exterior wall using insulation material and both sides of the wall to form a non-cavity composite insulation

structure, the fire resistance limit of the structure should conform to the relevant provisions of this specification; when the combustion performance of insulation material is Grade B1 or Grade B2, the wall on both sides of the insulation material should use the walls on both sides of the insulation material shall be made of non-combustible materials and shall not be less than 50mm in thickness.

4.【防火规】6.7.4 条：设置人员密集场所的建筑，其外墙外保温材料的燃烧性能应为 A 级。

4. [Code for fire protection design of buildings] 6.7.4: set up buildings in staff-intensive sites, its external wall insulation material combustion performance should be Grade A.

5.【防火规】6.7.5 条：与基层墙体、装饰层之间无空腔的建筑外墙外保温系统，其保温材料应符合下列规定。

5. [Code for fire protection design of buildings] 6.7.5: the building exterior wall insulation system without cavity between the base wall and decorative layer, its insulation materials should comply with the following provisions.

1）住宅建筑：

1）Residential buildings:

（1）建筑高度大于 100m 时，保温材料的燃烧性能应为 A 级；

（1）When the building height is greater than 100m, the combustion performance of the insulation material shall be Grade A;

（2）建筑高度大于 27m 但不大于 100m 时，保温材料的燃烧性能不应低于 B1 级；

（2）When the building height is greater than 27m, but not greater than 100m, the combustion performance of the insulation material should not be lower than the Grade B1;

（3）建筑高度不大于 27m 时，保温材料的燃烧性能不应低于 B2 级；

（3）When the building height is not greater than 27m, the combustion performance of the insulation material should not be lower than the Grade B2;

2）除住宅建筑和设置人员密集场所的建筑外，其他建筑：

2）In addition to residential buildings and buildings with densely populated premises, other buildings:

（1）建筑高度大于 50m 时，保温材料的燃烧性能应为 A 级；

（1）When the building height is greater than 50m, the combustion performance of the insulation material shall be Grade A;

（2）建筑高度大于 24m 但不大于 50m 时，保温材料的燃烧性能不应低于 B1 级；

（2）When the building height is greater than 24m, but not greater than 50m, the combustion performance of the insulation material should not be lower than the Grade B1;

（3）建筑高度不大于 24m 时，保温材料的燃烧性能不应低于 B2 级。

（3）When the building height is not greater than 24m, the combustion performance of the insulation material should not be lower than the Grade B2.

6.【防火规】6.7.6 条：除设置人员密集场所的建筑外，与基层墙体、装饰层之间有空腔的建筑外墙外保温系统，其保温材料应符合下列规定。

6. [Code for fire protection design of buildings] 6.7.6: In addition to the establishment of personnel intensive premises, with the base wall, decorative layer between the building exterior wall insulation system, its insulation materials should comply with the following provisions.

1）建筑高度大于 24m 时，保温材料的燃烧性能应为 A 级；

1）When the building height is greater than 24m, the combustion performance of the insulation material should be Grade A;

2）建筑高度不大于 24m 时，保温材料的燃烧性能不应低于 B1 级。

2）When the building height is not greater than 24m, the combustion performance of the insulation material should not be lower than the Grade B1.

7.【防火规】6.7.7 条：除本规范第 6.7.3 条规定的情况外，当建筑的外墙外保温系统按本规范第 6.7 节规定采用燃烧性能为 B1、B2 级的保温材料时，应符合下列规定。

7. [Code for fire protection design of buildings] 6.7.7: In addition to the circumstances stipulated in clause 6.7.3 of this code, when the exterior wall insulation system of the building adopts the insulation material with combustion performance of Grades B1 and B2 in accordance with the section 6.7 of this specification, the following provisions shall be met.

1）除采用 B1 级保温材料且建筑高度不大于 24m 的公共建筑或采用 B1 级保温材料且建筑高度不大于 27m 的住宅建筑外，建筑外墙上门、窗的耐火完整性不应低于 0.50h；

1）In addition to the use of Grade B1 insulation materials and building height is not greater than 24m of public buildings or the use of Grade B1 insulation materials and building height is not greater than 27m residential buildings, building exterior wall door, window fire-resistant integrity should not be less than 0.50h;

2）应在保温系统中每层设置水平防火隔离带。防火隔离带应采用燃烧性能为 A 级的材料，防火隔离带的高度不应小于 300mm。

2）A horizontal fire isolation belt should be arranged in each layer of the insulation system. Fire isolation belt should use combustion performance of Grade A of materials, the height of a fire isolation belt should not be less than 300mm.

8.【防火规】6.7.7 条：建筑的外墙外保温系统应采用不燃材料在其表面设置防护层，防护层应将保温材料完全包覆。除本规范第 6.7.3 条规定的情况外，当按本规范第 6.7 节规定采用 B1、B2 保温材料时，防护层厚度首层不应小于 15mm，其他层不应小于 5mm。

8. [Code for fire protection design of buildings] 6.7.7: Building exterior wall insulation

system should use non-combustible materials on its surface to set up a protective layer, protective layer should be completely coated insulation materials.Except as provided in article 6th. 7.3 of this specification, when B1 and B2 insulation materials are used in accordance with section 6.7 of this specification, the first layer of protective layer thickness shall not be less than 15mm, and the other layers shall not be less than 5mm.

9.【防火规】6.7.8 条：建筑外墙外保温系统与基层墙体、装饰层之间的空腔，应在每层楼板处采用防火封堵材料封堵。

9. [Code for fire protection design of buildings] 6.7.8: The building exterior wall insulation system and the base wall, decorative layer between the cavity, should be used in each floor of the floor fire blocking material plugging.

10.【防火规】6.7.10 条：建筑的屋面外保温系统，当屋面板的耐火极限不低于1.00h 时，保温材料的燃烧性能不应低于 B2 级；当屋面板的耐火极限低于1.00h 时，不应低于 B1 级。采用 B1、B2 级保温材料的外保温系统应采用不燃材料作防护层，防护层的厚度不应小于 10mm。

10. [Code for fire protection design of buildings] 6.7.10: Building of the roof external insulation system, when the fire resistance limit of the roof plate is not less than 1.00h, the combustion performance of the insulation material should not be lower than the Grade B2, when the fire resistance limit of the roof plate is less than 1.00h, should not be lower than the Grade B1. The external insulation system using B1 and Grade B2 insulation materials should use non-combustible materials as protective layer, and the thickness of the protective layer should not be less than 10mm.

当建筑的屋面和外墙外保温系统均采用 B1、B2 级保温材料时，屋面与外墙之间应采用宽度不小于 500mm 的不燃材料设置防火隔离带进行分隔。

When the roof and exterior wall insulation system of the building are using Grades B1 and B2 insulation materials, the roof and the exterior wall should be separated by a non-flammable material with a width of not less than 500mm.

11.【防火规】6.7.10 条：电气线路不应穿越或敷设在燃烧性能为 B1 或 B2 级的保温材料中；确需穿越或敷设时，应采取穿金属管并在金属管周围采用不燃隔热材料进行防火隔离等防火保护措施。设置开关、插座等电器配件的部位周围应采取不燃隔热材料进行防火隔离等防火保护措施。

11. [Code for fire protection design of buildings] 6.7.10: The electrical wiring should not be crossed or laid in insulation materials with combustion performance of Grade B1 or Grade B2, and when it is necessary to cross or lay, it should take fire protection measures such as wearing metal tubes and using non-combustible insulating materials around metal tubes for fire prevention and isolation.

Set switches, sockets and other electrical accessories around the site should take non-combustible insulation materials for fire protection and other fire prevention measures.

12.【防火规】6.7.12 条：建筑外墙的装饰层应采用燃烧性能为 A 级的材料，但建筑高度不大于 50m 时，可采用 B1 级材料。

12. [Code for fire protection design of buildings] 6.7.12：The decorative layer of the exterior wall of the building shall use a material with Grade A combustion performance, but the Grade B1 material may be used when the building height is not greater than 50m.

8.3.4 【防火规】6.2.5 条：建筑外墙上、下层开口之间应设置高度不小于 1.2m 的实体墙或挑出宽度不小于 1.0m、长度不小于开口宽度的防火挑檐；当室内设置自动喷水灭火系统时，上、下层开口之间的实体墙高度不应小于 0.8m。当上、下层开口之间设置实体墙确有困难时，可设置防火玻璃墙，但高层建筑的防火玻璃墙的耐火完整性不应低于 1.00h，单、多层建筑的防火玻璃墙的耐火完整性不应低于 0.50h。外窗的耐火完整性不应低于防火玻璃墙的耐火完整性要求。

8.3.4 [Code for fire protection design of buildings] 6.2.5: The solid walls with the height not less than 1.2m or fireproof overhanging eaves with the overhanging width not less than 1.0m and the length not less than the width of the openings shall be arranged between the openings on the upper and lower floors of exterior walls of buildings; When the automatic sprinkler systems are arranged indoors, the height of the solid walls between the openings of the upper and lower floors shall be not less than 0.8m. When the arrangement of the solid walls between the openings on the upper and lower floors is difficult, the fireproof glass walls can be arranged. The fire integrity of the fireproof glass walls of the high-rise buildings shall be not less than 1.00 hour, and the fire integrity of the fireproof glass walls of the single-storey and multi-storey buildings shall be not less than 0.50 hour. The fire integrity of exterior windows shall be not less than that in the requirements for the fire integrity of the fireproof glass walls.

住宅建筑外墙上相邻户开口之间的墙体宽度不应小于 1.0m；小于 1.0m 时，应在开口之间设置突出外墙不小于 0.6m 的隔板。

The wall width between adjacent openings on exterior walls of residential buildings shall be not less than 1.0m; When the wall width is less than 1.0m, the partition plates protruding out of exterior walls by not less than 0.6m shall be arranged between the openings.

实体墙、防火挑檐和隔板的耐火极限和燃烧性能，均不应低于相应耐火等级建筑外墙的要求。

The fire endurance and combustion performance of the solid walls, fireproof overhanging eaves and partition plates shall be not less than that in the requirements for exterior walls of buildings with the corresponding fire resistance ratings.

8.3.5 【防火规】6.2.6条：建筑幕墙应在每层楼板外沿处采取符合本规范第6.2.5条规定的防火措施，幕墙与每层楼板、隔墙处的缝隙应采用防火封堵材料封堵。

8.3.5 [Code for fire protection design of buildings] 6.2.6: For the curtain walls of the buildings at the outer edges of the floor slabs on each floor, the fire prevention measures complying with the provisions of Clause 6.2.5 of the Code shall be adopted, and the gaps between the curtain walls and the floor slabs and partition walls of each floor shall be blocked with fireproof blocking materials.

8.4 【隔声规】4条：墙体隔声

8.4 [Code for design of sound insulation of civil buildings] 4: Wall Sound Insulation

8.4.1 【隔声规】4条：建筑物的隔声减噪设计标准等级应按其实际使用要求确定，并应符合本规范的规定。

8.4.1 [Code for design of sound insulation of civil buildings] 4: The standard Grade of sound insulation and noise reduction design of buildings should be determined according to their actual use requirements, and should conform to the provisions of this code.

8.4.2 围护结构（隔墙和外墙）空气声隔声标准见表8.4.2。

8.4.2 For the airborne sound insulation standards for enclosure structures (partition walls and exterior walls), see Tab. 8.4.2.

【隔声规】Tab. 4.5.2：围护结构（隔墙）空气声隔声标准　　表8.4.2

[Code for design of sound insulation of civil buildings] Tab. 4.5.2: Airborne Sound Insulation Standards for Enclosure Structures (Partition Walls)　　Tab. 8.4.2

建筑类别 Building category	部 位 Position	隔声量（dB） Sound insulation volume (dB)	
		高要求标准 Standards with high requirements	一般标准 General standards
住宅 Residential buildings	分户墙 Household dividing walls	$R_w + C > 50$	$R_w + C > 45$

续表

建筑类别 Building category	部位 Position	隔声量（dB） Sound insulation volume (dB)	
		高要求标准 Standards with high requirements	一般标准 General standards
住宅 Residential buildings	外墙 Exterior walls	$R_w + C_{TR} \geq 45$	
	户内卧室墙 Indoor bedroom walls	$R_w + C \geq 35$	
	户内其他分室墙 Walls of other rooms indoor	$R_w + C \geq 30$	
学校 School	外墙 Exterior walls	$R_w + C_{TR} \geq 45$	
	语言教室、阅览教室的隔墙 Partition walls between language classrooms and reading rooms	$R_w + C > 50$	
	普通教室与各种产生噪声的房间之间的隔墙 Partition walls between ordinary classrooms and various rooms in which noise is generated	$R_w + C > 50$	
	普通教室之间的隔墙 Partition walls between ordinary classrooms	$R_w + C > 45$	
	音乐教室、琴房之间的隔墙 Partition walls between music classrooms and piano rooms	$R_w + C > 45$	
医院 Hospitals	病房与产生噪声的房间之间的隔墙 Partition walls between wards and rooms in which noise is generated	$R_w + C_{TR} > 55$	$R_w + C_{TR} > 50$
	手术室与产生噪声的房间之间的隔墙 Partition walls between operating rooms and rooms in which noise is generated	$R_w + C_{TR} > 50$	$R_w + C_{TR} > 45$
	病房之间及病房、手术室与普通房间之间的隔墙 Partition walls between wards, and between wards and operating rooms and ordinary rooms	$R_w + C > 50$	$R_w + C > 45$
	诊室之间的隔墙 Partition walls between consulting rooms	$R_w + C > 45$	$R_w + C > 40$
	听力测听室的隔墙 Partition walls of hearing detection rooms	—	$R_w + C > 50$
	体外震波碎石室、核磁共振室的隔墙 Partition walls of extracorporeal shock wave lithotripsy rooms and magnetic resonance imaging rooms	—	$R_w + C_{TR} > 50$
	外墙 Exterior wall	$R_w + C_{TR} \geq 45$	
旅馆 Hotels	等级 Grade	特级（五星） Special grade (five-star)	一级（三、四星） Grade I (three-star and four-star)	二级（其他） Grade II (others)

续表

建筑类别 Building category	部 位 Position	隔声量（dB） Sound insulation volume（dB）		
		高要求标准 Standards with high requirements	一般标准 General standards	
旅馆 Hotels	客房之间的隔墙 Partition walls between guest rooms	$R_w + C > 50$	$R_w + C > 45$	$R_w + C > 40$
	客房与走廊之间隔墙 Partition walls between guest rooms and corridors	$R_w + C > 45$	$R_w + C > 45$	$R_w + C > 40$
	客房外墙 Exterior walls of guest rooms	$R_w + C_{TR} > 40$	$R_w + C_{TR} > 35$	$R_w + C_{TR} > 30$
办公 Offices	办公室、会议室与产生噪声的房间之间的隔墙 Partition walls between offices, and between offices and rooms in which noise is generated	$R_w + C_{TR} > 50$	$R_w + C_{TR} > 45$	
	办公室、会议室与普通房间之间的隔墙 Partition walls between offices, and between offices and ordinary rooms	$R_w + C > 50$	$R_w + C > 45$	
	外墙 Exterior walls	$R_w + C \geq 45$		
商业 Commercial Buildings 【隔声规】9.3.1条 [Code for design of sound insulation of civil buildings] 9.3.1	健身房中心、娱乐场所等与噪声敏感房间之间隔墙 Partition walls between fitness centers, entertainment venues, etc., and noise sensitive rooms	$R_w + C_{TR} > 60$	$R_w + C_{TR} > 55$	
	购物中心、餐厅、会展中心等噪声敏感房间之间的隔墙 Partition walls between noise-sensitive rooms such as shopping centers, restaurants and exhibition centers, etc.	$R_w + C_{TR} > 50$	$R_w + C_{TR} > 45$	

注：$R_w + C$——计权隔声量+频谱修正量（A计权粉红噪声）；$R_w + C_{TR}$——计权隔声量+频谱修正量（A计权交通噪声）。
Note: $R_w + C$—Weighted sound insulation volume + spectrum correction amount（A weighted pink noise）; $R_w + C_{TR}$—Weighted sound insulation volume + spectrum correction amount（A weighted traffic noise）.

8.4.3 【住设规】7.3.4条：住宅建筑的体形、朝向和平面布置应有利于噪声控制。在住宅平面设计时，当卧室、起居室（厅）布置在噪声源一侧时，外窗应采取隔声降噪措施；当居住空间与可能产生噪声的房间相邻时，分隔墙和分隔楼板应采取隔声降噪措施；当内天井、凹天井中设置相邻户窗口时，宜采取隔声降噪措施。

8.4.3 [Design code of residential buildings] 7.3.4: The shape, orientation and plan layout of the residential building shall be conducive to noise control. In the graphic design for residential buildings, when bedrooms and living rooms are arranged on one side of the noise source, the sound insulation and noise reduction measures shall be taken for exterior windows;

When the residential space is adjacent to a room in which noises may be made, sound insulation and noise reduction measures shall be taken for the partition walls and partition floors; When the windows of the adjacent household are installed in the inner patio and sunken patio, the sound insulation and noise reduction measures shall be adopted.

8.4.4 【住设规】7.3.5 条：起居室（厅）不宜紧邻电梯布置。受条件限制起居室（厅）紧邻电梯布置时，必须采取有效的隔声和减振措施。

8.4.4 [Design code of residential buildings] 7.3.5: Living rooms (halls) shall not be arranged next to the elevators. If the conditions are limited, and living rooms (halls) are placed in close proximity to the elevator, effective sound insulation and vibration reduction measures must be taken.

9

幕墙、采光顶
Curtain Wall, Lighting Roof

9.1 【幕墙规】2条：常用类型

9.1 [Technical code for metal and stone curtain walls engineering] 2: Commonly Used Type

9.1.1 【幕墙规】3.1条

9.1.1 [Technical code for metal and stone curtain walls engineering] 3.1

1.【幕墙规】3.1.2条：玻璃幕墙应选用耐气候性的材料。金属材料和金属零配件除不锈钢及耐候钢外，钢材应进行表面热浸镀锌处理、无机富锌涂料处理或采取其他有效的防腐措施，铝合金材料应进行表面阳极氧化、电泳涂漆，粉末喷涂或氟碳漆喷涂处理。

1. [Technical code for metal and stone curtain walls engineering] 3.1.2: Glass curtain wall should be selected for weather-resistant materials. Metallic materials and metal parts in addition to stainless steel and weather-resistant steel, steel should be surface hot dip galvanizing treatment, inorganic zinc-rich coating treatment or take other effective anti-corrosion measures, aluminum alloy materials should be surface anodic oxidation, electrophoresis paint, powder spraying or fluorine paint spraying treatment.

2.【幕墙规】3.1.3条：玻璃幕墙材料宜采用不燃性材料或难燃性材料；防火密封构造应采用防火密封材料。

2. [Technical code for metal and stone curtain walls engineering] 3.1.3: Glass curtain wall materials should be made of non-flammable materials or refractory materials, fire sealing structure should be used fireproof sealing materials.

3.【幕墙规】3.1.4条：隐框和半隐框玻璃幕墙，其玻璃与铝型材的粘结必须采用中性硅酮结构密封胶；全玻幕墙和点支承慕墙采用镀膜玻璃时，不应采用酸性硅酮结构密封胶粘结。

3. [Technical code for metal and stone curtain walls engineering] 3.1.4: The hidden frame and semi-hidden frame glass curtain wall, its glass and aluminum profile bonding must use neutral silicone structure sealant, the whole glass curtain wall and point support wall using coated glass, should not use acidic silicone structure sealant bonding.

9.1.2 采光顶

9.1.2 Lighting roof

1.【采光顶】4.1.1条：按支承结构分为钢结构、索杆结构、铝合金结构、玻璃梁结构。

1. [Technical requirements of building glass skylight system] 4.1.1: It is divided into steel structure, cable rod structure, aluminum alloy structure, glass beam structure according to the support structure.

2.【采光顶】4.1.2条：按开合分为非开合、可开合形式。

2. [Technical requirements of building glass skylight system] 4.1.2: It is divided into non-open and open forms by opening and closing.

9.2 【幕墙规】3条：主要材料及选用要点

9.2 [Technical code for metal and stone curtain walls engineering] 3: Main Materials and Selection Points

建筑幕墙、采光顶常用材料主要分饰面材料、骨架材料、密封材料、五金件等。

Building curtain wall, lighting roof commonly used materials are mainly decorative surface materials, skeleton materials, sealing materials, hardware and so on.

9.2.1 【玻幕规】3.4.1条：幕墙玻璃的外观质量和性能应符合下列现行国家标准、行业标准的规定：《钢化玻璃》GB/T 9963、《幕墙用钢化玻璃与半钢化玻璃》GB/T 17841、《夹层玻璃》GB 9962、《中空玻璃》GB/T 11944、《浮法玻璃》GB 11614、《建筑用安全玻璃 防火玻璃》GB 15763.1《着色玻璃》GB/T 18701、《镀膜玻璃 第1部分：阳光控制镀膜玻璃》GB/T 18915.1、《镀膜玻璃 第2部分：低辐射镀膜玻璃》GB/T 18915.2。

9.2.1 [Technical code for glass curtain wall engineering] 3.4.1: The appearance quality and performance of curtain wall glass shall conform to the following current national standards and industrial standards: *Tempered Glass* GB/T 9963, *Curtain Wall with Tempered Glass and Semi-tempered Glass* GB/T 17841, *Laminated Glass* GB 9962, *Hollow Glass* GB/T 11944, *Float Glass* GB 11614, *Building Safety Glass Fireproof Glass* GB 15763.1、*Coloring Glass* GB/T 18701, *Coated Glass Part I Sun-controlled Coated Glass* GB/T 18915.1, *Coated Glass Part II Low-radiation Coated Glass* GB/T 18915.2.

9.2.2 【幕墙规】3.4.1条：建筑幕墙、采光顶用钢材选用要点。

9.2.2 [Technical code for metal and stone curtain walls engineering] 3.4.1: Building curtain wall, lighting used steel selection key points.

1.【幕墙规】3.3.1条：玻璃幕墙用碳素结构钢和低合金结构钢的钢种。

1. [Technical code for metal and stone curtain walls engineering] 3.3.1: Steel with carbon

structural steel and low alloy structural steel for glass curtain wall.

2.【幕墙规】3.4.3条：玻璃幕墙采用中空玻璃层厚度不应小于9mm。

2. [Technical code for metal and stone curtain walls engineering] 3.4.3: glass curtain wall using hollow glass layer thickness should not be less than 9mm.

9.2.3【幕墙规】3.4.3条：建筑幕墙、采光顶用密封材料选用要点。

9.2.3 [Technical code for metal and stone curtain walls engineering] 3.4.3: Selection key points of building curtain wall, lighting use sealing material.

1.【幕墙规】3.3.1条：玻璃幕墙用碳素结构钢和低合金结构钢的钢种。

1. [Technical code for metal and stone curtain walls engineering] 3.3.1: Steel type of carbon structural steel and low alloy structural steel for glass curtain wall.

2.【幕墙规】3.3.2条：玻璃幕墙用不锈钢材宜采用奥氏体不锈钢，且含镍量不应小于8%。

2. [Technical code for metal and stone curtain walls engineering] 3.3.2: Glass curtain wall with stainless steel should be used austenitic stainless steel, and nickel content should not be less than 8%.

3.【幕墙规】3.3.5条：支承结构用碳素钢和低合金高强度结构钢采用氟碳漆喷涂或聚氨酯漆喷涂时，涂膜的厚度不宜小于35μm；在空气污染严重及海滨地区，涂膜厚度不宜小于45μm。

3. [Technical code for metal and stone curtain walls engineering] 3.3.5: support structure with carbon steel and low alloy high-strength structural steel using fluorinated paint spraying or polyurethane paint spraying, the thickness of the coating film should not be less than 35μm; in the serious air pollution and coastal areas, the thickness of the film should not be less than 45μm.

4.【幕墙规】6.1.1条：支承玻璃幕墙单片玻璃的厚度不应小于6mm，夹层玻璃的单片厚度不宜小于5mm。夹层玻璃和中空玻璃的单片玻璃厚度相差不宜大于3mm。

4. [Technical code for metal and stone curtain walls engineering] 6.1.1: The thickness of monolithic glass of the supporting glass curtain wall should not be less than 6mm, laminated glass sheet thickness should not be less than 5mm. The thickness difference between laminated glass and hollow glass should not be greater than 3mm.

5.【幕墙规】7.1.1条：全玻幕墙玻璃高度大于表9.2.3限值的全玻幕墙应悬挂在主体结构上。

5. [Technical code for metal and stone curtain walls engineering] 7.1.1: Full-glass curtain wall with height greater than the limits given in Tab. 9.2.3 should be suspended on the main structure.

【幕墙规】表 7.1.1：下端支撑全玻幕墙的最大高度　　　　表 9.2.3

[Technical code for metal and stone curtain walls engineering] Tab. 7.1.1:
The Maximum Height of the Full-glass Curtain Wall with Lower End Supports　　Tab. 9.2.3

玻璃厚度（mm） Glass thickness（mm）	10、12	15	19
最大高度（m） Maximum height（m）	4	5	6

6.【幕墙规】8 条：点支承玻璃幕墙结构设计。

6. [Technical code for metal and stone curtain walls engineering] 8: point support glass curtain wall structure design.

玻璃面板：

Glass panel：

1）四边形玻璃面板可采用四点支承，有依据时也可采用六点支承；三角形玻璃面板可采用三点支承。玻璃面板支承孔边与板边的距离不宜小于 70mm。

1）Quadrilateral glass panel can use four point support, there is a basis can also be used six point support, triangular glass panel can use three point support. The distance between the glass panel support hole edge and the plate edge should not be less than 70mm.

2）采用浮头式连接件的幕墙玻璃厚度不应小于 6mm；采用沉头式连接件的幕墙玻璃厚度不应小于 8mm。

2）The thickness of curtain wall glass using floating head exchanger connector should not be less than 6mm, and the thickness of curtain wall glass using sunken head exchanger connector should not be less than 8mm.

3）玻璃之间的空隙宽度不应小于 10mm，且应采用硅酮建筑密封胶嵌缝。

3）The width of the gap between glass should not be less than 10mm, and silicone building sealant seam should be used.

9.2.4 【幕墙规】3.2.5 条：建筑幕墙用石材。

9.2.4 [Technical code for metal and stone curtain walls engineering] 3.2.5: Stone for building curtain wall.

1.【幕墙规】3.2.5 条：宜用花岗石、可选用大理石、石灰石、石英砂岩等。

1. [Technical code for metal and stone curtain walls engineering] 3.2.5: it should use granite, can choose marble, limestone, quartz sandstone etc.

2.【幕墙规】3.1.2 条：石材面板的性能应满足建筑物所在地的地理、气候、环境及幕墙功能的要求。

2. [Technical code for metal and stone curtain walls engineering] 3.1.2: The performance of

the stone panel should meet the geographical, climatic, environmental and curtain wall function requirements of the building site.

3.【幕墙规】3.1.4 条：石材的放射性应符合《建筑材料放射性核素限量》GB 6566 中 A 级、B 级、C 级的要求。

3. [Technical code for metal and stone curtain walls engineering] 3.1.4: The radioactivity of the stone shall conform to the requirements of Grades A, B and C in the *Radionuclide Limit for Building Materials*, GB 6566.

4.【幕墙规】3.2.2 条：花岗石板材的弯曲强度应经法定检测机构检测确定，其弯曲强度不应小于 8.0MPa。

4. [Technical code for metal and stone curtain walls engineering] 3.2.2: The bending strength of granite plate should be determined by the legal testing mechanism, its bending strength should not be less than 8.0MPa.

9.2.5【幕墙规】3.3.1 条：建筑幕墙用金属饰面板材料。

9.2.5 [Technical code for metal and stone curtain walls engineering] 3.3.1: Metal decorative panel materials used for building curtain wall with.

1. 常用材料：单层铝板、蜂窝铝板、彩色钢板、搪瓷涂层钢板、不锈钢板、锌合金板、钛合金板、铜合金板等。

1. Commonly used materials: single-layer aluminum plate, honeycomb aluminum plate, color steel plate, enamel coated steel plate, stainless steel plate, zinc alloy plate, titanium alloy plate, copper alloy plate and so on.

2.【幕墙规】3.3.9 条~3.3.12 条：板材常用厚度。

2. [Technical code for metal and stone curtain walls engineering] 3.3.9 ~ 3.3.12: Plate commonly used thickness.

单层铝板：2.5mm、3.0mm、4.0mm。

Single-layer aluminum plate: 2.5mm, 3.0mm, 4.0mm.

蜂窝铝板：10mm、15mm、20mm、25mm。

Honeycomb Aluminum Plate: 10mm, 15mm, 20mm, 25mm.

10

室内装修工程设计
Interior Decoration Engineering Design

10.1 【通则】6.15 条：一般规定

10.1 [Code for design of civil buildings] 6.15: General Provisions

10.1.1 【通则】6.15.1 条：基本要求。

10.1.1 [Code for design of civil buildings] 6.15.1: Basic requirement.

1. 室内外装修严禁破坏建筑结构安全；

1. Room interior and exterior decoration is strictly forbidden to destroy the safety of building structure;

2. 室内外装修应采用节能、环保型建筑材料；

2. Room interior and exterior decoration should use energy-saving, environment-friendly building materials;

3. 室内外装修工程应根据不同使用要求，采用防火、防污染、防潮、防水和控制有害气体和射线的装修材料和辅料；

3. Room interior and exterior decoration works according to different use requirements, the use of fire prevention, pollution prevention, moisture-proof, waterproof and control of harmful gases and rays of decoration materials and accessories;

4. 保护性建筑的内外装修尚应符合有关保护建筑条例的规定。

4. The interior and exterior decoration of a protective building shall comply with the provisions relating to the protection of Building regulations.

10.1.2 【通则】6.15.2 条：室内装修应符合下列规定。

10.1.2 [Code for design of civil buildings] 6.15.2: Interior decoration shall comply with the following provisions.

1. 室内装修不得遮挡消防设施标志、疏散指示标志及安全出口，并不得影响消防设施。

1. Interior decoration shall not block the signs of fire protection facilities, evacuation signs and safe exits, and shall not affect fire fighting facilities.

2. 室内如需要重新装修时，不得随意改变原有设施、设备管线系统。

2. If interior redecoration is needed, do not arbitrarily change the original facilities, equipment pipeline system.

10.1.3 【通则】6.15.3 条：室外装修应符合下列规定。

10.1.3 [Code for design of civil buildings] 6.15.3: Outdoor decoration shall comply with the following provisions.

1. 外墙装修必须与主体结构连接牢靠；

1. Exterior wall decoration must be securely connected with the main structure;

2. 外墙外保温材料应与主体结构和外墙饰面连接牢固，并应防开裂、防水、防冻、防腐蚀、防风化和防脱落；

2. Exterior wall insulation materials should be firmly connected with the main structure and exterior wall finishes, and should be anti-cracking, waterproof, antifreeze, anti-corrosion, windproof and anti-shedding;

3. 外墙装修应防止污染环境的强烈反光。

3. Exterior wall decoration should prevent the strong reflection of polluting the environment.

10.2 【装修防火规范】3 条：装修材料的分类和分级

10.2 [Code for fire prevention in design of interior decoration of buildings] 3: Classification and Grading of Decoration Materials

10.2.1 【装修防火规范】3.0.1 条：装修材料按其使用部位和功能，可划分为顶棚装修材料、墙面装修材料、地面装修材料、隔断装修材料、固定家具、装饰织物、其他装饰材料七类。

10.2.1 [Code for fire prevention in design of interior decoration of buildings] 3.0.1: Decoration materials according to its use of parts and functions, can be divided into ceiling decoration materials, wall decoration materials, ground decoration materials, partition decoration materials, fixed furniture, decorative fabrics, other decorative materials seven categories.

注：其他装饰材料系指楼梯扶手、挂镜线、踢脚板、窗帘盘、暖气罩等。
Note: Other decorative materials refer to stair handrails, hanging mirror line, kicking feet, curtain plates, heating covers etc.

10.2.2 【装修防火规范】3.0.2 条：装修材料按其燃烧性能应划分为四级，并应符合表 10.2.2 的规定。

10.2.2 [Code for fire prevention in design of interior decoration of buildings] 3.0.2: The decorative materials can be divided into four grades as per the combustion performance, and in conformity with the provisions in Tab. 10.2.2.

【装修防火规范】表3.0.2：装修材料燃烧性能等级　　　　表10.2.2

[Code for fire prevention in design of interior decoration of buildings] Tab. 3.02: Combustion Performance Levels of Decorative Materials　　Tab. 10.2.2

等级 Grade	装修材料燃烧性能 Combustion performance of decorative materials
A	不燃性 Incombustibility
B1	难燃性 Flame retardancy
B2	可燃性 Combustibility
B3	易燃性 Flammability

10.2.3　【装修防火规范】3.0.3条：装修材料的燃烧性能等级应按现行国家标准《建筑材料及制品燃烧性能分级》GB 8624的有关规定，经检测确定。

10.2.3　[Code for fire prevention in design of interior decoration of buildings] 3.0.3: The combustion performance rating of the decorative materials shall be determined by testing as per the relevant provisions in the current national standard *Classification for Burning Behavior of Building Materials and Products* GB 8624.

10.2.4　【装修防火规范】3.0.4条：安装在钢龙骨上燃烧性能达到B1级的纸面石膏板、矿棉吸声板，可作为A级装修材料使用。

10.2.4　[Code for fire prevention in design of interior decoration of buildings] 3.0.4: The gypsum plaster board and mineral wool sound absorbing board with Grade B1 combustion performance installed on the steel keel can be used as Grade A decorative materials.

10.2.5　【装修防火规范】3.0.5条：单位面积质量小于300g/m² 的纸质、布质壁纸，当直接粘贴在A级基材上时，可作为B1级装修材料使用。

10.2.5　[Code for fire prevention in design of interior decoration of buildings] 3.0.5: The papery and cloth wallpapers with the mass per area less than $300g/m^2$ can be used as grade B1 decorative materials when they are pasted directly on Grade A base materials.

10.2.6　【装修防火规范】3.0.6条：施涂于A级基材上的无机装饰涂料，可作为A级装修材料使用；施涂于A级基材上，湿涂覆比小于1.5kg/m²，且基层干膜厚度大于1.00mm的有机装饰涂料，可作为B1级装修材料使用。

10.2.6　[Code for fire prevention in design of interior decoration of buildings] 3.0.6: The inorganic decorative coating applied on grade A base materials can be used as grade A decorative materials ; the organic decorative coating on grade A base material and with the wet coating rate less than 1.5kg/m² and the thickness of the dry film on the base layer greater than 1.00 mm can be used as Grade B1 decorative materials.

10.2.7【装修防火规范】3.0.7条：当使用多层装修材料时，各层装修材料的燃烧性能等级均应符合建筑内部装修设计防火规范的规定。复合型装修材料的燃烧性能等级应进行整体检测确定。

10.2.7 [Code for fire prevention in design of interior decoration of buildings] 3.0.7: If the multi-layer decorative materials are used, the combustion performance ratings of decorative materials for each layer shall conform to the provisions in Code for fire prevention in design of interior decoration of buildings. The combustion performance rating of composite decorative materials shall be determined through overall testing.

10.3 【装修防火规范】4条：特别场所

10.3 [Code for fire prevention in design of interior decoration of buildings] 4: Special Places

10.3.1【装修防火规范】4.0.1条：建筑内部装修不应擅自减少、改动、拆除、遮挡消防设施、疏散指示标志、安全出口、疏散走道和防火分区、防烟分区等。

10.3.1 [Code for fire prevention in design of interior decoration of buildings] 4.0.1: For the interior decoration of buildings, fire facilities, evacuation indicators, safety exits, evacuation exits, evacuation walkways, fire compartments, smoke compartments, etc., shall not be reduced, changed, removed and shielded.

10.3.2【装修防火规范】4.0.2条：建筑内部消防栓箱门不应被装饰无遮掩，消防栓箱门四周的装修材料颜色应与消防栓箱门的颜色有明显区别或在消防栓箱门表面设置发光标志。

10.3.2 [Code for fire prevention in design of interior decoration of buildings] 4.0.2: The internal fire hydrant box doors in the buildings shall not be covered by decorations. The color of decorative materials around the fire hydrant box door shall be clearly different from the color of the fire hydrant box door, or luminous signs shall be set on the surface of the fire hydrant box door.

10.3.3【装修防火规范】4.0.3条：疏散走道和安全出口的顶棚、墙面不应采用影响人员安全疏散的镜面反光材料。

10.3.3 [Code for fire prevention in design of interior decoration of buildings] 4.0.3:

For the ceilings and walls of the evacuation corridors and safety exits, the specular reflective materials affecting safety evacuation of personnel shall not be adopted.

10.3.4 【装修防火规范】4.0.4 条：地上建筑的水平疏散走道和安全出口的门厅，其顶棚应采用 A 级装修材料，其他部位采用不低于 B1 级的装修材料；地下民用建筑的疏散走道和安全出口的门厅，其顶棚、墙面和地面均应采用 A 级装修材料。

10.3.4 [Code for fire prevention in design of interior decoration of buildings] 4.0.4: For the ceilings of the horizontal evacuation corridors and halls of the safety exits of the ground buildings, the Grade A decorative materials shall be adopted, and the decorative materials with the combustion performance not less than Grade B1 shall be adopted at the other positions; For the ceilings, walls and floors of the evacuation corridors and halls of the safety exits of the underground civil buildings, Grade A decorative materials shall be adopted.

10.3.5 【装修防火规范】4.0.5 条：疏散楼梯间和前室的顶棚、墙面和地面均应采用 A 级装修材料。

10.3.5 [Code for fire prevention in design of interior decoration of buildings] 4.0.5: For the ceilings, walls and floors of the evacuation staircases and anterooms, Grade A decorative materials shall be adopted.

10.3.6 【装修防火规范】4.0.6 条：建筑内设有上下层连通的中庭、走马廊、开敞楼梯、自动扶梯时，其连通部位的顶棚、墙面和地面均应采用 A 级装修材料，其他部位应采用不低于 B1 级的装修材料。

10.3.6 [Code for fire prevention in design of interior decoration of buildings] 4.0.6: When there are atriums, galleries over main halls, open staircases and escalators connecting upper and lower floors in buildings, for the ceilings, walls and floors of the connecting places, Grade A decorative materials shall be used, and for others positions, decorative materials with the combustion performance not less than Grade B1 shall be used.

10.3.7 【装修防火规范】4.0.7 条：建筑内部变形缝（包括沉降缝、伸缩缝、防震缝等）两侧基层的表面装修不应采用低于 B1 级的装修材料。

10.3.7 [Code for fire prevention in design of interior decoration of buildings] 4.0.7: For surface decoration of the bases on both sides of the deformation joints (including settlement joint, expansion joint, aseismatic joint, etc.) inside of buildings, decorative materials with the combustion performance less than Grade B1 shall not be used.

10.3.8 【装修防火规范】4.0.8 条：无窗房间内部装修材料的燃烧性能等级除 A 级外，应在【装修防火规范】表 5.1.1、表 5.2.1、表 5.3.1、表 6.0.1、表 6.0.5 规定的基础上提高一级。

10.3.8 [Code for fire prevention in design of interior decoration of building] 4.0.8: The combustion performance level of interior decoration materials of windowless rooms shall be improved by one level on the basis of Tab. 5.1.1, Tab. 5.2.1, Tab. 5.3.1, Tab. 6.0.1 and Tab. 6.0.5 of [Code for fire prevention in design of interior decoration of building].

10.3.9 【装修防火规范】4.0.9 条：消防水泵房、机械加压送风排风机房、固定灭火系统钢瓶间、配电室、变压器室、发电机房、储油间、通风和空调机房等，其内部所有装修均应采用 A 级装修材料。

10.3.9 [Code for fire prevention in design of interior decoration of buildings] 4.0.9: For the interior decoration of all fire pump rooms, mechanical pressurized blower and exhauster rooms, fixed extinguishing system cylinder rooms, power distribution rooms, transformer rooms, generator rooms, oil storage rooms, ventilator rooms, air conditioning machine rooms, etc., Class A decorative materials shall be used.

10.3.10 【装修防火规范】4.0.10 条：消防控制室等重要房间，其顶棚、墙面均应采用 A 级装修材料，地面及其他装修不应采用低于 B1 级的装修材料。

10.3.10 [Code for fire prevention in design of interior decoration of buildings] 4.0.10: For the ceilings and walls in important rooms such as fire control rooms, Grade A decorative materials shall be adopted, and for the decoration of floors and other positions, decorative materials with the combustion performance greater than or equal to Grade B1 shall be adopted.

10.3.11 【装修防火规范】4.0.11 条：建筑内部的厨房，其顶棚、墙面和地面均应采用 A 级装修材料。

10.3.11 [Code for fire prevention in design of interior decoration of buildings] 4.0.11: For the ceilings, walls and grounds of kitchens in buildings, Grade A decorative materials shall be adopted.

10.3.12 【装修防火规范】4.0.12 条：经常使用明火器具的餐厅、科研实验室，其材料的燃烧性能等级除 A 级外，应在【装修防火规范】表 5.1.1 及【装修防火规范】表 5.1.1、表 5.2.1、表 5.3.1、表 6.0.1、表 6.0.5 规定的基础上提高一级。

10.3.12 [Code for fire prevention in design of interior decoration of buildings] 4.0.12: In addition that Grade A of the combustion performance level of the materials shall be adopted for restaurants and research laboratories where open flame appliances are often used, the materials shall be raised by one grade on the basis of the provisions specified in [Code for Fire Prevention in Design of Interior Decoration of Buildings] Tab. 5.1.1, and [Code for Fire Prevention in Design of Interior Decoration of Buildings] Tab. 5.1.1, Tab. 5.2.1, Tab. 5.3.1, Tab. 6.0.1 and Tab. 6.0.5.

10.3.13 【装修防火规范】4.0.13 条：民用建筑内的厨房或贮藏间，其内部所有装修除应符合相应场所规定外，且应采用不低于 B1 级的装修材料。

10.3.13 [Code for fire prevention in design of interior decoration of buildings] 4.0.13:

In addition that the materials for all the interior decoration in kitchens or storage rooms in civil buildings shall conform to the provisions for the corresponding sites, and the decorative materials decorative materials with the combustion performance greater than or equal to Grade B1 shall be adopted.

10.3.14 【装修防火规范】4.0.14条：展览性场所装修设计应符合下列规定。

10.3.14 [Code for fire prevention in design of interior decoration of buildings] 4.0.14: The decoration and design of exhibition places shall conform to the following provisions.

1. 展台材料应采用不低于B1级的装修材料。

1. For the materials of exhibition stands, the decorative materials with the combustion performance not less than Grade B1 shall be adopted.

2. 在展厅内设置电加热设备的餐饮操作区内，与电加热设备贴邻的墙面、操作台均应采用A级装修材料。

2. In the catering operation areas in exhibition halls provided with the electric heating equipment, for all the walls and operation platforms near the electric heating equipment, the Grade A decorative materials shall be adopted.

3. 展台与卤钨灯高温照明灯具贴邻部位的材料应采用A级装修材料。

3. For materials adjacent to the position of the tungsten halogen lamp high-temperature lighting fixtures on the exhibition booth, the Grade A decorative materials shall be adopted.

10.3.15 【装修防火规范】4.0.15条：住宅建筑装修设计上应符合下列规定。

10.3.15 [Code for fire prevention in design of interior decoration of buildings] 4.0.15: The decoration design for residential buildings shall conform to the following provisions.

1. 不应改动住宅内部烟道、风道。

1. The interior flues and air ducts in the building shall not be modified.

2. 厨房内的固定橱柜宜采用不低于B1级的装修材料。

2. For fixed cabinets in the kitchen, the decorative materials with the combustion performance not less than Grade B1 shall be appropriately adopted.

3. 卫生间顶棚宜采用A级装修材料。

3. For ceilings in toilets, the Grade A decorative materials shall be appropriately adopted.

4. 阳台装修宜采用不低于B1级的装修材料。

4. In decoration of balconies, the decorative materials with the combustion performance not less than Grade B1 shall be adopted.

10.3.16 【装修防火规范】4.0.16条：照明灯具及电气设备、线路的高温部位，当靠近非A级装修材料或构件时，应采取隔热、散热等防火保护措施，与窗帘、帷幕、幕布、软包等装修材料的距离不应小于500mm；灯饰应采用不低于B1级的装修材料。

10.3.16 [Code for fire prevention in design of interior decoration of buildings] 4.0.16: At the high-temperature positions of lighting fixtures, electrical equipment and lines, when they are close to non-Grade A decorative materials or components, the fire protective measures of heat shielding, heat dissipation, etc., shall be adopted, and the distances to the decorative materials of curtain, screen, soft decoration, etc., shall be not less than 500 mm; For decorative materials for lamps, the decorative materials with the combustion performance not less than Grade B1 shall be adopt

10.3.17 【装修防火规范】4.0.17 条：建筑内部的配电箱、控制面板、接线盒、开关、插座等不应直接安装在低于 B1 级的装修材料上；用于顶棚和墙面装修的木质类板材，当内部含有电器、电线等物体时，应采用不低于 B1 级的装修材料。

10.3.17 [Code for fire prevention in design of interior decoration of buildings] 4.0.17: The distribution boxes, control panels, junction boxes, switches, sockets, etc., inside buildings shall not be installed on the decorative materials with the combustion performance less than Grade B1; when objects of electric appliances, electric wires, etc., are installed in the wooden boards in ceiling and wall decoration, the decorative materials with the combustion performance not less than Grade B1 shall be adopted.

10.3.18 【装修防火规范】4.0.18 条：当室内顶棚、墙面、地面和隔断装修材料内部安装电加热供暖系统时，室内采用的装修材料和绝热材料的燃烧性能等级应为 A 级。当室内顶棚、墙面、地面和隔断装修材料内部安装水暖（或蒸汽）供暖系统时，其顶棚采用的装修材料和绝缘材料的燃烧性能等级应为 A 级。其他部位的装修材料和绝热材料的燃烧性能等级不应低于 B1 级，尚应符合本规范有关公共场所的规定。

10.3.18 [Code for fire prevention in design of interior decoration of buildings] 4.0.18: When the electric heating systems are installed in the decorative materials for ceilings, walls, floors and partitions in the room, the combustion performance ratings of the decorative materials and the heat insulation materials used in the rooms shall be Grade A. When water（or steam）-heating systems are installed in the decorative materials for ceilings, walls, floors and partitions in the room, the combustion performance ratings of the decorative materials and insulation materials used in the ceilings shall be Grade A. The combustion performance ratings of the decorative materials and thermal insulation materials used in other positions shall be not less than Grade B1, and shall conform to the provisions relevant to public places in the Code.

10.3.19 【装修防火规范】4.0.19 条：建筑内部不宜设置采用 B3 级装饰材料制成的壁挂、布艺等，当需要设置时，不应靠近电气线路、火源或热源，应采取隔离措施。

10.3.19 [Code for fire prevention in design of interior decoration of buildings] 4.0.19: Wall hangings, fabrics, etc., made of Grade B3 decorative materials shall not be arranged inside buildings. If necessary, these hangings, fabrics, etc., shall not be arranged near the electrical circuits, fire sources or heat sources, or the isolation measures shall be adopted.

10.4 【装修防火规范】5条：民用建筑

10.4 [Code for fire prevention in design of interior decoration of buildings] 5: Civil Buildings

10.4.1 【装修防火规范】5.1条：单层、多层民用建筑

10.4.1 [Code for fire prevention in design of interior decoration of buildings] 5.1: Single-storey and Multi-storey Civil Buildings

1. 单层、多层民用建筑内部各部位装修材料的燃烧性能等级，不应低于本规范10.4.1中表10.4.1的规定。

1. The combustion performance ratings of the decorative materials for any part in single-storey and multi-storey civil buildings shall be not less than those specified in Tab. 10.4.1 of this code.

【装修防火规范】表5.1.1：单层和多层民用建筑不同部位装饰材料燃烧性能等级　　表10.4.1

[Code for fire prevention in design of interior decoration of buildings] Tab. 5.1.1: Combustion Performance Ratings of Decorative Materials for Different Parts in Single-storey and Multi-storey Civil Buildings　　Tab. 10.4.1

序号 No.	建筑物及场所 Buildings and places	建筑规模、性质 Building scale and nature	装修材料的燃烧性能等级 Combustion performance rating of decorative materials					装饰织物 Decorative fabric		
			顶棚 Ceilings	墙面 Walls	地面 Ground	隔断 Partition	固定家具 Fixed furniture	窗帘 Curtain	帷幕 Parocheth	其他装饰材料 Other decorative material
1	候机楼的候机大厅、贵宾候机室、商店、餐饮场所等 Waiting halls, VIP lounges, shops, catering places, etc., in terminal buildings	—	A	A	B1	B1	B1	B1	—	B1

续表

序号 No.	建筑物及场所 Buildings and places	建筑规模、性质 Building scale and nature	装修材料的燃烧性能等级 Combustion performance rating of decorative materials							
			顶棚 Ceilings	墙面 Walls	地面 Ground	隔断 Partition	固定家具 Fixed furniture	装饰织物 Decorative fabric		其他装修装饰材料 Other decorative material
								窗帘 Curtain	帷幕 Parocheth	
2	汽车站、火车站、轮船客运站的候车（船）室、商店、餐饮场所等 Waiting halls, shops, catering places, etc., at the bus station, railway station and ship passenger station	建筑面积＞10000m² Building area ＞ 10000m²	A	A	B1	B1	B1	B1	—	B2
		建筑面积≤10000m² Building area ≤ 10000 m²	A	B1	B1	B1	B1	B1	B1	B1
3	观众厅、会议厅、多功能厅、等候厅等 Auditorium, conference hall, multi-purpose hall, waiting hall, etc.	每个厅建筑面积＞400m² Building area of each hall ＞ 400m²	A	A	B1	B1	B1	B1	B1	B1
		每个厅建筑面积≤400m² Building area of each hall ≤ 400m²	A	B1	B1	B1	B2	B1	B1	B2
4	体育馆 Gymnasium	＞3000 座位 ＞ 3000 seats	A	A	B1	B1	B1	B1	B1	B1
		≤3000 座位 ≤ 3000 seats	A	B1	B1	B1	B2	B2	B1	B2
5	商店的营业厅 Business halls of stores	每层建筑面积＞1500m² 或总建筑面积＞3000m² Building area of each floor ＞ 1,500 m² or total building area ＞ 3000m²	A	B1	B1	B1	B1	B1	—	B2
6	宾馆、饭店的客房及公共活动用房等 Guest rooms of hotels and restaurants, and public activity places, etc.	设置送回风道（管）的集中空气调节系统 Centralized air conditioning system with air supply/return duct（pipe）	A	B1	B1	B1	B2	B2	—	B2
		其他 Others	B1	B2	B2	B2	B2	B2	—	—
7	养老院、托儿所、幼儿园的居住及活动场所 Residence and activity places of the nursing home, nursery and kindergarten	—	A	A	B1	B1	B2	B1	—	B2
8	医院的病房区、诊疗区、手术区 Ward area, diagnosis and treatment area and surgery area in hospital	—	A	A	B1	B1	B2	B1	—	B2
9	教学场所、教学实验场所 Teaching places and teaching and testing places	—	A	B1	B2	B2	B2	B2	B2	B2

续表

序号 No.	建筑物及场所 Buildings and places	建筑规模、性质 Building scale and nature	装修材料的燃烧性能等级 Combustion performance rating of decorative materials							
			顶棚 Ceilings	墙面 Walls	地面 Ground	隔断 Partition	固定家具 Fixed furniture	装饰织物 Decorative fabric		其他装修装饰材料 Other decorative material
								窗帘 Curtain	帷幕 Parocheth	
10	纪念馆、展览馆、博物馆、图书馆、档案馆、资料馆等的公众活动场所 Public activity places such as memorial halls, exhibition halls, museums, libraries, archives, etc.	—	A	B1	B1	B1	B2	B1	—	B2
11	存放文物、纪念展览物品、重要图书、档案、资料的场所 Places for the storage of cultural relics and commemoration of exhibition goods, important books, archives and materials	—	A	A	B1	B1	B2	B1	—	B2
12	歌舞娱乐游艺场所 Karaoke bars, dancing halls, entertainment rooms, video halls and amusement halls	—	A	B1	B1	B1	B1	B1	B1	B1
13	A、B级电子信息系统机房及装有重要机器、仪器的房间 Grades A and B electronic information system rooms and the rooms with important machines and instruments	—	A	A	B1	B1	B1	B1	B1	B1
14	餐饮场所 Catering places	营业面积＞100m² Business area ＞ 100m²	A	B1	B1	B1	B2	B1	—	B2
		营业面积≤100m² Business area ≤ 100m²	B1	B1	B1	B2	B2	B2	—	B2
15	办公场所 Places for offices	设置送回风道（管）的集中空气调节系统 Centralized air conditioning system with air supply/return duct（pipe）	A	B1	B1	B1	B2	B2	—	B2
		其他 Others	B1	B2	B2	B2	B2	—	—	—
16	其他公共场所 Other public places	—	B1	B2	B2	B2	B2	—	—	—
17	住宅 Residential buildings	—	B1	B1	B1	B1	B2	B2	—	B2

11

屋面和吊顶
Roof and Suspended Ceiling

11.1 【屋面技规】3.0.2 条：屋面类型及基本构造层次

11.1 [Technical code for roof engineering] 3.0.2: Roof Types and Basic Structure Levels

屋面类型及基本构造层次见表 11.1。
Roof types and basic structure levels see Tab.11.1.

【屋面技规】表 3.0.2：屋面的基本构造层次　　　　表 11.1
[Technical code for roof engineering] Tab. 3.0.2: Basic Structural Layers of Roof　　Tab. 11.1

屋面类型 Roof type	基本构造层次（自上至下） Basic structural layer (from top to bottom)
卷材、涂膜屋面 Coiled material and coated roof	保护层、隔离层、防水层、找平层、保温层、找平层、找坡层、结构层 Protective layer, isolation layer, waterproof layer, leveling layer, insulation layer, leveling layer, sloping layer, and structural layer
	保护层、保温层、防水层、找平层、找坡层、结构层 Protective layer, insulation layer, waterproof layer, leveling layer, sloping layer, and structural layer
	种植 隔热层、保护层、耐根穿刺防水层、防水层、找平层、找坡层结构层 Planting heat shielding layer, protective layer, root-penetration-resistant waterproof layer, waterproof layer, leveling layer, sloping layer, and structural layer
	加空隔热层、防水层、找平层、保温层、找平层、找坡层、结构层 Overhead heat shielding layer, waterproof layer, leveling layer, insulation layer, leveling layer, sloping layer, and structural layer
	蓄水隔热层、隔离层、防水层、找平层、保温层、找平层、找坡层、结构层 Water storage and heat shielding layer, isolation layer, waterproof layer, leveling layer, insulation layer, leveling layer, sloping layer, and structural layer
瓦屋面 Tile roof	块瓦、挂瓦条、顺水条、持钉层、防水层或防水垫层、保温层、结构层 Block tile, tile hanging batten, counter batten, nail-supporting layer, waterproof layer or waterproof cushion layer, insulation layer, and structural layer
	沥青瓦、防水层或防水垫层、保温层、结构层 Asphalt tile, waterproof layer or waterproof cushion layer, insulation layer, and structural layer
金属板屋面 Metal board roof	压型金属板、防水垫层、保温层、承托网、支撑结构 Profiled metal board, waterproof cushion layer, insulation layer, bearing mesh, and support structure
	上层压型金属板、防水垫层、保温层、底层压型金属板、支撑结构 Upper profiled metal board, waterproof cushion layer, insulation layer, bottom profiled metal board and support structure
	金属面绝热夹芯板、支撑结构 Metal-surface heat shielding sandwich board and support structure

续表

屋面类型 Roof type	基本构造层次（自上至下） Basic structural layer (from top to bottom)
玻璃采光顶 Glass daylighting roof	玻璃面板、金属框架、支撑结构 Glass panel, metal frame and support structure
	玻璃面板、点支撑结构、支撑结构 Glass panel, point-supported structure and support structure

注：1. 表中结构层包括混凝土基层和木基层；防水层包括卷材和涂膜防水层；保温层包括块体材料、水泥砂浆、细石混凝土保护层。
Notes: 1. The structural layer in the table includes the concrete base layer and wood base layer; the waterproof layer includes the coiled material and coated film waterproof layers; the insulation layer includes the block material, cement mortar and fine-aggregated concrete protective layers.
2. 有隔汽要求的屋面，应在保温层与结构层之间设隔汽层。
2. For roofs with vapor barrier requirements, a vapor barrier layer shall be set between the insulation layer and structural layer.

11.2 【屋面技规】3.0.5 条：屋面防水等级和设防要求

11.2 [Technical code for roof engineering] 3.0.5: Waterproof Level and Fortification Requirements of Roofs

屋面工程应根据建筑物的性质、重要程度、使用功能要求确定防水等级，并按相应等级进行防水设防；对防水有特殊要求的建筑屋面，应进行专项防水设计。屋面防水等级和设防要求应符合表 11.2 的规定。

For roof works, the waterproof grade shall be determined in accordance with the properties, importance degree and use function requirements of the building, and the waterproof fortification shall be designed in accordance with the corresponding grade; for building roofs with special requirements for waterproof, special waterproof design shall be carried out. The waterproof grades and fortification requirements of roofs shall conform to the provisions in Tab. 11.2.

【屋面技规】表 3.0.5：屋面防水等级和设防要求　　　　表 11.2
[Technical code for roof engineering] 3.0.5: Waterproof Grades and Fortification Requirements of Roofs　　Tab. 11.2

防水等级 Waterproof grade	建筑类别 Building category	设防要求 Fortification requirements
I 级 Grade I	重要建筑和高层建筑 Important buildings and high-rise buildings	两道防水设防 Waterproof fortification for two rounds
II 级 Grade II	一般建筑 General buildings	一道防水设防 Waterproof fortification for one round

11.3 屋面的找坡层和找平层

11.3 Sloping Layer and Leveling Layer of Roof

1.【屋面技规】4.3.1 条：混凝土结构层宜采取结构找坡，坡度不应小于 3%；当采用材料找坡时，宜采用质量轻，吸水率低和一定强度的材料，坡度宜为 2%。

1. [Technical code for roof engineering] 4.3.1: The structural sloping shall be adopted for the concrete structural layer, and the gradient shall be not less than 3%; if the material sloping is adopted, materials with light weight, low water absorption rate and certain strength shall be appropriately adopted, and the gradient shall be appropriately 2%.

2.【屋面技规】4.3.2 条：卷材、涂膜的基层宜设找平层。找平层厚度和技术要求应符合表 11.3 的规定。

2. [Technical code for roof engineering] 4.3.2: The leveling layer shall be appropriately arranged on the base layers of the coiled material and coated film. The thickness and technical requirements for leveling layers shall conform to the provisions in Tab. 11.3.

【屋面技规】表 4.3.2：找平层厚度和技术要求　　　表 11.3

[Technical code for roof engineering] Tab. 4.3.2: Thickness and Technical Requirements for Leveling Layers　　Tab. 11.3

找平层分类 Classification of leveling layers	使用的基层 Base layers used	厚度（mm） Thickness（mm）	技术要求 Technical requirements
水泥砂浆 Cement mortar	整体现浇混凝土板 Integral cast-in-situ concrete boards	15 ~ 20	M10 砂浆 M10 mortar
	整体材料保温层 Integral material insulation layers	20 ~ 25	
细石混凝土 Fine-aggregated concrete	装配式混凝土板 Assembly type concrete boards	30 ~ 35	C20 混凝土，宜加钢筋网 C20 concrete, appropriately added with reinforcement meshes
	块状材料保温层 Block-material insulation layer		C20 混凝土 C20 concrete

11.4 【屋面技规】4.2 条：屋面的排水设计

11.4 [Technical code for roof engineering] 4.2: Drainage Design for Roofs

1.【屋面技规】4.2.2 条：屋面的排水方式可分为有组织排水和无组织排水。有组织排水时，宜采用雨水收集系统。

1. [Technical code for roof engineering] 4.2.2: The drainage modes of roof include organized drainage and unorganized drainage. The rainwater collection system shall be adopted in the organized drainage.

2.【屋面技规】4.2.3 条：高层建筑屋面宜采用内排水；多层建筑屋面宜采用有组织的外排水；底层建筑及檐高小于 10m 的屋面，可采用无组织排水，多跨及汇水面积较大的屋面宜采用天沟排水，天沟找坡较长时，宜采用中间内排水和两端外排水。

2. [Technical code for roof engineering] 4.2.3: The internal drainage mode shall be adopted for the roofs of high-rise buildings; the organized external drainage shall be adopted for the roofs of multi-storey buildings; the unorganized drainage mode can be adopted for the roofs of low-rise buildings with eaves height less than 10m, gutters can be used for drainage for multi-span roofs and roofs with large catchment areas, and when the slopes of the gutters are long, middle internal drainage and both-end external drainage mode shall be adopted.

3.【屋面技规】4.2.4 条：屋面排水系统设计采用的雨水流量、暴雨强度、降雨历时、屋面回水面积等参数，应符合现行国家标准《建筑给水排水设计规范》GB 50015 的有关规定。

3. [Technical code for roof engineering] 4.2.4: When the rainwater flow, rainfall intensity and rainfall duration are adopted in the design of drainage systems, the parameters of backwater areas, etc., of the roofs shall conform to the relevant provisions in the current national standard, *Code for Design of Building Water Supply and Drainage* GB 50015.

4.【屋面技规】4.2.6 条：采用重力式排水时，屋面每个汇水面积内，雨水排水立管不宜少于 2 根；水落口和水落管的位置，应根据建筑物的造型要求和屋面汇水情况等因素确定。

4. [Technical code for roof engineering] 4.2.6: When the gravity-type drainage mode is adopted, at least 2 rainwater drainage riser pipes shall be provided in each roof catchment area; the positions of the downspout port and downpipe shall be confirmed in accordance with the factors of the modeling requirements for the building and roof catchment condition, etc.

5.【屋面技规】4.2.7条：高跨屋面为无组织排水时，其低跨屋面受雨水冲刷的部位应附加一层卷材，并应设40～50mm厚、300～500mm宽的C20细石混凝土保护层；高跨屋面为有组织排水时，水落管下应加设水簸箕。

5. [Technical code for roof engineering] 4.2.7: When the unorganized drainage mode is adopted for high-span roofs, one layer of coiled materials and C20 fine-aggregated concrete protective layer (40～50mm thick and 300～500mm wide) shall be arranged at the rainwash positions of the low-span roofs. When the organized drainage mode is adopted for high-span roofs, a water hopper shall be arranged below the downpipe.

6.【屋面技规】4.2.8条：暴雨强度较大的地区的大型屋面，宜采用虹吸式屋面雨水排水系统。

6. [Technical code for roof engineering] 4.2.8: For the large roofs in heavy rainstorm areas, the siphonic roof rainwater drainage system shall be appropriately adopted.

7.【屋面技规】4.2.9条：严寒地区应采用内排水，寒冷地区宜采用内排水。

7. [Technical code for roof engineering] 4.2.9: The internal drainage shall be adopted in severely cold regions, and the internal drainage shall be appropriately adopted in cold regions.

8.【屋面技规】4.2.10条：湿陷性黄土地区宜采用有组织排水，并应将雨雪水直接排至排水管网。

8. [Technical code for roof engineering] 4.2.10: The organized drainage shall be adopted in collapsible loess areas, and the rain and snow water shall be directly drained into the drainage pipeline network.

9.【屋面技规】4.2.11条：檐沟、天沟的过水断面，应根据汇水面积的水流量经计算确定。钢筋混凝土檐沟、天沟净宽不应小于300mm；分水线处最小深度不应小于100mm；沟内纵向坡度不应小于1%，沟底水落差不得超过200mm；檐沟、天沟排水不得流经变形缝和防火墙。

9. [Technical code for roof engineering] 4.2.11: Crossing sections of eaves and gutters shall be calculated on the basis of the flow of catchment area. The net width of the steel reinforced concrete eaves and gutters shall not be less than 300mm. The minimum depth at watershed shall not be less than 100mm; the longitudinal slope in ditch shall not be less than 1% and the water drop at the bottom of gully shall not exceed 200mm; drainage of eaves and gutters shall not flow through expansion joints and fire rated walls.

10.【屋面技规】4.2.12条：金属檐沟、天沟的纵向坡度宜为0.5%。

10. [Technical code for roof engineering] 4.2.12: The longitudinal gradient of the metal trenches and gutters shall be appropriately 0.5%.

11.【屋面技规】4.2.13条：坡屋面檐口宜采用有组织排水，檐口和水落斗可采用金属或塑料成品。

11. [Technical code for roof engineering] 4.2.13: The organized drainage shall be appropriately adopted for the cornices of the slope roofs, and the finished metal or plastic products can be adopted for the cornices and leader heads.

11.5 保温层和隔热层设计
11.5 Design of Insulation Layers and Heat Shielding Layers

11.5.1 【屋面技规】4.4.1 条：保温层应根据屋面所需传热系数或热阻选择轻质、高效的保温材料，保温层及其保温材料应符合表 11.5.1 的规定。

11.5.1 [Technical code for roof engineering] 4.4.1: For insulation layers, the high-efficiency light-weight insulation materials shall be selected as per the heat transfer coefficient or thermal resistance required on the roof, and the insulation layers and the insulation materials shall conform to the provisions in Tab. 11.5.1.

【屋面技规】表 4.4.1：保温层及其保温材料　　　　表 11.5.1
[Technical Code for Roof Engineering] Tab. 4.4.1: Insulation Layers and Insulation Materials　　Tab. 11.5.1

保温层 Insulation layer	保温材料 Insulation material
块状材料保温层 Block-material insulation layer	聚苯乙烯泡沫塑料，硬质聚氨酯泡沫塑料，膨胀珍珠岩制品，泡沫玻璃制品，加气混凝土砌块，泡沫混凝土砌块 Pow lystyrene foam plastics, rigid polyurethane foam plastics, expanded perlite product, foam glass products, aerated concrete block, and foam concrete block
纤维材料保温层 Insulation layer of fiber material	玻璃棉制品，岩棉，矿渣棉织品 Glass wool products, rock wool, and slag cotton fabric
整体材料保温层 Integral material insulation layers	喷涂硬泡聚氨酯，现浇泡沫混凝土 Sprayed hard foam polyurethane, and cast-in-situ foam concrete

11.5.2 【屋面技规】4.4.2 条：保温层设计应符合下列规定。

11.5.2 [Technical code for roof engineerings] 4.4.2: The design of the insulation layers shall conform to the following provisions.

1. 保温层宜选用吸水率低、密度和导热系数小，并有一定强度的保温材料；

1. For the insulation layers, the insulation materia with low water absorption rate, low density thermal conductivity coefficient and certain strength shall be selected;

2. 保温层厚度应根据所在地区现行节能设计标准，经计算确定；

2. The thickness of the insulation layers shall be determined through calculation in accordance with the current energy-saving design standards in the area where the project is located;

3. 保温层的含水率，应相当于该材料在当地自然风干状态下平衡含水率；

3. The moisture contents of the insulation layers shall be equivalent to the equilibrium moisture contents of the materials in the local natural air-dried state;

4. 屋面为停车场等高荷载情况时，应根据计算确定保温材料的强度；

4. If roofs are the heavy load areas such as parking areas, etc., the strength of the insulation materials shall be determined through calculation;

5. 纤维材料做保温层时，应采取防止压缩的措施；

5. When the fiber materials are used for the insulation layers, the compression prevention measures shall be adopted;

6. 屋面坡度较大时，采取防滑措施；

6. When the gradients of the roofs are large, the slip prevention measures shall be adopted;

7. 封闭式保温层或保温层干燥有困难的卷材屋面，宜采取排汽构造措施。

7. For the coiled material roofs with closed-type insulation layers or insulation layers difficult to be dried, the exhaust structure measures shall be adopted.

11.5.3 【屋面技规】4.4.6 条：倒置式屋面保温层设计应符合下列规定。

11.5.3 [Technical code for roof engineering] 4.4.6: The design of the insulation layer of the inverted roof shall conform to the following provisions.

1. 倒置式屋面的坡度宜为 3%；

1. The gradient of the inverted roof shall be 3%;

2. 保温层应采用吸水率低且长期浸水不变质的保温材料；

2. The insulation materials with low water absorption and no deterioration after long-term immersion in water shall be used for the insulation layers;

3. 板状材料的下部纵向边缘应设排水凹缝；

3. The concave drainage joints shall be arranged on the longitudinal edges of the lower parts of the slab materials;

4. 保温层与防水层所用材料应相容匹配；

4. The materials for the insulation layers shall match the materials for the waterproof layers;

5. 保温层上面宜采用块状材料或细石混凝土做保护层；

5. The bulk material or fine-aggregated concrete protective layers shall be used on the insulation layers;

6. 檐沟、水落口部位应采用现浇混凝土堵头或砖砌堵头，并应做好保温层排水处理。

6. The trenches and downspout port shall be blocked with cast-in-situ concrete or bricks, and the drainage treatment of the insulation layers shall be carried out.

11.5.4 【屋面技规】4.4.8 条：种植隔热层的设计应符合下列规定。

11.5.4 [Technical code for roof engineering] 4.4.8: The design of the planting heat shielding layer shall conform to the following provisions.

1. 种植隔热层的构造层次应包括植被层、种植土层、过滤层和排水层等。

1. The structural layers of the planting heat shielding layer shall include the vegetation layer, planting soil layer, filter layer and drainage layer, etc.

2. 种植隔热层所有材料及植物等应与当地的气候条件相适应，并应符合环境保护要求。

2. All the materials and plants for the planting heat shielding layer shall be compatible with the local climatic conditions and shall comply with the environmental protection requirements.

3. 种植隔热层宜根据植物种类及环境布局的需要进行分区布置，分区布置应设挡墙或挡板。

3. The planting heat shielding layer shall be arranged in partitions in accordance with the requirements for plants species and environmental layout, and retaining walls or baffles shall be set up in the partition arrangement.

4. 排水层材料应根据屋面功能及环境、经济条件进行选择；过滤层宜采用 200~400g/m² 的土工布，过滤层应沿种植土周边向上铺设至种植土高度。

4. The material for the drainage layer shall be selected in accordance with the functions of roof and the environmental and economic conditions; for the filter layer, $200 \sim 400 g/m^2$ geotextiles shall be used, and the filter layer shall be laid up to the height of the planting soil along the periphery of the planting soil.

5. 种植土四周应设挡墙，挡墙下部应设排水孔，并应与排水出口连通。

5. Retaining walls shall be arranged on the periphery of the planting soil, and drainage holes shall be set on the lower parts of the walls and be connected with the drainage outlet.

6. 种植土应根据种植植物的要求选择综合性能好的材料；种植土厚度应根据不同种植土和植物种类等确定。

6. For the planting soil, materials with good integrated performance shall be selected as per the requirements of planting plants; and the thickness of the planting soil shall be determined in accordance with the type of the soil, species of the plant, etc.

7. 种植隔热层的屋面坡度大于 20% 时，其排水层、种植土应采取防滑措施。

7. When the gradient of the roof of the planting heat shielding layer is more than 20%, the slip prevention measures shall be taken for the drainage layer and planting soil.

11.5.5 【屋面技规】4.4.9 条：架空隔热层的设计应符合下列规定。

11.5.5 [Technical code for roof engineerings] 4.4.9: The design of the overhead heat shielding layer shall conform to the following provisions.

1. 架空隔热层宜在屋顶有良好通风的建筑物上采用，不宜在寒冷地区采用；

1. The overhead heat shielding layers shall be used in building with the roof with good ventilation, and can not be used in cold regions;

2. 当采用钢筋混凝土板架空隔热层时，屋面坡度不宜大于5%；

2. When the overhead heat shielding layer of the reinforced concrete slab is used, the roof gradient shall be not greater than 5%;

3. 架空层的层间高度宜为180～300mm，架空板与女儿墙的距离不宜小于250mm；

3. The interlayer height of the overhead layer shall be 180 ～ 300 mm, and the distance between the overhead board and the parapet wall shall be not less than 250 mm;

4. 屋面宽度大于10m时，架空隔热层中部应设通风屋脊；

4. When the width of a roof is more than 10m, the ventilation roof ridge shall be arranged in the middle of the overhead heat shielding layer;

5. 架空隔热层的进风口，宜设置在当地炎热季节最大频率风向的正压区，出口宜设置在负压区。

5. The over-head insulated layer shall be arranged in the local positive pressure zone of the direction of the highest frequency wind in the hot season, and the outlet shall be arranged in the negative pressure zone.

11.5.6 【屋面技规】4.4.10 条：蓄水隔热层的设计应符合下列规定。

11.5.6 [Technical code for roof engineering] 4.4.10: The design of the water storage and heat shielding layer shall conform to the following provisions.

1. 蓄水隔热层不宜在寒冷地区、地震地区或振动较大的建筑物上使用。

1. The water storage and heat shielding layers shall not be used in cold regions, seismic zones and buildings with large vibration.

2. 蓄水隔热层的蓄水池应采用强度不低于C25、抗渗等级不低于P6的现浇混凝土，蓄水池内宜采用20mm厚防水砂浆抹面。

2. The cast-in-situ concrete with the strength not lower than C25 and the impermeability grade not less than P6 shall be adopted for the water storage tank of the water storage and heat shielding layer, and the waterproof mortar plastering with the thickness of 20mm shall be adopted for the inner side of the water storage tank.

3. 蓄水隔热层的排水坡度不宜大于0.5%。

3. The drainage gradient of the water storage and heat shielding layer shall be appropriately not greater than 0.5%.

4. 蓄水隔热层应划分若干蓄水区，每区的边长不宜大于10m，在变形缝的两侧应分成两个互不连通的蓄水区。长度超过40m的蓄水隔热层应分仓设置，分仓隔墙可采用现浇混凝土或砌体。

4. The water storage and heat shielding layer shall be divided into several water storage areas, and the side length of each area shall be appropriately not greater than 10m, and both sides of the deformation joint shall be divided into two disconnected water storage areas. The water storage and heat shielding layer with the length exceeding 40m shall be arranged in chambers, and the cast-in-situ concrete or masonry can be adopted for the partition walls.

5. 蓄水池应设溢水口、排水管和排水出口连通。

5. The overflow hole and drainage pipe shall be set in the water storage tank and be connected with the drainage outlet.

6. 蓄水池的蓄水深度宜为150～200mm。

6. The water storage depth of the tank shall be appropriately 150～200mm.

7. 蓄水池溢水口距分仓墙顶面的高度不得小于100mm。

7. The distance from the overflow hole of the water storage tank to the top surface of the partition wall shall be not less than 100mm.

8. 蓄水池应设人行通道。

8. For water storage tanks, pedestrian passages shall be arranged.

11.6 卷材及涂膜防水层设计

11.6 Design of Coiled Materials and Coated Film Waterproof Layers

11.6.1 【屋面技规】4.5.1条：卷材及涂膜防水等级和防水做法应符合表11.6.1的规定。

11.6.1 [Technical code for roof engineering] 4.5.1: The waterproof grades and waterproof methods for coiled materials and coated films shall conform to the provisions in Tab. 11.6.1.

【屋面技规】表 4.5.1：卷材及涂膜防水等级和防水做法　　表 11.6.1
[Technical code for roof engineering] Tab. 4.5.1:
Waterproof grades and Waterproof Methods for Coiled Materials and Coated Films　　Tab. 11.6.1

防水等级 Waterproof grade	防水做法 Waterproof methods
I 级 Grade I	卷材防水层和卷材防水层、卷材防水层和涂膜防水层、复合防水层 Coiled material waterproof layers and coiled material waterproof layers, coiled material waterproof layers and coated film waterproof layers, and composite waterproof layers
II 级 Grade II	卷材防水层、涂膜防水层、复合防水层 Coiled material waterproof layers, coated film waterproof layers and composite waterproof layers

注：在 I 级屋面防水做法中，防水层仅作单层卷材时，应符合有关单层防水卷材屋面技术的规定。
Note: In the Grade I waterproof methods of roofs, when the single-layer coiled materials are only made for the waterproof layers, the technical provisions for the single-layer waterproof coiled roofing shall be met.

11.6.2 【屋面技规】4.5.2 条：防水卷材的选择应符合下列规定。

11.6.2 [Technical code for roof engineering] 4.5.2: The selection of waterproof coiled materials shall conform to the following provisions.

1. 防水卷材可按合成高分子防水卷材和高聚物改性沥青防水卷材选用。其外观质量和品种、规格应符合国家现行有关材料标注的规定。

1. The waterproof coiled materials can be selected as per the synthetic polymer waterproof coiled materials and high-polymer modified asphalt waterproof coiled materials. The appearance quality, variety and specifications shall conform to the current national specifications relevant to material labels.

2. 根据当地历年最高气温、最低气温、屋面坡度和使用条件等因素，选择耐热度、低温柔性相适应的卷材。

2. In accordance with the factors of local maximum temperature and minimum temperature over the years, gradient of the roof, use conditions, etc., the coiled materials with adaptive heat resistance and low temperature flexibility shall be selected.

3. 应根据地基变形程度、结构形式、当地年温差、日温差和振动等因素，选择拉伸性能相适应的卷材。

3. In accordance with the factors of deformation degree of the foundation, structural form, year temperature difference, daily temperature difference, vibration, etc., the coiled materials with appropriate tensile properties shall be adopted.

4. 应根据屋面卷材暴露程度，选择耐紫外线、耐老化、耐霉烂相适应的材料。

4. In accordance with the exposure degree of the coiled materials for roof, the materials with adaptive ultraviolet resistance, aging resistance and mildew resistance shall be adopted.

5. 种植隔热屋面的防水层应选择耐穿刺防水材料。

5. For the waterproof layers of the planted heat shielding roofs, the root-penetration-resistant waterproof materials shall be adopted.

11.6.3 【屋面技规】4.5.3条：防水涂料的选择应符合下列规定。

11.6.3 [Technical code for roof engineering] 4.5.3: The selection of waterproof coatings shall conform to the following provisions.

1. 防水涂料可按合成高分子防水涂料、聚合物水泥防水涂料和高聚物改性沥青防水涂料选用，其外观质量和品种、型号应符合国家现行有关材料标注的规定；

1. The waterproof coatings can be selected as per the synthetic polymer waterproof coatings, polymer cement waterproof coatings and high-polymer modified asphalt waterproof coatings, and the appearance quality, varieties and models shall conform to the current national specifications relevant to material label;

2. 根据当地历年最高气温、最低气温、屋面坡度和使用条件等因素，选择耐热度、低温柔性相适应的涂料；

2. The coatings with adaptive heat resistance and low temperature flexibility shall be selected in accordance with the factors of local maximum temperature and minimum temperature over the years, gradient of the roofs, use conditions, etc.;

3. 应根据地基变形程度、结构形式、当地年温差、日温差和振动等因素，选择拉伸性能相适应的涂料；

3. The coatings with adaptive tensile properties shall be adopted in accordance with the factors of deformation degrees of foundations, structural forms, local year temperature difference, daily temperature difference, vibration, etc.;

4. 应根据屋面卷材暴露程度，选择耐紫外线、耐老化相适应的材料；

4. The materials with adaptive ultraviolet resistance and aging resistance shall be adopted in accordance with the exposure degrees of the coiled materials for the roofs;

5. 屋面坡度大于25%时。应选择成膜时间较短的涂料。

5. If the gradient of the roofs is greater than 25%, the coatings with short film-formation time shall be adopted.

11.6.4 【屋面技规】4.5.4条：复合防水层的设计应符合下列规定。

11.6.4 [Technical code for roof engineering] 4.5.4: The design of the composite waterproof layers shall conform to the following provisions.

1. 选择的防水卷材和防水涂料应相容；

1. The selected waterproof coiled materials and waterproof coatings shall be compatible;

2. 防水涂料宜设置在防水卷材的下面；

2. The waterproof coatings shall be arranged under the waterproof coiled materials;

3. 挥发固化型防水涂料不得作为防水卷材粘结材料使用；

3. The volatile curing type waterproof coatings shall not be used as the bonding materials of waterproof coiled materials;

4. 水溶型或合成高分子类防水涂膜上面，不得采用热熔型防水卷材；

4. The hot-melt waterproof coiled materials shall not be adopted on the water-soluble or synthetic polymer waterproof coating films;

5. 水溶型或水泥基类防水涂料，应待涂膜实干后再采用冷粘铺贴卷材。

5. For the water-soluble or cement-based waterproof coatings, the coiled materials shall be paved in the cold-bonding method after the coated films are totally dried.

11.6.5 【屋面技规】4.5.5 条：每道卷材防水层最小厚度应符合表 11.6.5 的规定。

11.6.5 [Technical code for roof engineering] 4.5.5: The minimum thickness of each coiled material waterproof layer shall conform to the provisions in Tab. 11.6.5.

【屋面技规】表 4.5.5：每道卷材防水层最小厚度（mm）　　　表 11.6.5

[Technical code for roof engineering] Tab. 4.5.5:
Minimum Thickness of Each Coiled Material Waterproof Layer (mm)　　Tab. 11.6.5

防水等级 Waterproof grade	合成高分子防水卷材 Synthetic polymer waterproof coiled materials	高聚物改性沥青防水卷材 High-polymer modified asphalt waterproof coiled material		
		聚酯胎、玻纤胎、聚乙烯胎 Polyester, fiberglass, and polyethylene	自粘聚酯胎 Self-adhesive polyester	自粘无胎 Self-adhesive and non-carcass
I 级 Grade I	1.2	3.0	2.0	1.5
II 级 Grade II	1.5	4.0	3.0	2.0

11.6.6 【屋面技规】4.5.6 条：每道涂膜防水层最小厚度应符合表 11.6.6 的规定。

11.6.6 [Technical code for roof engineering] 4.5.6: The minimum thickness of each coated film waterproof layer shall conform to the provisions in Tab. 11.6.6.

【屋面技规】表 4.5.6：每道涂膜防水层最小厚度（mm）　　　表 11.6.6

[Technical code for roof engineering] Tab. 4.5.6:
Minimum Thickness of Each Coated Film Waterproof Layer (mm)　　Tab. 11.6.6

防水等级 Waterproof grade	合成高分子防水涂膜 Synthetic polymer waterproof coated film	聚合物水泥防水涂膜 Polymer cement waterproof coated film	高聚物改性沥青防水涂膜 High-polymer modified asphalt waterproof coated film
I 级 Grade I	1.5	1.5	2.0

续表

防水等级 Waterproof grade	合成高分子防水涂膜 Synthetic polymer waterproof coated film	聚合物水泥防水涂膜 Polymer cement waterproof coated film	高聚物改性沥青防水涂膜 High-polymer modified asphalt waterproof coated film
Ⅱ 级 Grade Ⅱ	2.0	2.0	3.0

11.6.7 【屋面技规】4.5.7 条：复合防水层最小厚度应符合表 11.6.7 的规定。

11.6.7 [Technical code for roof engineering] 4.5.7: The minimum thickness of the composite waterproof layer shall conform to the provisions in Tab. 11.6.7.

【屋面技规】表 4.5.7：复合防水层最小厚度（mm） 表 11.6.7
[Technical code for roof engineering] Tab. 4.5.7:
Minimum Thickness of Composite Waterproof Layer (mm)   Tab. 11.6.7

防水等级 Waterproof grade	合成高分子防水卷材 + 合成高分子防水涂膜 Synthetic polymer waterproof coiled material + synthetic polymer waterproof coated film	自粘聚合物改性沥青防水卷材（无胎）+ 合成高分子防水涂膜 Self-adhesive polymer modified asphalt waterproof coiled material (carcass-free) + synthetic polymer waterproof coated film	高聚物改性沥青防水卷材 + 高聚物改性沥青防水涂膜 High-polymer modified asphalt waterproof coiled material + high-polymer modified asphalt waterproof coated film	聚乙烯丙纶卷材 + 聚合物水泥防水胶结材料 Polypropylene coiled material + polymer cement waterwproof cementing material
Ⅰ 级 Grade Ⅰ	1.2+1.5	1.5+1.5	3.0+2.0	（0.75+1.3）×2
Ⅱ 级 Grade Ⅱ	1.0+1.0	1.2+1.0	3.0+1.2	0.7+1.3

11.6.8 【屋面技规】4.5.8 条：下列情况不得作为屋面的一道防水设防。

11.6.8 [Technical code for roof engineering] 4.5.8: The following cases shall not be used as one of the roof waterproof fortifications.

1. 混凝土结构层；

1. Concrete structural layer;

2. Ⅰ 型喷涂硬泡聚氨酯保温层；

2. I-shaped hard foam polyurethane sprayed insulation layer;

3. 装饰瓦以及不搭接瓦的屋面；

3. Decorative tile and the roof not overlapping tile;

4. 隔汽层；

4. Vapor barrier layer;

5. 细石混凝土层；

5. Fine-aggregated concrete layer;

6. 卷材或涂膜厚度不符合本规范规定的防水层。

6. Waterproof layer the thickness of which coiled material or coated film does not conform to the provisions in the Code.

11.7 【屋面技规】4.7 条：保护层和隔离层设计

11.7 [Technical code for roof engineering] 4.7: Design of Protective Layers and Isolation Layers

11.7.1 【屋面技规】4.7.1 条：上人屋面保护层可采用块体材料、细石混凝土等材料，不上人屋面保护层可采用浅色涂料、铝箔、矿物粒料、水泥砂浆等材料。保护层材料的适用范围和技术要求应符合表 11.7.1 的规定。

11.7.1 [Technical code for roof engineering] 4.7.1: For protective layers of accessible roofs, materials such as block material, fine-aggregated concrete, etc., can be used, but for protective layers of inaccessible roofs, materials of light-color coating, aluminum foil, mineral aggregate, cement mortar, etc., can be used. And the application scopes and technical requirements of materials for protective layers shall conform to the provisions in Tab. 11.7.1.

【屋面技规】表 4.7.1：保护层材料的适用范围和技术要求　　表 11.7.1

[Technical code for roof engineering] Tab. 4.7.1: Application Scopes and Technical Requirements of Materials for Protective Layers　　Tab. 11.7.1

保护层材料 Materials for protective layers	适用范围 Application scope	技术要求 Technical requirements
浅色涂料 Light-color coatings	不上人屋面 Inaccessible roofs	丙烯酸系反射涂料 Reflective crylic acid coatings
铝箔 Aluminum foils	不上人屋面 Inaccessible roofs	0.05mm 厚铝箔反射膜 0.05 mm thick reflective aluminum foil films
矿物粒料 Mineral aggregates	不上人屋面 Inaccessible roofs	不透明的矿物颗粒 Non-transparent mineral particles
水泥砂浆 Cement mortar	上人屋面 Accessible roofs	20mm 厚 M15 水泥砂浆 20 mm thick M15 cement mortar
块体材料 Block materials	上人屋面 Accessible roofs	地砖或 30mm 厚 C20 细石混凝土预制板 Floor tiles or 30 mm thick C20 precast fine-aggregated concrete boards
细石混凝土 Fine-aggregated concrete	上人屋面 Accessible roofs	40mm 厚 C20 细石混凝土或 50mm 厚 C20 细石混凝土内配 $\phi 4 @ 100$ 双向钢筋网 40 mm thick C20 fine-aggregated concrete or 50 mm thick C20 fine-aggregated concrete, equipped with $\phi 4 @ 100$ bi-directional reinforcement meshes

11.7.2 【屋面技规】4.7.2 条：采用块体材料做保护层时，宜设分仓缝，其纵横间距不宜大于 10m，分格缝宽度宜为 20mm，并应用密封材料嵌填。

11.7.2 [Technical code for roof engineering] 4.7.2: If the block materials are adopted as protective layers, the partition joints shall be arranged; the vertical and horizontal spacings shall be appropriately not greater than 10m; the width of the dividing joints shall be appropriately 20 mm; and the joints shall be filled with sealing materials.

11.7.3 【屋面技规】4.7.3 条：采用水泥砂浆做保护层时，表面应抹平压光，并应设表面分格缝，分格缝面积 1m²。

11.7.3 [Technical code for roof engineering] 4.7.3: When the cement mortar protective layers are adopted, the surfaces shall be smoothed and polished. The dividing joints on the surfaces shall be arranged, and the area of the dividing joints shall be 1m².

11.7.4 【屋面技规】4.7.4 条：采用细石混凝土做保护层时，表面应抹平压光，并应设表面分格缝，其纵横间距不宜大于 6m，分格缝宽度宜为 10～20mm，并应用密封材料嵌填。

11.7.4 [Technical code for roof engineering] 4.7.4: When the fine-aggregated concrete protective layer is used, the surface shall be smoothed and polished. The dividing joints on the surface shall be arranged, the vertical and horizontal spacings shall be not greater than 6m; the width of the dividing joint shall be 10～20mm, and the joints shall be filled with sealing materials.

11.8 瓦屋面

11.8 Tile Roofs

11.8.1 【屋面技规】4.8.1 条：瓦屋面防水等级和防水做法应符合表 11.8.1 的规定。

11.8.1 [Technical code for roof engineering] 4.8.1: The waterproof grades and waterproof methods for the tile roofs shall conform to the provisions in Tab. 11.8.1.

【屋面技规】表 4.8.1：瓦屋面防水等级和防水做法　　表 11.8.1
[Technical code for roof engineering] Tab. 4.8.1:
Waterproof grades and Waterproof Methods for Tile Roofs　　Tab. 11.8.1

防水等级 Waterproof grade	防水做法 Waterproof methods
I 级 Grade I	瓦＋防水层 Tiles + waterproof layers

续表

防水等级 Waterproof grade	防水做法 Waterproof methods
II 级 Grade II	瓦 + 防水垫层 Tiles + waterproof cushion layers

注：防水层厚度应符合《屋面工程技术规范》GB 50345—2012 第 4.5.5 或 4.5.6 条 II 级防暑的规定。
Note: The thickness of waterproof layers shall conform to the Grade II waterproof provisions in clause 4.5.5 or 4.5.6 of *Technical code for roof engineering* GB 50345—2012.

11.8.2 【屋面技规】4.8.2 条：瓦屋面应根据瓦的类型和基层种类采取相应的构造做法。

11.8.2 [Technical code for roof engineering] 4.8.2: For tile roofs, the corresponding construction methods shall be adopted as per the types of the tiles and base layers.

11.8.3 【屋面技规】4.8.3 条：瓦屋面与山墙及突出屋面结构的交接处，均应做不小于 250mm 高的泛水处理。

11.8.3 [Technical code for roof engineering] 4.8.3: At the junctions of the tile roofs with gable walls and the structures protruding the roofs, the flashing treatment with the height not less than 250mm shall be made.

11.8.4 【屋面技规】4.8.4 条：在大风及地震设防地区或屋面坡度大于 100% 时，瓦片应采取固定的加固措施。

11.8.4 [Technical code for roof engineering] 4.8.4: For tiles in the gale and earthquake fortification areas or areas where roof gradients are greater than 100%, fixed reinforcement measures shall be adopted.

11.8.5 【屋面技规】4.8.5 条：严寒及寒冷地区瓦屋面，檐口部位应采取防止冰雪融化下坠和冰坝形成等措施。

11.8.5 [Technical code for roof engineering] 4.8.5: Measures to prevent the falling of melted snow and ice and the formation of ice dams shall be adopted at the cornices of the tile roofs in the severely cold and cold regions.

11.8.6 【屋面技规】4.8.6 条：防水垫层宜采用自粘聚合物沥青防水垫层、聚合物改性沥青防水垫层，其最小厚度和搭接宽度应符合表 11.8.6 的规定。

11.8.6 [Technical code for roof engineering] 4.8.6: For waterproof cushion layers, the self-adhesive polymer asphalt waterproof cushion layers and polymer modified asphalt waterproof cushion layers shall be adopted, and the minimum thickness and overlapping width shall conform to the provisions in Tab. 11.8.6.

防水垫层的最小厚度和搭接宽度（mm） 表 11.8.6
Minimum Thickness and Overlapping Width of the Waterproof Cushion Layers（mm） Tab. 11.8.6

防水垫层种类 Type of the waterproof cushion layers	最小厚度 Minimum thickness	搭接宽度 Overlapping width
自粘聚合物沥青防水垫层 Self-adhesive polymer asphalt waterproof cushion layers	1.0	80
聚合物改性沥青防水垫层 Polymer modified asphalt waterproof cushion layers	2.0	100

11.8.7 【屋面技规】4.8.7 条：在满足屋面荷载的前提下，瓦屋面持钉层厚度应符合表 11.8.7 的规定。

11.8.7 [Technical code for roof engineering] 4.8.7: Under the premise of meeting the roof loads, the thickness of the nail-supporting layers of tile roofs shall conform to the provisions in Tab. 11.8.7.

瓦屋面持钉层厚度（mm） 表 11.8.7
Thickness of Nail-supporting Layers of Tile Roofs（mm） Tab. 11.8.7

持钉层类别 Category of nail-supporting layers	持钉层厚度 Thickness of nail-supporting layers
木板 Boards	20
人造板 Artificial boards	16
细石混凝土 Fine-aggregated concrete	35

11.8.8 烧结瓦、混凝土瓦屋面要求应符合表 11.8.8 的规定。

11.8.8 The requirements for roofs with sintered tiles and concrete tiles shall conform to the provisions in Tab. 11.8.8.

【屋面技规】4.8.9～12：烧结瓦、混凝土瓦屋面要求 表 11.8.8
[Technical code for roof engineering] 4.8.9～12: Requirements for Roofs with Sintered Tiles and Concrete Tiles Tab. 11.8.8

类别 Category	技术要求 Technical requirement
屋面坡度 Roof gradient	不应小于 30% Not less than 30%

续表

类别 Category		技术要求 Technical requirement
基层及挂瓦类型 Base and tile-hanging types	木基层、顺水条、挂瓦条 Wood base, counter batten, and tile hanging batten	防腐、防火和防蛀 Corrosion prevention, fire prevention and moth proofing
	金属顺水条、挂瓦条 Metal counter batten and tile hanging batten	防腐、防锈蚀 Corrosion prevention and rust prevention
	烧结瓦、混凝土瓦 Sintered tile and concrete tile	干法挂瓦屋面应固定牢靠 The dry tile-hanging roofs shall be securely fixed
烧结瓦、混凝土瓦铺装相关尺寸 Relevant sizes to the pavement of sintered tile and concrete tile		
瓦屋面檐口挑出墙面 Overhanging of tile roof cornices out of walls		不宜小于 300mm Appropriately not less than 300mm
脊瓦在两坡面瓦上的搭盖宽度 Width of the overlapping cover of the ridge tiles on the two-slope tiles		每边不应小于 40mm Not less than 40mm for each side
脊瓦下端距坡面瓦的高度 Height from the lower ends of ridge tiles to slope tiles		不宜大于 80mm Appropriately not greater than 80mm
瓦头深入檐沟、天沟内的长度 Lengths of tiles into trenches and gutters		宜为 50~70mm To be appropriately 50~70mm
金属檐沟、天沟深入瓦内的宽度 Width of metal trenches and gutters into tiles		不应小于 150mm Not less than 150mm
瓦头挑出檐口的长度 Length of tiles overhanging cornices		宜为 50~70mm To be appropriately 50~70mm
挑出屋面结构的侧面瓦深入泛水的宽度 Width of side tiles into the flashing which overhang the roof structures		不应小于 50mm Not less than 50mm

11.8.9 沥青瓦屋面：

11.8.9 Asphalt tile roofs:

1.【屋面技规】4.8.13 条：沥青瓦屋面的坡度不应小于 20%。

1. [Technical code for roof engineering] 4.8.13: The gradients of roofs with asphalt tiles shall be not less than 20%.

2.【屋面技规】4.8.14 条：沥青瓦应具有自粘胶带或相互搭接的连锁构造。矿物粒料或片料覆面沥青瓦的厚度不应小于 2.6mm，金属箔面沥青瓦的厚度不应小于 2mm。

2. [Technical code for roof engineering] 4.8.14: The asphalt tiles shall be provided with self-adhesive tapes or interlocked structures with the tiles overlapped with each other. The thickness of mineral aggregate or flake covered asphalt tiles shall be not less than 2.6mm, and the thickness of the asphalt tiles with the metal foil surfaces shall be not less than 2mm.

3.【屋面技规】4.8.15 条：沥青瓦的固定方式应以钉为主、粘结为辅。每张瓦片上不得少于 4 个固定钉；在大风地区或屋面坡度大于 100% 时，每张瓦片不得少于 6 个固定钉。

3. [Technical code for roof engineering] 4.8.15: For the fixing methods for asphalt tiles, nailing shall be dominant and bonding shall be subordinate. At least 4 fixing nails are required on each tile; in the gale areas or areas where roof gradients are greater than 100%, at least 6 fixing nails are required on each tile.

4.【屋面技规】4.8.16 条：天沟部位铺设的沥青瓦可采用搭接式、编织式、敞开式。搭接式、编织式铺设时，沥青瓦下应设置不小于 1000mm 宽的附加层；敞开铺设时，在防水层或防水垫层上应铺设厚度不小于 0.45mm 的金属板材，沥青瓦与金属板材应用沥青基胶结材料粘结，其搭接宽度不应小于 100mm。

4. [Technical code for roof engineering] 4.8.16: Asphalt tiles can be laid in the gutters in an overlapped, woven or open way. When the asphalt tiles are laid in an overlapped or woven way, the additional layers with the width not less than 1000mm shall be arranged under the asphalt tiles; during exposed laying, the metal boards with the thickness not less than 0.45mm shall be laid on the waterproof layers or waterproof cushion layers, the asphalt tiles and metal boards shall be bonded with asphalt-based cementing materials, and the overlapping width shall be not less than 100mm.

5. 沥青瓦铺装的尺寸要求应符合表 11.8.9 的规定。

5. The dimensional requirements for the pavement of asphalt tiles shall conform to the provisions in Tab. 11.8.9.

【屋面技规】表 4.8.17：沥青瓦铺装的尺寸要求　　　　　表 11.8.9
[Technical code for roof engineering] Tab. 4.8.17:
Dimensional Requirements for the Pavement of Asphalt Tiles　　Tab. 11.8.9

沥青瓦铺装形式 Paving forms of asphalt tiles	搭接宽度（mm） Overlapping width（mm）
脊瓦在两坡面瓦上的搭盖宽度 Width of the overlapping cover of the ridge tiles on the two-slope tiles	每边不应少于 150 Not less than 150 for each side
脊瓦与脊瓦的压盖面 Cover between two ridge tiles	不应小于脊瓦面积的 1/2 Not less than 1/2 of the areas of ridge tiles
沥青瓦挑出檐口的长度 Length of asphalt tiles overhanging cornices	宜为 10～20 To be appropriately 10～20
金属泛水板与沥青瓦的搭盖宽度 Width of the overlapping cover of metal flashing boards and asphalt tiles	不应小于 100 Not less than 100
金属泛水板与突出屋面墙体的搭接高度 Overlapping height of metal flashing boards and walls protruding roofs	不应小于 250 Not less than 250
金属泛水板深入沥青瓦下的宽度 Width of the metal flashing boards under the asphalt tiles	不应小于 80 Not less than 80

11.9 金属屋面设计

11.9 Design for Metal Board Roofs

11.9.1 【屋面技规】4.9.1 条：金属屋面防水等级和防水做法应符合表 11.9.1 的规定。

11.9.1 [Technical code for roof engineering] 4.9.1: The waterproof grades and waterproof methods for metal roofs shall conform to the provisions in Tab. 11.9.1.

【屋面技规】表 4.9.1：金属屋面防水等级和防水做法　　　　表 11.9.1

[Technical code for roof engineering] Tab. 4.9.1:
Waterproof Grades and Waterproof Methods for Metal Roofs　　　Tab. 11.9.1

防水等级 Waterproof grade	防水做法 Waterproof methods
I 级 Grade I	压型金属板 + 防水垫层 Profiled metal boards + waterproof cushion layers
II 级 Grade II	压型金属板。金属面绝热夹芯板 Profiled metal boards Metal-surface heat shielding sandwich boards

注：1. 当防水等级为 I 级时，压型铝合金板基板厚度不应小于 0.9mm，压型钢板基板厚度不应小于 0.6mm。
Notes: 1. For Grade I waterproof, the thickness of the profiled aluminum alloy substrate shall be not less than 0.9mm, and the thickness of the profiled steel substrate shall be not less than 0.6mm.
2. 当防水等级为 I 级时，压型铝合金板应采用 360°咬口锁边连接方式。
2. For Grade I waterproof, for profiled aluminum alloy plates, the 360° seam lapping and overlocking connection mode shall be adopted.
3. 在 I 级屋面防水做法中，仅作压型金属板时，应符合相关技术规定。
3. If only profiled metal boards are adopted in Grade I roof waterproof, the boards shall conform to the relevant technical specifications.

11.9.2 【屋面技规】4.9.2 条：金属板屋面可按建筑设计要求，选用镀层钢板、涂层钢板、铝合金板、不锈钢板和钛锌板等金属板材。金属板材及其配套的紧固件、密封材料，其材料的品种、规格和性能等应符合现行国家有关材料标准的规定。

11.9.2 [Technical code for roof engineering] 4.9.2: For metal board roofs, the metal boards of clad steel plate, coated steel plate, aluminum alloy plate, stainless steel plate and titanium-zinc plate, etc., shall be selected as per the requirements of building design. The types, specifications, performance, etc., of the metal boards and the supporting fasteners and sealing materials shall conform to the relevant requirements.

11.9.3 【屋面技规】4.9.3 条：金属板屋面按围护结构设计，并应具有相应的承载力、刚度、稳定性和变形能力。

11.9.3 [Technical code for roof engineering] 4.9.3: The metal board roofs shall be provided with the corresponding bearing capacity, stiffness, stability and deformation capacity as per the design for the enclosure structures.

11.9.4 【屋面技规】4.9.4 条：金属板屋面应根据当地风荷载、结构性体、热工性能、屋面坡度等情况，采用相应的压型金属板板型及构造系统。

11.9.4 [Technical code for roof engineering] 4.9.4: The corresponding profiled metal boards and structure systems shall be adopted in accordance with the conditions of local wind load, structure, thermal performance, roof gradient, etc.

11.9.5 【屋面技规】4.9.5 条：金属板屋面在保温层下面宜设置隔汽层，在保温层上面宜设置防水透汽膜。

11.9.5 [Technical code for roof engineering] 4.9.5: The vapor barrier layers shall be arranged under the insulation layer of metal board roofs, and the waterproof and breathable films shall be arranged above the insulation layers.

11.9.6 【屋面技规】4.9.6 条：金属板屋面的防结露设计，应符合现行国家标准《民用建筑热工设计规范》的规定。

11.9.6 [Technical code for roof engineering] 4.9.6: The anti-condensation design of metal sheet roof shall conform to the provisions in the current national standard *Thermal Design Code for Civil Building.*

11.9.7 【屋面技规】4.9.7 条：压型金属板采用咬口锁边连接时，屋面的排水坡度不宜小于 5%；压型金属板采用紧固件连接时，屋面的排水坡度不宜小于 10%。

11.9.7 [Technical code for roof engineering] 4.9.7: When the seam lapping and overlocking connection are adopted for profiled metal boards, the drainage gradient of the roof shall be not less than 5%; When the profiled metal boards are connected with fasteners, the drainage gradient of the roof shall be not less than 10%.

11.9.8 【屋面技规】4.9.8 条：金属檐沟、天沟的伸缩缝间距不宜大于 30m，内檐沟及内天沟应设置溢水口或溢流系统，沟内宜按 0.5% 找坡。

11.9.8 [Technical code for roof engineering] 4.9.8: The spacings of expansion joints of metal trench and gutter shall be not greater than 30m, the overflow port or overflow system shall be arranged in the inner trench and inner gutter, and 0.5% slope shall be made in the trench.

11.9.9 【屋面技规】4.9.9 条：金属板的伸缩缝变形除应满足咬口锁边连接或紧固件连接的要求外，还应满足檩条、檐口及天沟等使用要求，且金属板最大伸缩变形量不应超过 100mm。

11.9.9 [Technical code for roof engineering] 4.9.9: The deformation of expansion joints in metal boards shall conform to the application requirements for purlins, cornices, gutters, etc., in addition to the requirements for seam lapping and overlocking connection or fastener connection, and the maximum expansion deformation amount of the metal board shall not exceed 100 mm.

11.9.10 【屋面技规】4.9.10条：金属板在主体结构的变形处宜断开，变形缝上部加扣带伸缩的金属盖板。

11.9.10 [Technical code for roof engineering] 4.9.10: The metal boards shall be disconnected at the deformation positions of the main structures, and metal boards with expansion shall be additionally provided at the upper parts of the deformation joints.

11.9.11 【屋面技规】4.9.12条：压型金属板采用咬口锁边连接的构造应符合下列规定。

11.9.11 [Technical code for roof engineering] 4.9.12: The structure of seam lapping and overlocking connection for the profiled metal board shall conform to the following provisions.

1. 在檩条上应设置与压型金属板波形相配套的专用固定支座，并应用自攻螺钉与檩条连接；

1. The approved fixing supports matching with the wavy shape of the profiled metal boards shall be arranged on the purlins, and connected to the purlins with self-tapping screws;

2. 压型金属板应搁置在固定的支座上，两片金属板的侧边应确保在风吸力等因素作用下扣合或咬合连接可靠；

2. The profiled metal boards shall be arranged on the fixed supports, and for the sides of two metal boards, the reliable buckle or occlusion connection under the effect of wind suction and other factors shall be ensured;

3. 在大风地区或高度大于30m的屋面，压型金属板应采用360°咬口锁边连接；

3. In windy areas or roof with a height of more than 30m, the profiled metal plates shall be connected by 360° locking;

4. 大面积屋面和弧状或组合弧状屋面，压型金属板的立边咬口宜采用暗扣直立锁边屋面系统；

4. On the roofs with big area or the curving or combined curving roofs, the hidden-buckle vertical overlocking roof system shall be adopted for the seam lapping connection on the vertical edges of the profiled metal boards;

5. 单坡尺寸过长或环境温差过大的屋面，压型金属板宜采用滑动式支座的360°咬口锁边连接。

5. On the roofs with overlong single slopes or excessive environment temperature difference, the 360° seam lapping and overlocking connection for the sliding supports shall be adopted for the profiled metal boards.

11.9.12 【屋面技规】4.9.13条：压型金属板采用紧固件连接的构造应符合下列规定。

11.9.12 [Technical code for roof engineering] 4.9.13: The structures in which fastener connection is adopted for profiled metal boards shall conform to the following provisions.

1. 铺设高波压型金属板时，在檩条上应设置固定支架，固定支架应采用自攻螺钉与檩条连接，连接件宜每波设置一个。

1. When the high-wave profiled metal boards are laid, fixing brackets shall be arranged on the purlins, connected to the purlins with self-tapping screws, and one connecting component shall be arranged at each wave.

2. 铺设低波压型金属板时，可不设固定支架，应在波峰处采用带防水密封胶垫的自攻螺钉与檩条连接，连接件可每波或隔波设置一个，但每块板不得少于3个。

2. When the low-wave profiled metal boards are laid, fixed brackets may not be arranged; however, the low-wave profiled metal boards shall be connected to the purlins with self-tapping screws with waterproof sealing gaskets at the wave peaks; one connecting component can be arranged at each wave or every other wave, and at least 3 components shall be arranged for each plate.

3. 压型金属板的纵向搭接应位于檩条处，搭接端处与檩条有可靠的连接，搭接部位应设置防水密封胶带。压型金属板的纵向搭接长度应符合表11.9.12的规定。

3. The longitudinal overlapping of profiled metal boards shall be carried out at the purlins, the overlapping ends shall be reliably connected with the purlins, and the overlapping positions shall be provided with the waterproof sealing tapes. The longitudinal overlapping length of profiled metal boards shall conform to the provisions in Tab. 11.9.12.

【屋面技规】表4.9.13：压型金属板的纵向搭接长度（mm） 表11.9.12

[Technical code for roof engineering] Tab. 4.9.13: Longitudinal Overlapping Length of Profiled Metal Boards（mm） Tab. 11.9.12

压型金属板 Profiled metal boards		纵向最小搭接长度 Minimum longitudinal overlapping length
高波压型金属板 High-wave profiled metal boards		350
低波压型金属板 Low-wave profiled metal boards	屋面坡度≤10% Roof gradient ≤ 10%	250
	屋面坡度＞10% Roof gradient ＞ 10%	200

4. 压型金属板的横向搭接方向宜与主导风向一致，搭接不应小于一个波，搭接部位应设置防水密封胶带。搭接处用连接件紧固时，连接件应采用防水密封胶垫的自攻螺钉

设置在波峰上。

4. The transverse overlapping direction of profiled metal boards shall be consistent with the direction of the dominant wind, the overlapping size shall be not less than one wave, and the waterproof sealing tapes shall be adopted at the overlapping positions. When the overlapping positions are fastened with connectors, the connectors shall be arranged on the wave peaks with self-tapping screws with the waterproof sealing gaskets.

11.10 玻璃采光顶设计
11.10 Design for Glass Daylighting Roofs

11.10.1 【屋面技规】4.10.1 条：玻璃采光顶设计应根据建筑物的屋面形式、使用功能和美观要求，选择结构类型、材料和细部构造。

11.10.1 [Technical code for roof engineering] 4.10.1: The type and material of the structure and the detail structure shall be selected in the design for glass daylighting roofs in accordance with the requirements for the form, function and aesthetic quality of the building roof.

11.10.2 【屋面技规】4.10.2 条：玻璃采光顶的物理性能等级，应根据建筑物的类别、高度、体形、功能以及建筑物所在的地理位置、气候和环境条件进行设计。玻璃采光顶的物理性能指标，应符合现行行业标准的有关规定。

11.10.2 [Technical code for roof engineering] 4.10.2: The physical performance grade of the glass daylighting roofs shall be designed in accordance with the category, height, shape and function of the building and the geographical conditions, climate conditions and environmental conditions of the building location. The physical performance index of the glass daylighting roof shall conform to the relevant provisions in the current industry standard.

12

楼梯、台阶、坡道
Staircases, Steps and Ramps

12.1 一般规定

12.1 General Provisions

12.1.1 【通则】6.7.1 条：楼梯的数量、位置、宽度和楼梯间形式应满足使用方便和安全疏散的要求。

12.1.1 [Code for design of civil buildings] 6.7.1: The number, location, width and stairwell form of the staircase should meet the requirements of ease of use and safe evacuation.

12.1.2 【通则】6.7.2 条：墙面至扶手中心线或扶手中心线之间的水平距离即楼梯梯段宽度除应符合防火规范的规定外，供日常主要交通用的楼梯的梯段宽度应根据建筑物使用特征，按每股人流为0.55+(0~0.15)m的人流股数确定，并不应少于两股人流。0~0.15m为人流在行进中人体的摆幅，公共建筑人流众多的场所应取上限值（图12.1.2）。

12.1.2 [Code for design of civil buildings] 6.7.2: The horizontal distance between walls and handrail centerlines or between handrail centerlines is equal to the width of the stair flight. In addition to the conformity with the provisions in the Code for Fire Protection Design of Buildings, the width of the stair flight of the staircase for the main daily traffic shall be determined in accordance with the use characteristics of the building at the number of streams of 0.55+(0~0.15)m per stream of people, and can not be less than two streams of people flows. 0 ~ 0.15 m is the swing amplitude of people in movement, and the upper limit value shall be adopted at places with crowded people in public buildings (see Fig.12.1.2).

12.1.3 【通则】6.7.6 条：楼梯应至少于一侧设扶手，梯段净宽达三股人流时应两侧设扶手，达四股人流时宜加设中间扶手 [图12.1.2（b）、(c)]。

12.1.3 [Code for design of civil buildings] 6.7.6: Staircase should be set up handrail at least one side; if net width of the ladder section is up to three strands of flow, handrails should be set on both sides; and an additional middle handrail should be set up there are four strands of flow[See Fig. 12.1.2（b）,(c)].

12.1.4 【通则】6.7.3 条：梯段改变方向时，扶手转向端处的平台最小宽度不应小于梯段宽度，并不得小于1.20m，当有搬运大型物件需要时应适量加宽。

12.1.4 [Code for design of civil buildings] 6.7.3: When the ladder changes direction, the minimum width of the platform at the end of the handrail shall not be less than the width of the ladder segment and shall not be less than 1.20m should be widened appropriately when it is

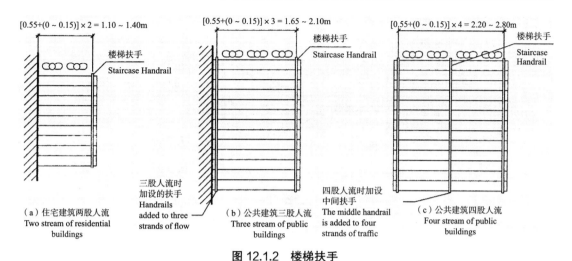

图 12.1.2 楼梯扶手
Fig. 12.1.2 Staircase Handrails

necessary to carry large objects.

12.1.5 【通则】6.7.5 条：楼梯平台上部及下部过道处的净高不应小于2m，梯段净高不宜小于2.20m。

12.1.5 [Code for design of civil buildings] 6.7.5: The net height of the upper and lower sidewalks of the stair platform should not be less than 2m, and the net height of the ladder section should not be less than 2.20m.

注：梯段净高为自踏步前缘（包括最低和最高一级踏步前缘线以外0.30m范围内）量至上方突出物下缘间的垂直高度（图 12.1.5）。

Note: The net height of the ladder is vertical height from the self-stepping front edge (including in the 0.30m range outside of the lowest and highest level of the steps) to the lower edge of the protruding object (Fig. 12.1.5).

12.1.6 【通则】6.7.7 条：室内楼梯扶手高度自踏步前缘线量起不宜小于0.90m。靠楼梯井一侧水平扶手长度超过0.50m时，其高度不应小于1.05m。

12.1.6 [Code for design of civil buildings] 6.7.7: Indoor staircase handrail height from the pedal front line amount should not be less than 0.90m. If the length of the horizontal handrail on one side of the stairwell is more than 0.50m, its height should not be less than 1.05m.

12.1.7 【通则】6.7.9 条：托儿所、幼儿园、中小学及少年儿童专用活动场所的楼梯，梯井净宽大于0.20m时，必须采取防止少年儿童攀滑的措施，楼梯栏杆应采取不易攀登的构造，当采用垂直杆件做栏杆时，其杆件净距不应大于0.11m。

12.1.7 [Code for design of civil buildings] 6.7.9: The staircase of the nursery, kindergarten, primary and secondary schools and children's special activity places is greater than 0.20m for the ladder well, measures must be taken to prevent children from climbing, stair railings should take the construction of not easy to climb, when the use of vertical rod to make railings, the net distance of the rod should not be greater than 0.11m.

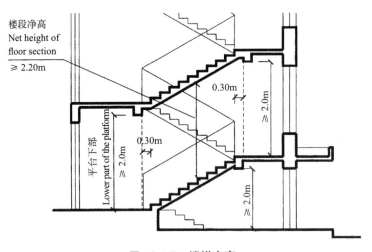

图 12.1.5 楼梯净高
Fig. 12.1.5 Staircase Net Height

12.1.8 【通则】6.7.4 条：每个梯段的踏步不应超过 18 级，亦不应少于 3 级。

12.1.8 [Code for design of civil buildings] 6.7.4: The step of each ladder shall not exceed 18 steps, nor shall it be less than 3 steps.

12.1.9 【通则】6.7.8 条：踏步应采取防滑措施。

12.1.9 [Code for design of civil buildings] 6.7.8: Steps should take anti-skid measures.

12.2 楼梯、楼梯间设计

12.2 Design of Staircase and Staircase Room

12.2.1 常用建筑楼梯的踏步最小宽度和最大高度（表 12.2.1）。

12.2.1 The minimum and maximum height of the step of a common building stair （Tab.12.2.1）.

12.2.2 梯段设计。

12.2.2 Design of stair flight.

楼梯梯段基本要求见表 12.2.2。

The basic requirements of stair flights are as shown in Tab. 12.2.2.

12 楼梯、台阶、坡道 Staircases, Steps and Ramps

楼梯踏步最小宽度和最大高度（mm） 表 12.2.1
The Minimum Width and Maximum Height of Staircase Steps (mm) Tab. 12.2.1

楼梯类型 Staircase type	最小宽度 Minimum width	最大高度 Maximum height	备注 Note
住宅共用楼梯 Residential shared staircase	260	175	【住设规】6.3.2 条 [Design code of residential buildings] 6.3.2
住宅套内楼梯 Residential set inside Staircase	220	200	【住设规】5.7.4 条 [Design code of residential buildings] 5.7.4
幼儿园、托儿所幼儿使用的楼梯 Stairs used by kindergarten and nursery children	260	130	【托规】4.1.11 条 [Code for design of nursery and kindergarten buildings] 4.1.11
小学楼梯 Elementary school Staircase	260	150	【校规】8.7.3 条 [Code for design of school] 8.7.3
电影院、剧场、体育馆、医院、旅馆、大中学校楼梯 Cinemas, theatres, gymnasiums, hospitals, hotels, grand school Stairs	280	160	【通则】6.7.10 条、【校规】8.7.3 条 [Gode for design of civil buildings] 6.7.10、[Code for design of school] 8.7.3
宿舍楼梯 Dormitory staircase	270	165	【宿舍规】4.5.1 条 [Code for design of dormitory building] 4.5.1
商店建筑营业区公用楼梯 Shop building business Area public staircase	280	160	【商店建规】4.1.6 条 [Code for design of store buildings] 4.1.6
商店建筑专用疏散楼梯 Shop Building dedicated evacuation staircase	260	170	【商店建规】4.1.6 条 [Code for design of store buildings] 4.1.6
商店建筑室外楼梯 Shop Building Outdoor Staircase	300	150	【商店建规】4.1.6 条 [Code for design of store buildings] 4.1.6
其他建筑楼梯 Other building stairs	260	170	【通则】6.7.10 条 [Code for design of civil buildings] 6.7.10
专用疏散楼梯 Dedicated evacuation staircase	250	180	【通则】6.7.10 条 [Code for design of civil buildings] 6.7.10

注：无中柱螺旋楼梯和弧形楼梯离内侧扶手中心 0.25m 处的踏步宽度不应小于 0.22m。
Note: the step width of spiral staircase without middle column and curved staircase 0.25m from the inner handrail center should not be less than 0.22m.

常用建筑楼梯的梯段宽度（mm） 表 12.2.2
Staircase Flight Width of Staircase of Common Buildings (mm) Tab. 12.2.2

楼梯类型 Staircase type	限定条件或其他要求 Qualifications or other requirements	最小宽度 Minimum width	引自规范、标准 Cited in specifications, standards
住宅共用楼梯 Residential shared staircase	建筑高度＞18m Building height >18m	1100	【防火规】5.5.30 条【住设规】6.3.1 条 [Code for fire protection design of buildings] 5.5.30、[Design code of residential buildings] 6.3.1
	建筑高度≤18m 且一边设置栏杆 The building is highly ≤ 18m and the railing is set on one side	1000	

续表

楼梯类型 Staircase type	限定条件或其他要求 Qualifications or other requirements	最小宽度 Minimum width	引自规范、标准 Cited in specifications, standards
住宅套内楼梯 Residential set inside staircase	一边临空时 While in the air	750	【住设规】5.7.3 条 [Design code of residential buildings] 5.7.3
	两侧有墙，且其中一侧墙面设置扶手 There are walls on both sides, and one side of the wall sets the handrail	900	
中小学校教学用房 Teaching rooms in primary and secondary schools	应按 0.60m 的整数倍增加梯段宽度每个梯段可增加不超过 0.15m 的摆幅宽度 There are walls on both sides, and one side of the wall sets the handrail. The width of the ladder should be increased by an integer double of 0.60m, and the swing width of no more than 0.15m can be increased for each ladder segment	1200	【校规】8.7.2 条 [Code for design of school] 8.7.2
医院 Hospital	主楼梯 Main staircase	1650	【医规】5.1.5 条 [Code for design of general hospital] 5.1.5
	高层医院疏散楼梯 High-rise hospital evacuation staircase	1300	【防火规】5.5.18 条 [Code for fire protection design of buildings] 5.5.18
高层公建疏散楼梯 High-rise public building evacuation staircase		1200	【防火规】5.5.18 条 [Code for fire protection design of building] 5.5.18
商店建筑营业区公用楼梯 Shop building business area public staircase		1400	【商店建规】4.1.6 条 [Code for design of store buildings] 4.1.6
商店建筑专用疏散楼梯 Shop building dedicated evacuation staircase		1200	【商店建规】4.1.6 条 [Code for design of store buildings] 4.1.6
商店建筑室外楼梯 Shop building outdoor staircase		1400	【商店建规】4.1.6 条 [Code for design of store buildings] 4.1.6
室外疏散楼梯 Outdoor evacuation staircase		900	【防火规】6.4.5 条 [Code for fire protection design of building] 6.4.5
多层公建疏散楼梯 Multi-storey public construction evacuation staircase	最低要求，有特殊规定的除外 Minimum requirements, except where there are special provisions	1100	【防火规】5.5.18 条 [Code for fire protection design of building] 5.5.18

12.2.3 扶手、栏杆（板）的设计。

12.2.3 Design of handrail and railing (plate).

1.【建结规】5.5.2、【校规】8.1.6：栏杆（板）应用坚固、耐久的材料制作，并能承受荷载规范规定的水平荷载，栏杆顶部的水平荷载应符合表12.2.3的规定。

1. [Load code for design of the building structure] 5.5.2、[Code for design of school] 8.1.6: Railings (balustrades) shall be made of strong and durable materials and shall be able to bear the horizontal loads specified in the Code of load; and the horizontal loads of the railings at the top shall meet the provisions in Tab. 12.2.3.

【建结规】5.5.2、【校规】8.1.6：栏杆顶部水平荷载　　　　表12.2.3

[Load code for design of the building structures] 5.5.2、[Code for design of school] 8.1.6: Horizontal Loads of the Railings at the Top　　　Tab. 12.2.3

建筑类型 Building types	栏杆顶部水平荷载、最薄弱处水平推力（kN/m） Horizontal load at the top of the railing, weakest horizontal thrust (kN/m)
住宅、宿舍、办公楼、旅馆、医院、托儿所、幼儿园 Residential buildings, dormitories, office buildings, hotels, hospitals, nurseries and kindergartens	≥ 1.0
食堂、剧场、电影院、车站、礼堂、展览馆、体育馆、商场营业厅 Dining rooms, theaters, cinemas, stations, auditoriums, exhibition halls, gymnasiums, business halls	≥ 1.0
学校上人屋面、外廊、楼梯、平台、阳台等临空部位 Open places of accessible roofs, verandas, staircases, platforms, balconies, etc., in schools	≥ 1.5

2.【托规】4.1.12条：幼儿使用的楼梯，当楼梯井净宽度大于0.11m时，必须采取防止幼儿攀滑措施。楼梯栏杆应采取不易攀爬的构造，当采用垂直杆件做栏杆时，其杆件净距不应大于0.11m。

2. [Code for design of nursery and kindergarten buildings] 4.1.12: For young children use staircases, when the net width of the staircase well is greater than 0.11m, measures must be taken to prevent young children from climbing. Staircase railings should take a construction that is not easy to climb, when the use of vertical rod to make railings, the net distance of the rod should not be greater than 0.11m.

3.【托规】4.1.9条：托儿所、幼儿园的外廊、室内回廊、内天井、阳台、上人屋面、平台、看台及室外楼梯等临空处应设置防护栏杆，栏杆应以坚固、耐久的材料制作，防护栏杆水平承载能力应符合《建筑结构荷载规范》GB 50009的规定。防护栏杆的高度应从地面计算，且净高不应小于1.10m。防护栏杆必须采用防止幼儿攀登和穿过的构造，当采用垂直杆件做栏杆时，其杆件净距离不应大于 0.11m。

3. [Code for design of nursery and kindergarten buildings] 4.1.9: Nursery, kindergarten porch, indoor sidewalk, inner patio, balcony, roof, platform, grandstand and outdoor staircase and other empty space shall be provided with protective railings, railings shall be made of sturdy, durable materials, and the horizontal carrying capacity of protective railings shall conform to the provisions of the *Code for the load of building structures* GB 50009. The height of the protective railing should be calculated from the ground, and the net height should not be less than 1.10m. Protective railings must be constructed to prevent young children from climbing and passing through, and when vertical rods are used as railings, the net distance of their rods should not be greater than 0.11m.

4.【托规】4.1.11 条：托儿所、幼儿园的楼梯除设成人扶手外，应在梯段两侧设幼儿扶手，其高度宜为 0.60m。

4. [Code for design of nursery and kindergarten buildings] 4.1.11: Nursery, kindergarten staircase in addition to the establishment of adult handrails, should be on both sides of the ladder to set up children's handrails, its height should be 0.60m.

5.【住设规】6.3.2 条：住宅的楼梯栏杆垂直杆件间净空不应大于 0.11m。

5. [Design cade of residential buildings] 6.3.2: The clearance between vertical rods of the staircase railing of the house shall not be greater than 0.11m.

6.【住设规】6.3.5 条：住宅楼梯井净宽大于 0.11m 时，必须采取防止儿童攀滑的措施。

6. [Design cade of residential buildings] 6.3.5: The net width of the residential staircase well is greater than 0.11m, measures must be taken to prevent children from slipping.

7.【商店建规】4.1.6 条：商店的楼梯、室内回廊、内天井等临空处的栏杆应采用防攀爬的构造，当采用垂直杆件做栏杆时，其杆件净距不应大于 0.11m。

7. [Code for design of store buildings] 4.1.6: Shop staircase, indoor sidewalk, inner patio and other empty space railing should use anti-climbing structure, when the use of vertical rod to make railings, the net distance of the rod should not be greater than 0.11m.

12.2.4【老年人照料设施标】5.6.6 条：老年人照料设施中，老年人使用的楼梯严禁采用弧形楼梯和旋转楼梯。

12.2.4 [Standard for design of care facilities for the aged] 5.6.6: Elderly care facilities, the staircase used by the elderly is strictly forbidden to use curved staircase and revolving staircase.

12.2.5【托规】4.1.11 条：幼儿使用的楼梯不应采用扇形、螺旋形踏步。

12.2.5 [Code for design of nursery and kindergarten buildings] 4.1.11: The staircase used by young children should not be fan-shaped and spiral-shaped.

12.2.6 【校规】8.7.4 条：中小学校疏散楼梯不得采用螺旋楼梯和扇形踏步。

12.2.6 [Code for design of school] 8.7.4: The evacuation staircase of primary and secondary schools shall not use spiral stairs and fan-shaped stepping.

12.3 楼梯、楼梯间防火设计
12.3 Fire Protection Design for Stairs and Staircases

12.3.1 【防火规】6.4.1 条：疏散楼梯间应符合下列规定：

12.3.1 [Code for fire protection design of buildings] 6.4.1: Evacuation staircases shall conform to the following provisions

1. 楼梯间应能天然采光和自然通风，并宜靠外墙设置。靠外墙设置时，楼梯间、前室及合用前室外墙上的窗口与两侧门、窗、洞口最近边缘的水平距离不应小于 1.0m。

1. Staircases shall be provided with natural daylighting and natural ventilation, and shall be arranged by exterior walls. When a staircase is set by the exterior wall, the horizontal distances between the windows in the exterior walls of the staircase, anteroom and shared anteroom and the nearest edge of the door, window and opening on both sides shall be not less than 1.0m.

2. 楼梯间内不应设置烧水间、可燃材料储藏室、垃圾道。

2. In staircases, no water heater rooms, combustible material storage rooms and refuse chutes shall be arranged.

3. 楼梯间内不应有影响疏散的凸出物或其他障碍物。

3. There shall be no projections or other obstacles affecting evacuation in staircases.

4. 封闭楼梯间、防烟楼梯间及其前室，不应设置卷帘。

4. For the enclosed staircases, smoke-proof staircases and the anterooms, roller shutters shall not be arranged.

5. 楼梯间内不应设置甲、乙、丙类液体管道。

5. The pipelines for Classes A, B and C liquids shall not be arranged in staircases.

6. 封闭楼梯间、防烟楼梯间及其前室内禁止穿过或设置可燃气体管道。敞开楼梯间内不应设置可燃气体管道，当住宅建筑的敞开楼梯间内确需设置可燃气体管道和可燃气体计量表时，应采用金属管和设置切断气源的阀门。

6. The pipelines for combustible gases shall not pass through or be arranged in the enclosed staircases, smoke-proof staircases and anterooms. The pipelines for combustible gases shall

not be arranged in the open type staircases; and when the pipelines for combustible gases and combustible gas meters are required to be arranged in the open type staircases in residential buildings, the metal pipes shall be adopted and the air source shutoff valves shall be arranged.

12.3.2 【防火规】6.4.2条：封闭楼梯间除符合第12.3.1条规定外，尚应符合下列规定。

12.3.2 [Code for fire protection design of buildings] 6.4.2: Enclosed staircases shall conform to the provisions in Clause 12.3.1 and the following provisions additionally.

1. 不能自然通风或自然通风不能满足要求时，应设置机械加压送风系统或采用防烟楼梯间；

1. If natural ventilation can not be carried out or the natural ventilation can not meet the requirements, the mechanical pressurized air supply systems shall be arranged or the smoke-proof staircases shall be adopted;

2. 除楼梯间的出入口和外窗外，楼梯间的墙上不应开设其他门、窗、洞口；

2. Except for entrances and exits and exterior windows of the staircases, other doors, windows and openings shall not be arranged in the walls of staircases;

3. 高层建筑、人员密集的公共建筑、人员密集的多层丙类厂房、甲、乙类厂房，其封闭楼梯间的门应采用乙级防火门，并应向疏散方向开启；其他建筑，可采用双向弹簧门；

3. For the doors of the enclosed staircases of the high-rise buildings, crowded public buildings, crowded multi-storey Classes C, A and B plants, the Grade B fire doors shall be adopted, and opened to the evacuation directions; bi-directional spring doors can be adopted for other buildings;

4. 楼梯间的首层可将走道和门厅等包括在楼梯间内形成扩大的封闭楼梯间，但应采用乙级防火门等与其他走道和房间分隔。

4. On the first floor of the staircases, sidewalks, halls and others can be included into the staircases, to form the enlarged enclosed staircases, but the Grade B fire doors and others shall be adopted to separate it from other sidewalks and rooms.

12.3.3 【防火规】6.4.3条：防烟楼梯间除符合第12.3.1条规定外，尚应符合下列规定。

12.3.3 [Code for fire protection design of buildings] 6.4.3: In addition to the provisions in Clause 12.3.1, the smokeproof staircases shall conform to the following provisions.

1. 应设置防烟设施。

1. The smoke preventive facilities shall be arranged.

2. 前室可与消防电梯间前室合用。

2. The anteroom and the one of the fire elevator room can be shared.

3. 前室的使用面积：公共建筑、高层厂房（仓库），不应小于6.0m²；住宅建筑，不应小于4.5m²。

3. Usable floor area of the lobby: the area for public buildings and high-rise plants (warehouses) shall not be less than 6.0m²; the area for residential buildings shall not be less than 4.5m².

与消防电梯间前室合用时，合用前室的使用面积：公共建筑、高层厂房（仓库），不应小于10.0m²；住宅建筑，不应小于6.0m²。

When it is shared with the fire elevator room, the usable floor area of the anteroom shall not be less than 10.0m² for the public buildings and high-rise plants (warehouses); and not less than 6.0m² for residential buildings.

4. 疏散走道通向前室以及前室通向楼梯间的门应采用乙级防火门。

4. The Grade B fire doors shall be used as the doors of the evacuation sidewalks to the anterooms and the doors of the anterooms to the staircases.

5. 除住宅建筑的楼梯间前室外，防烟楼梯间和前室内的墙上不应开设除疏散门和送风口外的其他门、窗、洞口。

5. Except for the anteroom of the staircase of the residential building, doors, windows and openings other than evacuation doors and air supply inlets shall not be arranged in the walls of smoke-proof staircases and anterooms.

6. 楼梯间的首层可将走道和门厅等包括在楼梯间前室内形成扩大的前室，但应采用乙级防火门等与其他走道和房间分隔。

6. On the first floor of the staircase, sidewalks, halls and others can be included in the anterooms of staircases to form an expanded anteroom, which shall be separated from other sidewalks and rooms with the Grade B fire doors, etc.

12.3.4【防火规】6.4.4条：除通向避难层错位的疏散楼梯外，建筑内的疏散楼梯间在各层的平面位置不应改变。

12.3.4 [Code for fire protection design of buildings] 6.4.4: Except for dislocated evacuation staircases to refuge floors, the plane positions of the evacuation staircases on all floors of buildings shall not be changed.

除住宅建筑套内的自用楼梯外，地下或半地下建筑（室）的疏散楼梯间，应符合下列规定：

Except for the owner-occupied staircases in the residential building suites, the evacuation staircases of underground or semi-underground buildings (rooms) shall conform to the following provisions:

1. 室内地面与室外出入口地坪高差大于10m或3层及以上的地下、半地下建筑（室），其疏散楼梯应采用防烟楼梯间；其他地下或半地下建筑（室），其疏散楼梯应采用封闭楼梯间。

1. For the underground and semi-underground buildings (rooms) with the height difference between the indoor floor and the outdoor access floors greater than 10m or with three floors or above, the smoke-proof staircases shall be adopted for the evacuation staircases; the enclosed staircases shall be adopted for other underground or semi-underground buildings (rooms).

2. 应在首层采用耐火极限不低于 2.00h 的防火隔墙与其他部位分隔并应直通室外，确需在隔墙上开门时，应采用乙级防火门。

2. The fireproof partition walls with fire resistance not lower than 2.00h shall be adopted on the first floor for separation from other positions, the fireproof partition walls shall directly lead to outdoors, and the Grade B fire doors shall be adopted if doors are required to be arranged in the partition walls.

3. 建筑的地下或半地下部分与地上部分不应共用楼梯间，确需共用楼梯间时，应在首层采用耐火极限不低于 2.00h 的防火隔墙和乙级防火门将地下或半地下部分与地上部分的连通部位完全分隔，并应设置明显的标志。

3. The staircase shall not be shared between the underground or semi-underground and overground parts of buildings; if such sharing is required indeed, the fire partitions and Class B fire doors with the fire endurance not less than 2.00h shall be adopted on the first floor for full separation of the joints of the underground or semi-underground parts and overground parts, and clear marks shall be arranged.

12.3.5 各种疏散楼梯、楼梯间的适用范围见表 12.3.5-1、表 12.3.5-2。

12.3.5 For the application scope of various evacuation staircases and staircase, see Tab. 12.3.5-1 and Tab. 12.3.5-2.

各类疏散楼梯间定义表　　　　　　　　　　表 12.3.5-1
Definition Table of Various Evacuation Staircases　　　Tab. 12.3.5-1

类型 Type	术语 Terms	简图 Schematic diagram
敞开楼梯间 Open type staircases	楼梯的进出端敞开，其他三个面用满足耐火极限要求的隔墙封闭的楼梯间 Staircases with open access but other three sides enclosed by the partitions with appropriate fire endurance	外窗 Exterior window 走道 Walkway

续表

类型 Type	术语 Terms	简图 Schematic diagram
封闭楼梯间【防火规】 2.1.15 条 Enclosed staircases [Code for fire protection design of buildings] 2.1.15	在楼梯间入口处设置门，以防止火灾的烟和热气进入的楼梯间 Staircases with a door at the staircase entrance to prevent the entry of fire smoke and hot air	外窗 Exterior window FM 乙（乙级防火门） Grade B fire doors
防烟楼梯间【防火规】 2.1.16 条 Smokeproof staircases [Code for fire protection design of buildings] 2.1.16	在楼梯间入口处设置防烟的前室、开敞式阳台或凹廊（统称前室）等设施，且通向前室和楼梯间的门均为防火门，以防止火灾的烟和热气进入的楼梯间 The facilities of smoke-proof anteroom, open balcony or concave corridor (collectively referred to as the anteroom) etc., shall be arranged at the staircase entrance, fire doors shall be used as the doors to the anterooms and the doors of the staircases, to prevent the entry of fire smoke and hot air into the staircase	FM 乙 Grade B fire doors 前室 Anteroom FM 乙（乙级防火门） Grade B fire doors

各种疏散楼梯、楼梯间的规定和适用条件　　　　　　表 12.3.5-2

Provisions and Applicable Conditions for Various Evacuation Stairs and Staircases　Tab. 12.3.5-2

类型 Type		规定和适用条件 Provisions and applicable conditions	备注 Remarks
敞开楼梯间 Open type staircases	住宅建筑 Residential buildings	1. 建筑高度不大于 21m 的住宅建筑可采用敞开楼梯间；与电梯井相邻布置的疏散楼梯应采用封闭楼梯间，当户门采用乙级防火门时，仍可采用敞开楼梯间； 1. For the residential buildings with the building height not greater than 21m, open type staircases can be adopted; for the evacuation staircases arranged adjacent to the elevator shafts, enclosed staircases shall be adopted; and when Grade B fire doors are adopted, open type staircases can be also adopted; 2. 建筑高度大于 21m、不大于 33m 的住宅建筑应采用封闭楼梯间；当户门采用乙级防火门时，可采用敞开楼梯间 2. For residential buildings with the building height greater than 21m and not greater than 33m, enclosed staircases shall be adopted; when Grade B fire doors are adopted, open type staircases can be also adopted	【防火规】5.5.27 条 [Code for fire protection design of buildings] 5.5.27
	多层公共建筑 Multi-storey public buildings	下列多层公共建筑的疏散楼梯，除与敞开式外廊直接相连的楼梯间外，均应采用封闭楼梯间： Except for the staircases directly connected to the open verandas, the enclosed staircases shall be adopted for the following evacuation staircases of the multi-storey public buildings: 1. 医疗建筑、旅馆及类似使用功能的建筑； 1. Medical buildings, hotel, and buildings with the similar use functions; 2. 设置歌舞娱乐放映游艺场所的建筑； 2. Buildings in which karaoke bars, dancing halls, entertainment rooms, video halls, amusement halls, etc., are arranged; 3. 商店、图书馆、展览建筑、会议中心及类似使用功能的建筑； 3. Shops, libraries, exhibition buildings, conference centers, and buildings with the similar use functions; 4. 6 层及以上的其他建筑 4. Other buildings of 6 or more storeys	【防火规】5.5.13 条 [Code for Fire protection design of buildings] 5.5.13

续表

类型 Type		规定和适用条件 Provisions and applicable conditions	备注 Remarks
封闭楼梯间 Enclosed staircases	住宅建筑 Residential buildings	1. 建筑高度不大于21m的住宅建筑可采用敞开楼梯间；与电梯井相邻布置的疏散楼梯应采用封闭楼梯间； 1. For the residential buildings with the building height not greater than 21m, open type staircases can be adopted; for the evacuation staircases arranged adjacent to the elevator shafts, the enclosed staircases shall be adopted; 2. 建筑高度大于21m、不大于33m的住宅建筑应采用封闭楼梯间；当户门采用乙级防火门时，可采用敞开楼梯间 2. For residential buildings with the building height greater than 21m and not greater than 33m, enclosed staircases shall be adopted; when Grade B fire doors are adopted, open type staircases can be also adopted	【防火规】5.5.27.2【防火规】5.5.27.3 [Code for fire protection design of buildings] 5.5.27.2 [Code for fire protection design of buildings] 5.5.27.3
	公共建筑 Public buildings	下列多层公共建筑的疏散楼梯，除与敞开式外廊直接相连的楼梯间外，均应采用封闭楼梯间： Except for the staircases directly connected to the open verandas, the enclosed staircases shall be adopted for the following evacuation staircases of the multi-storey public buildings: 1. 医疗建筑、旅馆及类似使用功能的建筑； 1. Medical buildings, hotel, and buildings with the similar use functions; 2. 设置歌舞娱乐放映游艺场所的建筑； 2. Buildings in which karaoke bars, dancing halls, entertainment rooms, video halls, amusement halls and so on are arranged; 3. 商店、图书馆、展览建筑、会议中心及类似使用功能的建筑； 3. Shops, libraries, exhibition buildings, conference centers, and buildings with the similar use functions; 4. 6层及以上的其他建筑 4. Other buildings of 6 or more floors	【防火规】5.5.13条 [Code for fire protection design of buildings] 5.5.13
		裙房和建筑高度不大于32m的二类高层公共建筑，其疏散楼梯应采用封闭楼梯间。（注：当裙房与高层建筑主体之间设置防火墙时，裙房的疏散楼梯可按《建筑设计防火规范（2018年版）》GB 50016—2014有关单、多层建筑的要求确定） For the podiums of the Class I high-rise public buildings and the Class II high-rise public buildings with the height not less than 32 m, the enclosed staircases shall be adopted for the evacuation staircases. (Note: when the fire walls are arranged between the podiums and the main bodies of the high-rise buildings, the evacuation staircases in podiums can be determined in accordance with the relevant requirements for single-storey and multi-storey buildings in *Code for Fire Protection Design of Buildings* GB 50016—2014)	【防火规】5.5.12条 [Code for fire protection design of buildings] 5.5.12
	其他建筑 Other buildings	除住宅建筑套内的自用楼梯外，地下或半地下建筑（室）的疏散楼梯间：室内地面与室外出入口地坪高差不大于10m且2层及以下的地下、半地下建筑（室），其疏散楼梯应采用封闭楼梯间 Except for the owner-occupied staircases in the residential building suites, for the evacuation staircases of underground or semi-underground buildings (rooms) with the height difference between the indoor floors and floors at the outdoor entrances and exits less than 10 m or the evacuation staircases of the underground and semi-underground buildings (rooms) of 2 floors and below and other underground or semi-underground buildings (rooms), enclosed staircases shall be adopted	【防火规】6.4.4条 [Code for fire protection design of buildings] 6.4.4

续表

类型 Type		规定和适用条件 Provisions and applicable conditions	备注 Remarks
防烟楼梯间 Smokeproof staircases		建筑高度小于32m的高层汽车库、室内地面与室外出入口地坪的高差小于10m的地下汽车库 High-rise garages with the building height less than 32m and underground garages with the height difference between the indoor floors and the floors at the outdoor entrances and exits less than 10m	【汽防规】6.0.3.1条 [Code for fire protection design of garage, motor repair shop and parking area] 6.0.3.1
	住宅建筑 Residential buildings	1. 建筑高度大于33m的住宅建筑应采用防烟楼梯间。户门不宜直接开向前室，确有困难时，每层开向同一前室的户门不应大于3樘且应采用乙级防火门。 1.The smokeproof staircases shall be adopted for the residential buildings with the building height greater than 33m. The doors opened to the anterooms shall not be appropriately arranged; if it is difficult to arrange in this way, the doors on each floor opened to the same anteroom shall be not greater than 3 sets, and Grade B fire door shall be adopted. 2. 住宅建筑的疏散楼梯采用剪刀楼梯间时 2. If the evacuation staircase of residential buildings use scissor-shaped staircase rooms	【防火规】5.5.27.3条、 【防火规】5.5.28.1条 [Code for fire protection design of buildings] 5.5.27.3, [Code for fire protection design of building] 5.5.28.1
	公共建筑 Public buildings	1. 一类高层公共建筑； 1. Class I high-rise public buildings; 2. 建筑高度大于32m的二类高层公共建筑； 2. Class II high-rise public buildings with the building higher than 32m; 3. 高层公共建筑的疏散楼梯采用剪刀楼梯间时 3. When the scissors type staircases are adopted as the evacuation staircases of high-rise public buildings	【防火规】5.5.12条 [Code for fire protection design of buildings] 5.5.12
	汽车库 Garages	建筑高度大于32m的高层汽车库、室内地面与室外出入口地坪的高差大于10m的地下汽车库 High-Rise garages with the building higher than 32m and the underground garages with the height difference between the indoor floor and the ground floor at the outdoor access greater than 10m	【汽防规】6.0.3.1条 [Code for fire protection design of garage, motor repair shop and parking area] 6.0.3.1
	其他建筑 Other buildings	室内地面与室外出入口地坪高差大于10m或3层及以上的地下、半地下建筑（室） Underground and semi-underground buildings（rooms）with the height difference between the indoor floor and the outdoor entrance and exit floor greater than 10m or with three storeys	【防火规】6.4.4.1条 [Code for fire protection design of buildings] 6.4.4.1

12.3.6 剪刀楼梯、楼梯间的设计要求见表12.3.6。

12.3.6 Design requirements for scissors type stairs and staircases see Tab.12.3.6.

剪刀楼梯、楼梯间的设计要求　　表 12.3.6
Design Requirements for Scissors Type Stairs and Staircases　　Tab.12.3.6

类型 Type		适用范围 Application scope	简图 Schematic diagram
剪刀楼梯、楼梯间 Scissors type stairs and staircases	住宅建筑 Apartment building	【防火规】5.5.28 条：住宅单元的疏散楼梯，当分散设置确有困难且任一户门至最近疏散楼梯间入口的距离不大于 10m 时，可采用剪刀楼梯间，但应符合下列规定。 [Code for fire protection design of buildings] 5.5.28: For the evacuation staircase in the residential buildings, when there is difficulty in decentralized arrangement, and the distance from any door to the nearest evacuation staircase entrance is not more than 10m, the scissors type staircases shall be adopted and conform to the following provisions. 1. 应采用防烟楼梯间； 1. The smokeproof staircases shall be adopted; 2. 梯段之间应设置耐火极限不低于 1.00h 的防火隔墙； 2. The fire partitions with the fire endurance not less than 1.00 hour shall be arranged between the stair flights; 3. 楼梯间的前室不宜共用；共用时，前室的使用面积不应小于 6.0m²； 3. The anteroom of staircases shall not be shared; when the anteroom of staircases is shared; the usable floor area of the lobby shall be not less than 6.0m²; 4. 楼梯间的前室或共用前室不宜与消防电梯的前室合用；楼梯间的共用前室与消防电梯的前室合用时，合用前室的使用面积不应小于 12.0m²，且短边不应小于 2.4m 4. The staircase or shared anteroom and the fire elevator shall not share an anteroom; When the shared anteroom of the staircase is also used as the anteroom of the fire elevator, the usable floor area of the shared anteroom shall be not less than 12.0m², and the short side shall be not less than 2.4m	消防电梯 Fire elevator 合用前室 Shared anteroom 前室 Anteroom FH 丙 （丙级防火门） Grade C fire doors FM 乙 （乙级防火门） Grade B fire doors 加压送风道和加压送风口 Pressurized air supply duct and pressurized air supply inlet
	高层公共建筑 High-Rise public buildings	【防火规】5.5.10 条：高层公共建筑的疏散楼梯，当分散设置确有困难且从任一疏散门至最近疏散楼梯间入口的距离不大于 10m 时，可采用剪刀楼梯间，但应符合下列规定。 [Code for fire protection design of zbuildings] 5.5.10: For the evacuation staircase in the high-rise public buildings, when there is difficulty in decentralized arrangement, and the distance from any evacuation door to the nearest evacuation staircase entrance is not more than 10m, the scissors type staircases shall be adopted and conform to the following provisions. 1. 楼梯间应为防烟楼梯间； 1. The staircase shall be a smokeproof staircase; 2. 梯段之间应设置耐火极限不低于 1.00h 的防火隔墙； 2. The fire partitions with the fire endurance not less than 1.00 hour shall be arranged between the stair flights; 3. 楼梯间的前室应分别设置 3. The anterooms of the staircase shall be arranged separately	消防电梯 Fire elevator ≥ 5.0m 合用前室 Shared anteroom FH 丙 （丙级防火门） Grade C fire doors ≥ 2.40m 加压送风道和加压送风口 Pressurized air supply duct and pressurized air supply inlet

12.3.7 室外疏散楼梯（表 12.3.7）。

12.3.7 Outdoor evacuation staircases (See Tab.12.3.7).

室外疏散楼梯　　　　　　　　　　　　　　　表 12.3.7
Outdoor evacuation staircases　　　　　　　Tab.12.3.7

类型 Type	适用范围 Application scope	简图 Schematic diagram
室外疏散楼梯 Outdoor evacuation staircases	【防火规】6.4.5 条：室外疏散楼梯应符合下列规定。 [Code for fire protection design of buildings] 6.4.5: The outdoor evacuation staircases shall conform to the following provisions. 1. 栏杆扶手的高度不应小于 1.10m，楼梯的净宽度不应小于 0.90m。 1. The handrail height shall not be less than 1.10m, and the net width of the staircase shall not be less than 0.90m. 2. 倾斜角度不应大于 45°。 2. The inclination angle shall not be greater than 45°. 3. 梯段和平台均应采用不燃材料制作。平台的耐火极限不应低于 1.00h，梯段的耐火极限不应低于 0.25h。 3. The stair flights and platforms shall be made of incombustible material. The fire endurance of the platform shall not be less than 1.00h , and the fire endurance of the stair flight shall not be less than 0.25h . 4. 通向室外楼梯的门应采用乙级防火门，并应向外开启。 4. For the doors to the outdoor staircases, Grade B fire doors shall be adopted, and the doors shall be opened outward. 5. 除疏散门外，楼梯周围 2m 内的墙面上不应设置门、窗、洞口。疏散门不应正对梯段。 5. Except for evacuation doors, doors, windows and openings shall not be arranged in the walls 2m around the staircases. Evacuation doors shall not be exactly against the stair flights	

12.3.8 疏散楼梯宽度的设计。

12.3.8 Design of the width of evacuation staircases.

【防火规】5.5.21 条：除剧场、电影院、礼堂、体育馆外的其他公共建筑，其房间疏散门、安全出口、疏散走道和疏散楼梯的各自总净宽度，应符合下列规定。

[Code for fire protection design of buildings] 5.5.21: For other public buildings except theaters, cinemas, auditoriums and gymnasiums, the respective total net width of the evacuation doors, safety exits, evacuation corridors and evacuation staircases for the rooms shall conform to the following provisions.

1. 每层的房间疏散门、安全出口、疏散走道和疏散楼梯的各自总净宽度，应根据疏散人数按每 100 人的最小疏散净宽度不小于表 12.3.8-1 的规定计算确定。当每层疏散人数不等时，疏散楼梯的总净宽度可分层计算，地上建筑内下层楼梯的总净宽度应按该层及以上疏散人数最多一层的人数计算；地下建筑内上层楼梯的总净宽度应按该层及以下疏散人数最多一层的人数计算。

1. The respective total net width of the evacuation doors, safety exits, evacuation sidewalks and evacuation staircases for the rooms on each floor shall be determined through calculation in accordance with the number of evacuees (the minimum net evacuation width per 100 people shall not be less than that in Tab. 12.3.8-1). When the number of evacuees of each floor is unequal, the total net width of the evacuation staircases shall be calculated respectively, and the total net width of the staircases of the lower floors in overground buildings shall be calculated as per the maximum number of the evacuees on and above the floor in the buildings; the total net width of the staircases for upper floors in underground buildings shall be calculated as per the maximum number of the evacuees on and below the floor in the buildings.

【防火规】5.5.21-1: 每层的房间疏散门、安全出口、疏散走道和疏散楼梯的每 100 人最小疏散净宽度（m/100 人） 表 12.3.8-1

[Code for Fire Protection Design of Buildings] 5.5.21-1: Minimum Net evacuation width (m/100 People) Per 100 People of the Evacuation Doors, Safety Exits, Evacuation Corridors and Evacuation Staircases for the Rooms on Each Floor   Tab. 12.3.8-1

建筑层数 Building floors		建筑的耐火等级 Fire resistance rating of the building		
		一、二级 Grade I and Grade II	三级 Grade III	四级 Grade IV
地上楼层 Overground floors	1～2层 1 to 2 floors	0.65	0.75	1.00
	3层 3 floors	0.75	1.00	—
	≥ 4层 ≥ 4 floors	1.00	1.25	—
地下楼层 Underground floor	与地面出入口地面的高差 $\Delta H \leq 10m$ Height difference ΔH between the floors of the access ≤ 10m	0.75	—	—
	与地面出入口地面的高差 $\Delta H > 10m$ Height difference ΔH between the floors of the access > 10m	1.00	—	—

2. 地下或半地下人员密集的厅、室和歌舞娱乐放映游艺场所，其房间疏散门、安全出口、疏散走道和疏散楼梯的各自总净宽度，应根据疏散人数按每 100 人不小于 1.00m 计算确定。

2.The respective total net width of evacuation doors, safety exits, evacuation sidewalks and evacuation staircases for rooms of underground or semi-underground crowded halls, rooms and karaoke bars, dancing halls, entertainment rooms, video halls, amusement halls and others shall be determined in accordance with the number of evacuees as not less than 1.00m per 100 people.

3. 首层外门的总净宽度应按该建筑疏散人数最多一层的人数计算确定，不供其他楼层人员疏散的外门，可按本层的疏散人数计算确定。

3. The total net width of the first-floor exterior evacuation doors shall be calculated and determined as per the largest number of evacuees at the floor in the building, and the total net width of the interior doors not for the personnel evacuation of other floors can be calculated and determined as per the number of evacuees of the floor.

4. 歌舞娱乐放映游艺场所中录像厅的疏散人数，应根据厅、室的建筑面积按不小于 1.0 人 /m² 计算；其他歌舞娱乐放映游艺场所的疏散人数，应根据厅、室的建筑面积按不小于 0.5 人 /m² 计算。

4. The number of evacuees in the video halls in karaoke, dancing halls, entertainment rooms, video halls, amusement halls and others shall be calculated as at least 1.0 person/m²; the number of evacuees in other karaokes, dancing halls, entertainment rooms, video halls, amusement halls and others shall be calculated in accordance with the building area of the halls and rooms as not less than 0.5 person/m².

5. 有固定座位的场所，其疏散人数可按实际座位数的 1.1 倍计算。

5. The number of evacuees at places with fixed seats can be calculated as per 1.1 times of the actual quantity of the seats.

6. 展览厅的疏散人数应根据展览厅的建筑面积和人员密度计算，展览厅内的人员密度不宜小于 0.75 人 /m²。

6. The number of evacuees in the exhibition hall shall be calculated in accordance with the building area and personnel density of the exhibition hall, and the personnel density of the exhibition halls shall be not less than 0.75 person/m².

7. 商店的疏散人数应按每层营业厅的建筑面积乘以表 12.3.8-2 规定的人员密度计算。对于建材商店、家具和灯饰展示建筑，其人员密度可按表 12.3.8-2 规定值的 30% 确定。

7. The number of evacuees in the stores shall be calculated by multiplying the building area of the business hall in each floor by the personnel density specified in Tab. 12.3.8-2. The personnel density in the building material stores and furniture and lighting display buildings can be determined as per 30% of the specified values in Tab. 12.3.8-2.

商店营业厅内的人员密度（人 /m²） 表 12.3.8-2
Personnel Density of Shops and Business Halls (person/m²) Tab. 12.3.8-2

楼层位置 Positions of the floors	地下第二层 Second floor underground	地下第一层 First floor underground	地上第一、二层 First and second floors above the ground	地上第三层 Third floor above the ground	地上第四层及以上各层 Fourth floor above the ground and above
人员密度 Personnel density	0.56	0.60	0.43 ~ 0.60	0.39 ~ 0.54	0.30 ~ 0.42

12.3.9 楼梯、楼梯间材料的防火性能要求。

12.3.9 Requirements for the fireproof performance of materials for stairs and staircases.

1. 楼梯、楼梯间材料的燃烧性能和耐火极限见表12.3.9。

1. Combustion performance and fire endurance of materials for stairs and staircases see Tab. 12.3.9.

【防火规】表5.1.2：楼梯、楼梯间材料的燃烧性能和耐火极限（h） 表12.3.9

[Code for fire protection design of buildings] Tab. 5.1.2: Combustion Performance and Fire Endurance of Materials for Stairs and Staircases (h) Tab. 12.3.9

构件名称	耐火等级 Fire resistance rating			
	一级 Grade I	二级 Grade II	三级 Grade III	四级 Grade IV
疏散楼梯 Evacuation staircase	不燃性 1.50 Incombustibility 1.50	不燃性 1.00 Incombustibility 1.00	不燃 0.50 Incombustibility 0.50	可燃性 Combustibility
楼梯间的墙 Walls of the staircase	不燃性 2.00 Incombustibility 2.00	不燃性 2.00 Incombustibility 2.00	不燃 1.50 Incombustibility 1.50	难燃性 0.50 Flame retardancy 0.50

2.【装修防火规范】4.0.5条：疏散楼梯间和前室的顶棚、墙面和地面应采用A级装修材料。

2. [Code for fire protection in design of interior decoration of buildings] 4.0.5: Grade A non-combustible decorative materials shall be used for the ceilings, walls and floors of the evacuation staircases.

12.3.10 楼梯间的布局。

12.3.10 Layout of Staircases.

1. 公共建筑的安全疏散距离及相关措施，见本书6.3节。

1. See chapter 6.3 of this book for safe evacuation distances for public buildings and related measures.

2. 住宅建筑的安全疏散距离及相关措施，见本书6.3节。

2. See chapter 6.3 of this book for safe evacuation distances for residential buildings and related measures.

3.【防火规】5.5.30条：住宅建筑的户门、安全出口、疏散走道和疏散楼梯的各自总净宽度应经计算确定，且户门和安全出口的净宽度不应小于0.90m，疏散走道、疏散楼梯和首层疏散外门的净宽度不应小于1.10m。建筑高度不大于18m的住宅中一边设置栏杆的疏散楼梯，其净宽度不应小于1.0m。

3. [Code for fire protection design of buildings] 5.5.30: The total net widths of doors of residential buildings, safety exits, evacuation sidewalks and evacuation staircases shall be determined through calculation; the net width of the door and safety exit shall not be less than 0.90m; and the net widths of the evacuation sidewalk, evacuation staircase and exterior

evacuation door on the first floor shall not be less than 1.10m. For the evacuation staircase with one side provided with the railings in the residential building with the building height not greater than 18m, the net width shall not be less than 1.0m.

4. 公共建筑允许设置1个安全出口或1部疏散楼梯的条件：

4.Conditions in which one safety exit or one evacuation staircase is allowed to be arranged in the building:

【防火规】表5.5.8：公共建筑内每个防火分区或一个防火分区的每个楼层，其安全出口的数量应经计算确定，且不应少于2个。符合下列条件之一的公共建筑，可设置1个安全出口或1部疏散楼梯。

[Code for fire protection design of buildings] Tab. 5.5.8: The quantity of the safety exits of each fire compartment in the public buildings or each floor of a fire compartment shall be determined through calculation, and shall not be less than 2. For the public buildings conforming to one of the following conditions, one safety exit or one evacuation staircase can be arranged.

1）除托儿所、幼儿园外，建筑面积不大于200m²且人数不超过50人的单层公共建筑或多层公共建筑的首层；

1) The first floor of a single-storey public building or a multi-storey public building with the building area not greater than 200m² and the number of people not exceeding 50 except for nurseries and kindergartens;

2）除医疗建筑，老年人照料设施，托儿所、幼儿园的儿童用房，儿童游乐厅等儿童活动场所和歌舞娱乐放映游艺场所等外，符合表12.3.10-1规定的公共建筑。

2) Public buildings conforming to the provisions in Tab. 12.3.10-1 except for medical buildings, care facilities for the aged, children's houses in nurseries and kindergartens, activity places for children like children's play hall, and karaoke bars, dancing halls, entertainment rooms, video halls, amusement halls, etc.

【防火规】表5.5.8：可设置一部疏散楼梯的公共建筑　　　　表12.3.10-1

[Code for fire protection design of buildings] Tab. 5.5.8:
Public Buildings in Which One Evacuation Staircase can be Arranged　　Tab. 12.3.10-1

耐火等级 Fire resistance rating	最多层数 Maximum number of storeys	每层最大建筑面积（m²） Maximum building area of each floor（m²）	人数 Number of people
一、二级 Grade I and Grade II	3层 3 storeys	200	第二、第三层的人数之和不超过50人 The total number of people on second and third floors shall be not greater than 50
三级 Grade III	3层 3 storeys	200	第二、第三层的人数之和不超过25人 The total number of people on second and third floors shall be not greater than 25

续表

耐火等级 Fire resistance rating	最多层数 Maximum number of storeys	每层最大建筑面积（m²） Maximum building area of each floor (m²)	人数 Number of people
四级 Grade IV	2 层 2 storeys	200	第二层人数不超过 15 人 Number of people on second floors not greater than 15

5. 住宅建筑允许设置一个安全出口或一部疏散楼梯的条件符合表 12.3.10-2 规定的住宅建筑可设置 1 个安全出口或 1 部疏散楼梯。

5. Conditions in which one safety exit or one evacuation staircase is allowed to be arranged in the public building conform to the provisions in Tab. 12.3.10-2.

【防火规】5.5.25：住宅建筑设置一个安全出口或一部疏散楼梯的条件　　　表 12.3.10-2

[Code for fire protection design of buildings] 5.5.25: Conditions in Which One Safety Exit or One Evacuation Staircase is Allowed to be Arranged in the Residential Buildings　Tab. 12.3.10-2

建筑类别 Building Category	建筑高度 Building Height	条件 Conditions
住宅 Residential buildings	≤ 27	每个单元任一层的建筑面积不大于 650m²，且任一户门至最近安全出口的距离不大于 15m Building area of any floor of all the units is not greater than 650 m² and the distance from any door to the nearest safety exit is not greater than 15m
	27< H ≤ 54	每个单元任一层的建筑面积不大于 650m²，且任一户门至最近安全出口的距离不大于 10m Building area of any floor of all the units is not greater than 650 m² and the distance from any door to the nearest safety exit is not greater than 10m

6. 地下或半地下建筑（室）疏散楼梯设置要求，见表 12.3.10-3。

6. Requirements for the installation of evacuation staircases in underground or semi-underground buildings（rooms），see Tab.12.3.10-3.

地下或半地下建筑（室）疏散楼梯设置要求　　　表 12.3.10-3

Requirements for the Installation of Evacuation Staircases in Sewi-underground Building (Rooms)　Tab. 12.3.10-3

建筑类别 Building category	埋深（m） Burial depth (m)	使用人数（人） Number of users（person）	最大建筑面积或最大防火分面积（m²） Maximum building area or maximum fire compartment area (m²)	疏散距离 Evacuation distance	其他条件 Other conditions
除人员密集场所外的地下或半地下建筑（室） Underground or semi-underground building（room）except for crowded place	≤ 10	≤ 30	≤ 500	满足表 12.3.10-1 的要求 The requirements in Tab. 12.3.10-1 shall be met	还需另设一直通室外的金属竖向梯作为第二安全出口 Another vertical metal staircase directly to the outdoor shall be arranged as a second safety exit

续表

建筑类别 Building category	埋深（m） Burial depth(m)	使用人数（人） Number of users (person)	最大建筑面积或最大防火分区面积（m²） Maximum building area or maximum fire compartment area(m²)	疏散距离 Evacuation distance	其他条件 Other conditions
地下或半地下设备间 Underground or semi-underground mechanical room	—	—	防火分区建筑面积 ≤ 200m² Building area of the fire compartment ≤ 200m²	满足表12.3.10-1的要求 The requirements in Tab. 12.3.10-1 shall be met	
除歌舞娱乐放映游艺场所外地下或半地下建筑（室） Underground or semi-underground building (room) except for karaokes, dancing halls, entertainment rooms, video halls, amusement halls, etc.	—	15	防火分区建筑面积 ≤ 50m²且经常停留人数不超过15人 Building area of the fire compartment is not greater than 50 m², and no more than 15 people stay regularly		

12.3.11 楼梯出屋顶的相关设计规定。

12.3.11 Relevant design regulations for a staircase to roof.

1.【防火规】5.5.3条：建筑的楼梯间宜通至屋顶，通向屋面的门或窗应向外开启。

1. [Code for fire protection design of buildings] 5.5.3: The staircases of the building shall be connected to the roof, and the doors or windows to the roof shall be opened outward.

2.【防火规】5.5.26条：建筑高度大于27m，但不大于54m的住宅建筑，每个单元设置一座疏散楼梯时，疏散楼梯应通至屋面，且单元之间的疏散楼梯应能通过屋面连通，户门应采用乙级防火门。当不能通至屋面或不能通过屋面连通时，应设置2个安全出口。

2. [Code for fire orotection design of buildings] 5.5.26：When an evacuation staircase is arranged in each unit of the residential building with the building higher than 27m but not greater than 54m, the evacuation staircase shall be connected to the roof, the evacuation staircases between units shall be connected by roof, and the Grade B fire door shall be used. If the roof cannot be accessed or connected through, 2 safety exits should be arranged.

3.【商店建规】5.2.5条：大型百货商店、商场建筑物的营业层在五层以上时，应设置直通屋顶平台的疏散楼梯间不少于2座，屋顶平台上无障碍物的避难面积不宜小于最大营业层建筑面积的50%。

3. [Code for design of store buildings] 5.2.5: If the business floors of the large department store and mall buildings are above five, at least two evacuation staircases directly to the roof platform shall be arranged, and the area of the obstacle-free refuge on the roof platform shall be appropriately not less than 50% of the building area of the maximum business floor.

4.【防火规】6.4.9条：高度大于10m的三级耐火等级建筑应设置通至屋顶的室外消防梯。室外消防梯不应面对老虎窗，宽度不应小于0.6m，且宜从离地面3.0m高处设置。

4. [Code for fire protection design of buildings] 6.4.9: In the Grade III fire-resistant buildings with the height greater than 10m, the outdoor fire ladders to the roof shall be arranged. The outdoor fire ladders shall not face dormers, with the width not less than 0.6m, and shall be arranged at the height of 3.0m from the ground.

5.【防火规】5.5.11条：设置不少于2部疏散楼梯的一、二级耐火等级多层公共建筑，如顶层局部升高，当高出部分的层数不超过2层、人数之和不超过50人且每层建筑面积不大于200m²时，高出部分可设置1部疏散楼梯，但至少应另外设置1个直通建筑主体上人平屋面的安全出口，且上人屋面应符合人员安全疏散的要求。

5. [Code for fire protection design of buildings] 5.5.11: For Grade I and Grade II fire-resistant multi-storey public buildings with at least two evacuation staircases, if the top floors are partially raised, when the number of floors of the higher parts are not more than two, total number of people is not more than 50 and building area of each storey is not greater than 200m², and one evacuation staircase can be arranged for the higher parts, but at least one safety exit directly to accessible flat roofs of the main bodies of the buildings shall be additionally arranged, and the accessible roofs shall conform to the requirements for safe evacuation of personnel.

12.4　台阶、坡道

12.4　Step and Ramp

12.4.1　台阶设计应符合下列要求：

12.4.1　The design for steps shall conform to the following requirements:

1.【通则】6.6.1.1条：公共建筑室内外台阶踏步宽度不宜小于0.30m，踏步高度不宜大于0.15m，并不宜小于0.10m，踏步应防滑。室内台阶踏步数不应少于2级，当高差不足2级时，应按坡道设置。

1. [Code for design of civil buildings] 6.6.1.1: The width of the indoor and outdoor step treads of the public building shall be appropriately not less than 0.30m; the step height shall be appropriately not greater than 0.15m, and shall be appropriately not less than 0.10m; and the steps shall be anti-slip. The number of indoor step treads shall be not less than two, and when the height difference is less than two, the step treads shall be arranged as per ramps.

2.【通则】6.6.1.2条：人流密集的场所台阶高度超过0.70m并侧面临空时，应有防护设施。

2. [Code for design of civil buildings] 6.6.1.2: If the height of steps in crowded places exceeds 0.70m with the sides open, protective facilities shall be provided.

3.【办规】4.1.9.2条：高差不足两级踏步时，不应设置台阶，应设坡道，其坡度不宜大于1∶8。

3. [Design code for office building] 4.1.9.2: When the height difference is less than two steps, the steps shall not be arranged, and the ramps with the gradient not greater than 1∶8 shall be set up.

12.4.2 坡道设计应符合下列要求：

12.4.2 The ramp design shall conform to the following requirements:

1.【通则】6.6.2.1条：室内坡道坡度不宜大于1∶8，室外坡道坡度不宜大于1∶10。

1. [Code for design of civil buildings] 6.6.2.1: The gradient of indoor ramps shall be appropriately not greater than 1∶8, and the gradient of outdoor ramps shall be appropriately not greater than 1∶10.

2.【通则】6.6.2.2条：室内坡道水平投影长度超过15m时，应设休息平台，平台宽度应根据使用功能或设备尺寸所需缓冲空间而定。

2. [Code for design of civil buildings] 6.6.2.2: When the length of the horizontal projection of the indoor ramp exceeds 15m, the stair landing shall be arranged, and the platform width shall be determined in accordance with use function or buffer space required for the equipment dimensions.

3.【通则】6.6.2.3条：供轮椅使用的坡道不应大于1∶12，困难地段不应大于1∶8。

3. [Code for design of civil buildings] 6.6.2.3: The gradients of wheelchair ramps shall be not greater than 1∶12, and if difficult, the gradients shall be not greater than 1∶8.

4.【通则】6.6.2.4条：自行车推行坡道每段坡长不宜超过6m，坡度不宜大于1∶5（图12.4.2）。

4. [Code for design of civil buildings] 6.6.2.4: The length of each section of the ramp for pushing bicycles shall be not greater than 6m, and the gradient shall be appropriately not greater than 1∶5 (Fig. 12.4.2).

5.【车库建规】6.2.5条：非机动车库出入口宜采用直线形坡道，当坡道长度超过6.8m或转换方向时，应设休息平台，平台长度不应小于2.00m，并应能保持非机动车推行的连续性。【车库建规】6.2.6条：踏步式出入口推车斜坡的坡度不宜大于25%，单向净宽不应小于0.35m，总净宽度不应小于1.80m。坡道式出入口的斜坡坡度不宜大于15%，坡道宽度不应小于1.80m。

5. [Code for design of parking garage building] 6.2.5: non-motorized reservoir entrances should use linear ramps, when the length of the ramp exceeds 6.8m or the direction of

conversion, should be set up a rest platform, the length of the platform should not be less than 2.00m, and should be able to maintain the continuity of non-motor vehicle implementation. [Code for design of parking garage building] 6.2.6: the slope of the pedal entrance cart slope should not be greater than 25%, the one-way net width should not be less than 0.35m, the total net width should not be less than 1.80m. Ramp entrance and exit slope slope should not be greater than 15%, ramp width should not be less than 1.80m.

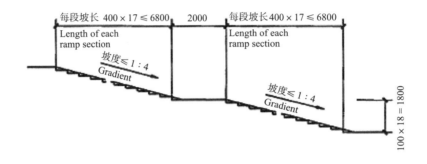

图 12.4.2 自行车推行坡道示意图

Fig. 12.4.2 Schematic Diagram of Ramp for Pushing Bicycle

6.【老年人照料设施标】6.1.3 条：老年人照料设施中，老年人使用的室内外交通空间，当地面有高差时，应设轮椅坡道连接，且坡度不应大于 1/12。当轮椅坡道的高度大于 0.10m 时，应同时设无障碍台阶。

6. [Standard for design of care facilities for the aged] 6.1.3: In care facilities for the elderly, the indoor and outdoor traffic space used by the elderly should be connected by a wheelchair ramp when there is a high difference on the ground, and the slope should not be greater than 1/12. When the height of the wheelchair ramp is greater than 0.10m, the barrier-free steps should be set at the same time.

7. 供机动车行驶的坡道参见本书 7.3.3 节地下汽车库。

7. Ramps for motor vehicles shall conform to the requirements in the underground garage in 7.3.3 of this book.

8.【通则】6.6.2.6 条：坡道应采取防滑措施。

8. [Code for design of civil buidings] 6.6.2.6: Anti-slip measures should be taken on the ramp.

9.【无规】3.4 条：轮椅坡道

9. [Code for accessibility design] 3.4: Wheelchair Ramps

1）【无规】3.4.1 条：轮椅坡道宜设计成直线型、直角形或折返型。

1）[Code for accessibility design] 3.4.1: The wheelchair ramps shall be designed to be linear, right angle or turning-back.

2）【无规】3.4.2 条：轮椅坡道的净宽度不应小于 1.00m，无障碍出入口的轮椅坡道净宽度不应小于 1.20m。

2）[Code for accessibility design] 3.4.2: The net width of wheelchair ramps shall be not less than 1.00m, and net width of the wheelchair ramps at barrier-free exits and entrances shall be not less than 1.20m.

3）【无规】3.4.3 条：轮椅坡道的高度超过 300mm 且坡度大于 1∶20 时，应在两侧设置扶手，坡道与休息平台的扶手应保持连贯。

3）[Code for accessibility design] 3.4.3: When the height of wheelchair ramps exceeds 300mm and the gradient is greater than 1∶20, handrails shall be arranged on both sides, and the handrails on the ramps and stair landing shall be consistent.

4）【无规】3.4.4 条：轮椅坡道的高度和水平长度应符合表 12.4.2 规定。

4）[Code for accessibility design] 3.4.4: The height and horizontal length of wheelchair ramps shall conform to Tab. 12.4.2.

【无规】表 3.4.4：轮椅坡道的高度和水平长度　　　　表 12.4.2

[Code for accessibility design] Tab. 3.4.4: Height and Horizontal Length of Wheelchair Ramps　　　Tab. 12.4.2

坡 度 Gradient	1∶20	1∶16	1∶12	1∶10	1∶8
最大高度（m） Maximum height（m）	1.20	0.90	0.75	0.60	0.30
水平长度（m） Horizontal length（m）	24.00	14.40	9.00	6.00	2.40

5）【无规】3.4.6 条：轮椅坡道起点、终点和中间休息平台的水平长度不应小于 1.50m。

5）[Code for accessibility design] 3.4.6: The horizontal length of the starting and ending points of wheelchair ramps and the middle stair landing shall be not less than 1.50m.

13

电梯、自动扶梯、自动人行道
Elevators, Escalators and Moving Walkways

13.1　一般规定

13.1　General Provisions

13.1.1【09措施】9.1.2条：按建筑使用功能要求和电梯类别、性质、特点合理选用和配置电梯。电梯类别、性质和特点见表13.1.1。

13.1.1 [Measures 2009] 9.1.2: Elevators shall be reasonably selected and conFig.d on the basis of the use function requirements of the building and the category, property and characteristics of elevators. For the categories, properties and characteristics of elevators, see Tab. 13.1.1.

【09措施】表6.9.1.1：电梯类别和性质、特点　　表 13.1.1

[Measures 2009] Tab. 6.9.1.1: Categories, Properties and Characteristics of Elevators　Tab. 13.1.1

类别 Category	名称 Name	性质、特点 Properties and characteristics	简称 Short description
Ⅰ类 Class I	乘客电梯 Passenger elevator	运送乘客的电梯 Elevator for transporting passengers	客梯 Hereinafter referred to as passenger elevator
Ⅱ类 Class II	客货电梯 Passenger-freight elevator	主要为运送乘客、同时也可运送货物的电梯 Elevators mainly used for transporting passengers as well as goods	客货梯 Hereinafter referred to as passenger-freight elevator
Ⅲ类 Class III	病床电梯 Sickbed elevator	运送病床（包括病人）和医疗设备的电梯 Elevators for the transport of sickbeds (including patients) and medical equipment	病床梯 Hereinafter referred to as sickbed elevators
Ⅳ类 Class IV	载货电梯 Freight elevators	运送通常有人伴随的货物的电梯 Elevators for the transport of goods usually accompanied by the people	货梯 Hereinafter referred to as freight elevators
Ⅴ类 Class V	杂物电梯 Sundries elevators	供运送图书、资料、文件、杂物、食品等的提升装置，由于结构形式和尺寸关系，轿厢内不能进人 Lifting devices for the transport of books, materials, documents, sundries, food, etc. People shall not enter the cars because of the structure forms and dimensions)	杂物梯 Hereinafter referred to as sundries elevators

注：1. 本表摘自国家标准《电梯主参数及轿厢、井道、机房的型式与尺寸》GB 7025。该标准等效采用国际标准《电梯的安装》ISO4190。

Notes:1. The table is extracted from the national standard *Lifts - Main Parameters and the Dimensions, Types of Its Cars, Wells and Machine Rooms* GB 7025. In this standard, the international standard *Installation of Elevators* ISO 4190 is adopted by equivalent.

2. I类、III类电梯与II类电梯的主要区别在于轿厢内的装修。
2. The main difference between the Classes I and III elevators and the Class II elevators is the decoration in the cars.
3. 住宅与非住宅用电梯都是乘客电梯，住宅用电梯宜采用II类电梯。
3. Elevators in residential buildings or non-residential buildings shall be passenger elevators, and for the elevators in residential buildings, the Class II elevators shall be adopted.
4. 立体车库中运送汽车的电梯为汽车电梯，此类电梯可并入IV类电梯中。
4. Elevators for the transport of vehicles in the stereo garages are car elevators, and these elevators can be incorporated into Class IV elevators.

13.1.2 电梯的设置及要求（以下均为最低要求，设计时可根据工程具体情况提高标准）：

13.1.2 Arrangement and requirements of elevators (the following criteria are the minimum requirements, and the criteria can be raised as per the specific situations of the project during design):

1.【住设规】6.4.1条：住宅建筑属下列情况之一，必须设置电梯。

1. [Design code of residential buildings] 6.4.1: For residential buildings in any of the following cases, elevators must be arranged.

1）七层及七层以上住宅或住户入口层楼面距室外设计地面的高度超过16m时；

1）Residential buildings of seven storeys or above or whose height between the floors of the entrance floors and the outdoor design ground is greater than 16m;

2）底层作为商店或其他用房的六层及六层以下住宅，其住户入口层楼面距该建筑物的室外设计地面高度超过16m时；

2）When the distance from the entrance floor of the residential buildings to the outdoor design floor of the residential buildings of six storeys or below with the ground floors as shops or in other use purposes exceed 16m;

3）底层做架空层或贮存空间的六层及六层以下住宅，其住户入口层楼面距该建筑物的室外设计地面高度超过16m时；

3）When the distances from the entrance floors of the residential buildings to the outdoor design floors of the residential buildings of six storeys or below with the ground floors as the overhead floors or storage spaces exceed 16m;

4）顶层为两层一套的跃层住宅时，跃层部分不计层数，其顶层住户入口层楼面距该建筑物室外设计地面的高度超过16m时。

4）When top floors are two-storey duplex apartment floors, the duplex parts are not counted into the number of storeys, and the height from the floors of the top entrance floors of the residential buildings to the outdoor design floors of the buildings exceed 16m.

2.【办规】4.1.3条：五层及以上的办公建筑应设电梯。

2. [Design code for office building] 4.1.3: Elevators shall be arranged in office buildings of five storeys or above.

3.【医规】5.1.4.1 条：二层医疗用房宜设电梯；三层及以上的医疗用房应设电梯，且不得少于 2 台。

3. [Code for design of general hospital] 5.1.4.1: Elevators shall be arranged in two-storey medical buildings; medical buildings of three storeys or above shall be equipped with at least two elevators.

4.【旅规】4.1.11.1 条：四级、五级旅馆建筑二层宜设乘客电梯，三层及三层以上应设乘客电梯。一级、二级、三级旅馆建筑三层宜设乘客电梯，四层及四层以上应设乘客电梯。

4. [Code for design of hotel building] 4.1.11.1: Grade IV and Grade V hotel buildings of two storeys are appropriate to be equipped with passenger elevators, and Grade IV and Grade V hotel buildings of three storeys or above shall be set with passenger elevators. Grade I, Grade II and Grade III hotel buildings with three floors are appropriate to be equipped with passenger elevators, and Grade I, Grade II and Grade III hotel buildings with three or more floors shall be set with passenger elevators.

5.【图规】4.1.4 条：图书馆的四层及四层以上设有阅览室时，应设置为读者服务的电梯，并应至少设一台无障碍电梯。

5. [Code for design of library buildings] 4.1.4: When reading rooms are arranged on the fourth floor or above of libraries, elevators shall be set for readers, and at least one barrier-free elevator shall be set.

6.【档规】4.1.4 条：四层及四层以上的对外服务用房、档案业务和技术用房应设电梯。两层或两层以上的档案库应设垂直运输设备。

6. [Code for design of archives buildings] 4.1.4: For external service buildings, and archives business and technical buildings of four storeys or above elevators shall be equipped. Archive storehouses of two storeys or above shall be set with vertical transport equipment.

7.【疗养院规】3.1.2 条：疗养院建筑不宜超过四层，若超过四层应设置电梯。

7. [Code for design of sanatoriums] 3.1.2: Sanatorium buildings are appropriately to be not more than four storeys; otherwise, elevators shall be provided for sanatorium buildings of more than four storeys.

8.【商店建规】4.1.7 条：大型和中型商店的营业区宜设乘客电梯、自动扶梯、自动人行道；多层商店宜设置货梯或提升机。

8. [Code for design of store buildings] 4.1.7: Foe the business areas of large and medium-sized shops, passenger elevators, escalators and moving walkways are appropriately to be arranged; and for multi-storey stores, the freight elevators or hoisters are to be arranged.

9.【老年人照料设施标】5.6.4 条：老年人照料设施中，二层及以上楼层、地下室、半地下室设置老年人用房时应设电梯，电梯应为无障碍电梯，且至少一台能容纳担架。

9. [Standard for design of care facilities for the aged] 5.6.4: In the elderly care facilities, the second floor and above floors, basements, semi-basements to set up elderly rooms should be equipped with elevators, elevators should be barrier-free elevators, and at least one can accommodate stretchers.

10.【宿舍规】4.5.4 条：六层及六层以上宿舍或居室最高入口层楼面距室外设计地面的高度大于 15m 时，宜设置电梯，高度大于 18m 时，应设置电梯；并宜有一部电梯供担架平入。

10. [Code for design of dormitory building] 4.5.4: For a six-storey or above dormitory or residential building, when the distance between the floor of the storey with the highest entrance and the outdoor design ground is greater than 15m, elevators are appropriately to be arranged; and when the height is greater than 18m, elevators shall be arranged; and one elevator are appropriately to be arranged for the access of stretchers in a flat way.

11.【饮食标】4.1.5 条：位于二层及二层以上的餐馆、饮品店和位于三层及三层以上的快餐店宜设置乘客电梯；位于二层及二层以上的大型和特大型食堂宜设置自动扶梯。

11. [Standard for design of dietetic buildings] 4.1.5: Passenger lifts are recommended for restaurants and beverage shops located on the second and above and fast food restaurants located on the third floor and above; escalators are recommended for large and extra-large dining halls located on the second and above.

13.2 电梯

13.2 Elevators

13.2.1 【09 措施】9.2.1 条：电梯的配置及要求。

13.2.1 [Measures 2009] 9.2.1: Configuration of elevators and the requirements.

1. 电梯应尽可能地集中在一个区域设置，以便乘客在同一个地方候梯，从而达到乘客对电梯的均匀化分布。

1. Elevators shall be concentrated in one area as much as possible so that passengers can wait at the same places and the even distribution of passengers for elevators can be achieved.

2. 以电梯为主要垂直交通的每幢建筑物或每个服务区，乘客电梯不应少于 2 台，以备高峰客流或轮流检修的需要。两台宜并排布置，以利故障时互救。

2. For each building or service area with elevators as the main vertical transport means, passenger elevators shall be not less than two sets for the requirements of the passenger flow

peak or inspection and maintenance in turns. The two sets shall be arranged side by side to facilitate mutual rescue in case of failure.

3. 当建筑物的出入口为两层或以上时，可用自动扶梯连接出入口层之间的交通，使始发站集中在一层，从而提高运输效率。

3. When the entrances and exits of buildings have two or more floors, escalators can be used to connect the traffic between the entrance and exit floors so that the starting stations can be concentrated on one floor, and the transport efficiency can be improved.

4. 对服务站或运行速度一致的电梯，应采用并联和群控管理。

4. For service stations or elevators with the same running speed, the parallel connection and group control management shall be adopted.

5. 对于主要需要局部运行的电梯的建筑物，为提高电梯运输能力，宜选择局部实效高的电梯而非一味考虑高额定速度。对高层建筑或超高层建筑，电梯应分层、分区设置。

5. For main buildings requiring for partial operation of elevators, in order to improve the transport capacity of the elevators, the partially efficient elevators shall be appropriately selected, and elevators with high rated speed shall not be considered all the way. For high-rise buildings or super high-rise buildings, elevators shall be in operation for division of floors and areas.

6. 建筑面积巨大，且工作、生活人数很多的超高层建筑，为提高输效率，可配置双层轿厢电梯。

6. For super high-rise buildings with large building area and large number of people working and living, double-deck elevators can be arranged to improve the transport efficiency.

7. 建筑面积较大，且建筑设计标准很高的办公建筑，货梯与客梯的比例大约为1：4的关系，大约建筑面积每20000m² 需配置1台额定载重量1000kg左右的货梯。若货梯数量较多，占核心筒面积太大时，可用较大载重量的货梯。

7. For office buildings with large building area and very high building design standards, the proportion of freight elevators to passenger elevators is about 1：4, and one freight elevator with the rated load of about 1000kg shall be arranged for the building area of 20000m². If the quantity of freight elevators is large, and the area in the elevator passageway is too large, freight elevators with larger loading capacity can be adopted.

13.2.2【09措施】表9.2.2：乘客电梯台数的确定，需要根据不同建筑类型、层数、每层面积、人数、电梯主要技术参数等因素综合考虑。方案设计阶段可参照表13.2.2。

13.2.2 [Measures 2009] Tab. 9.2.2: In determination of the number of passenger elevators, overall consideration shall be made as per factors such as different building types, number of floors, area of each floor, number of people, main technical parameters of the elevator, etc. For project design, see Tab. 13.2.2.

13 电梯、自动扶梯、自动人行道 Elevators, Escalators and Moving Walkways

【09 措施】表 9.2.2：电梯数量、主要技术参数　　　表 13.2.2
[Measures 2009] Tab. 9.2.2: Quantity and Main Technical Parameters of Elevators　Tab. 13.2.2

建筑类别 Building category		标准 Standard 数量 Quantity				额定载重量（kg）和乘客人数（人） Rated loading capacity (kg) and number of passengers (Person)					额定速度（m/s） Rated speed (m/s)
		经济型 Economical type	常用型 Common type	舒适型 Comfortable type	豪华型 Luxury type						
住宅 Residential buildings		90~100 户/台 90-100 households/set	60~90 户/台 60-90 households/set	30~60 户/台 30-60 households/set	<30 户/台 <30 households/set	400	630	1000			0.63, 1.00, 1.60, 2.50
						5	8	13			
旅馆 Hotel		120~140 客房/台 120-140 guest rooms/set	100~120 客房/台 100-120 guest rooms/set	70~100 客房/台 70-100 guest rooms/set	<70 客房/台 <70 guest rooms/set	630	800	1000	1250	1600	
办公 Offices	建筑面积 Building area	6000 m²/台 m²/set	5000 m²/台 m²/set	4000 m²/台 m²/set	<2000m²/台 <2000 m²/set	8	10	13	16	21	
	使用面积 Usable floor area	3000 m²/台 m²/set	2500 m²/台 m²/set	2000 m²/台 m²/set	<2000m²/台 <2000 m²/set						
	按人数 As per the number of people	350 人/台 person/set	300 人/台 person/set	250 人/台 person/set	<250 人/台 <250 people/set						
医院住院部 In-patient department of a hospital		200 床/台 sickbed/set	150 床/台 sickbed/set	100 床/台 sickbed/set	<100 床/台 <100 sickbeds/set	1600	200	2500			
						21	26	33			

注：1. 本表的电梯台数不包括消防和服务电梯。
Notes: 1. The number of fire and service elevators are not included in the number of elevators in the table.
2. 旅馆的工作、服务电梯台数等于 0.3~0.5 倍客梯数。住宅的消防电梯可与客梯合用。
2. The quantity of the working and service elevators in hotel equals 0.3~0.5 times the quantity of passenger elevators. The fire elevator of a residential building and passenger elevator can be shared.
3. 十二层及十二层以上的高层住宅，其电梯数不应少于 2 台。当每层居住 25 人，层数为 24 层以上时，应设 3 台电梯；每层居住 25 人，层数为 35 层以上时，应设 4 台电梯。
3. At least two elevators shall be arranged in the high-rise residential buildings of twelve storeys or above. three elevators shall be arranged on each floor when 25 people live on each floor of the building with 24 storeys or above; four elevators shall be arranged on each floor when 25 people live on each floor of the building with 35 storeys or above.
4. 医院住院部宜增设 1~2 台供医护人员专用的客梯。
4. One or two passenger elevator (s) for medical staff shall be arranged in the in-patient department of a hospital.
5. 超过 3 层的门诊楼应设 2 台以上的乘客电梯。

5. Two or more sets of passenger elevators shall be arranged in the out-patient building with more than three storeys.

6. 办公建筑的有效使用面积为总建筑面积的67%~73%,一般宜取70%。有效使用面积为总建筑面积扣除不能供人居住或办公的面积,如楼梯间、电梯间、公共走道、卫生间、设备间、结构面积等。

6. The effective usable floor area of a office building shall be 67%~73% (generally, 70% shall be taken) of the total building area. The effective usable area shall be the area of the total building area after deduction of the area (such as staircases, elevator hoistways, public corridors, toilets, mechanical rooms, structure area, etc.) which shall not be used for living or office.

7. 办公建筑中的使用人数可按4~10m²/人的使用面积估算。计算办公建筑的建筑面积,应将首层不使用电梯的建筑面积和裙房的建筑面积扣除。

7. The number of users in office buildings can be estimated as per the usable floor area of 4~10m²/person. For calculation of the building area of the office building, the building area without elevator on the first floor and area of the podium shall be deducted.

8. 在各类建筑物中,至少应配置1~2台能使轮椅使用者进出的无障碍电梯。

8. At least 1~2 barrier-free elevator (s) allowing access of wheelchair users shall be arranged in various buildings.

13.2.3 【09措施】9.2.5条:电梯井道底坑深度和顶层高度与额定速度和额定载重量有关,方案设计阶段可参照表13.2.3,具体工程设计时,应按供货厂提供的土建技术条件确定。

13.2.3 [Measures 2009] 9.2.5: The pit depth and top floor height of the elevator shafts are related to the rated speed and rated loading capacity. For project design stage, refer to Tab. 13.2.3. For specific design of building, , the pit depth and top floor height shall be determined as per the technical conditions for civil engineering provided by the supplier.

电梯井道底坑深度和顶层高度　　　　表 13.2.3
Pit Depth and Top Floor Height of Elevator Shafts　　　Tab. 13.2.3

额定速度 (m/s) Rated speed (m/s)	地坑深度 P 顶层盖度 Q (mm) Pit depth P and height Q (mm) of the top floor	乘客电梯额定载重量 Rated load capacity of passenger elevator (kg)					病床电梯额定载重量 Rated load capacities of sickbed elevators (kg)		
		630	800	1000	1200	1600	1600	2000	2500
0.63	P	1400	1400	1400	1600	1600	1600	1600	1800
	Q	3800	3800	3200	4400	4400	4400	4400	4600
1.00	P	1400	1400	1600	1600	1600	1700	1700	1900
	Q	3800	3800	4200	4400	4400	4400	4400	4600
1.60	P	1600	1600	1600	1600	1600	1900	1900	2100
	Q	4000	4000	4200	4400	4400	4400	4400	4600
2.50	P	—	2200	2200	2200	2200	2500	2500	2500
	Q	—	5000	5200	5400	5400	5400	5400	5600
		病床电梯额定载重量 Rated load capacities of sickbed elevators			载重电梯额定载重量 Rated loading capacities of loading elevators					
0.63	P	1600	1600	1800	—	—	—	—	—
	Q	4400	4400	4600	—	—	—	—	—

续表

额定速度 (m/s) Rated speed (m/s)	地坑深度 P 顶层盖度 Q (mm) Pit depth P and height Q (mm) of the top floor	乘客电梯额定载重量 Rated load capacity of passenger elevator (kg)					病床电梯额定载重量 Rated load capacities of sickbed elevators (kg)			
		630	800	1000	1200	1600	1600	2000	2500	
1.00	P	1700	1700	1900	—	—	—	—	1400	1600
	Q	4400	4400	4600	—	—	—	—	4300	4500
1.60	P	1900	1900	2100	1500	1500	1700	1700	—	—
	Q	4400	4400	4600	4100	4100	4100	4100	—	—
2.50	P	2500	2500	2500						
	Q	4400	4400	5600						

注：1. 本表摘自国家标准《电梯主参数及轿厢、井道、机房的型式与尺寸》GB 7025，该标准等效采用《电梯的安装》ISO 4190。
Notes:1. This table is extracted from the national standard *Lifts - Main Parameters and the Dimensions, Types of Its Cars, Wells and Machine Rooms* GB 7025, for which the standard *Installation of Elevators* ISO 4190 is adopted by equivalent.
2. 顶层高度为顶层层站至电梯井道顶板底的垂直距离。
2. The height of top floor refers to the vertical distance from the landing on the top floor to the bottom of the elevator shaft.

13.2.4【住设规】6.4 条：住宅电梯设置。

13.2.4 [Design code of residential buildings] 6.4: Setting of residential elevator.

1.【住设规】6.4.2 条：十二层及十二层以上的住宅，每栋楼设置电梯不应少于两台，其中应设置一台可容纳担架的电梯。

1. [Design code of residential buildings] 6.4.2: At least two elevators shall be arranged in each residential building with twelve storeys or above, and therein one elevator accommodating stretchers shall be arranged.

2.【住设规】6.4.3 条：十二层及十二层以上的住宅每单元只设置一部电梯时，从第十二层起应设置与相邻住宅单元联通的联系廊。联系廊可隔层设置，上下联系廊之间的间隔不应超过五层。联系廊的净宽不应小于 1.10m，局部净高不应低于 2.00m。

2. [Design code of residential buildings] 6.4.3: When one elevator is arranged in each unit of twelve-storey and higher residential buildings, the connection corridor connected to the adjacent residential building unit shall be arranged from Floor 12. One connection corridor can be set every other floor, and the partition between the upper and lower connection corridors shall not exceed five floors. The net width of the connection corridor shall be not less than 1.10m, and the partial net height shall not be less than 2.00m.

3.【住设规】6.4.4 条：十二层及十二层以上的住宅由二个及二个以上的住宅单元组成。且其中有一个或一个以上住宅单元未设置可容纳担架的电梯时，应从第十二层起设置与可容纳担架的电梯联通的联系廊。联系廊可隔层设置，上下联系廊之间的间隔不应超过五层。联系廊的净宽不应小于 1.10m，局部净高不应低于 2.00m。

3. [Design code of residential buildings] 6.4.4: The twelve-storey and higher residential buildings are composed of two or more residential building units. When the elevator accommodating stretchers is not arranged in one or more residential unit（s）, the connection corridor connected to the elevator accommodating stretchers shall be arranged from Floor 12. One connection corridor can be set every other floor, and the partition between the upper and lower connection corridors shall not exceed five floors. The net width of the connection corridor shall be not less than 1.10m, and the partial net height shall not be less than 2.00m.

4.【住设规】6.4.5 条：七层及七层以上住宅电梯应在设有户门和公共走廊的每层设站。住宅电梯宜成组集中布置。

4. [Design code of residential buildings] 6.4.5: The landings of elevators in residential buildings with seven and above storeys shall be set up on each floor with doors and public corridors. The elevators in residential buildings shall be arranged in groups.

5.【住设规】6.4.6 条：候梯厅深度不应小于多台电梯中最大轿箱的深度，且不应小于 1.50m。

5. [Design code of residential buildings] 6.4.6: The depth of elevator waiting halls shall be not less than the depth of the largest car in multiple elevators, and shall be not less than 1.50m.

6.【住设规】6.4.7 条：电梯不应紧邻卧室布置。当受条件限制，电梯不得不紧邻兼起居的卧室布置时，应采取隔声、减震的构造措施。

6. [Design code of residential buildings] 6.4.7: Elevators shall not be set close to bedrooms. When elevators have to be arranged close to a bedrooms due to limited conditions, structural measures for sound insulation and vibration reduction shall be taken.

13.2.5 无障碍电梯的设计要求，详见本书 15.2.7。

13.2.5 Refer to 15.2.7 of this Book for Design requirements for barrier-free elevators.

13.2.6 【医规】5.1.4 条：电梯的设置应符合下列规定。

13.2.6 [Code for design of general hospital] 5.1.4: The arrangement of elevators shall conform to the following provisions.

1. 二层医疗用房宜设电梯；三层及三层以上的医疗用房应设电梯，且不得少于 2 台。

1. Two-storey medical buildings are advisable to be equipped with elevators; medical buildings of three storeys or above shall be equipped with at least 2 elevators.

2. 供患者使用的电梯和污物梯，应采用病床梯。

2. For elevators for patients and elevators for medical wastes, sickbed elevators shall be adopted.

3. 医院住院部宜增设供医护人员专用的客梯、送餐和污物专用货梯。

3. The elevators for medical staff and elevators for medical wastes and meal delivery services shall be arranged in the in-patient department of a hospital.

4. 电梯井道不应与有安静要求的用房贴邻。

4. The elevator shafts shall not be set close to the rooms with requirements for quietness.

13.2.7 【09 措施】9.2.13 条：货梯的设置应满足下列要求。

13.2.7 [Measures 2009] 9.2.13: The arrangement of freight elevators shall meet the following requirements.

1. 载货电梯应靠近货流出入口，以免水平运输距离过长。货流、人流宜分开，尽量减少交叉。前后开门的轿厢只能在工艺流程必需时采用，以减少安全隐患、节省造价。

1. Freight elevators shall be set close to the entrances and exits of goods flow to avoid overlong horizontal transport distance. The goods flow and the people flow shall be separated to minimize the crossing. The car with doors opened in the front and rear can only be adopted when the technological process requires, to reduce hidden safety risks and save construction costs.

2. 地下室仓库和地下停车库应有货梯或客货梯下达，以便充分利用地下空间。仓库设在人防区内时，电梯厅应布置在防毒通道的防护密闭门之外。

2. Freight elevators or passenger-freight elevators shall be able to reach the basement warehouses and underground parking garages for full utilization of the underground space. When the warehouse is arranged in the civil air defence zone, the elevator hall shall be arranged outside the protective closed door for gas defence passages.

3. 必须考虑不同功能、对象的人、货在运送时的隔绝，以符合卫生防疫规定，如生熟食品的隔绝、医疗机械和污净制品的隔绝，以及高级宾馆客人和工作人员、杂物分流等。为节省造价，运生食电梯可与其他货梯兼用。服务用梯宜采用客货梯。

3. The isolation of people and objects of different functions during transport must be considered to conform to the regulations for hygiene and epidemic prevention, such as the isolation of raw and cooked food, isolation of medical equipment and polluted and clean products, isolation of very important guests and staff, and the isolation of sundries. To save the construction costs, the elevator for raw food can be also used as the elevators for other freights. The passenger-freight elevators can be used as service elevators.

4. 货梯的载重量为运送货物的最大重量、随行人员及运输工具和司机的总重量。

4. The loading capacities of freight elevators are the total weight of transported goods with maximum weight, accompanying people, means of transport and drivers.

13.2.8 【09 措施】9.2.14 条：建筑高度超过 75m 和层数为 25 层及以上的高层公共建筑的乘客电梯宜分层（奇数、偶数层）停靠或宜按低区、中区、高区、分区运行。超高层建筑的乘客电梯应分层、分区停靠，见图 13.2.8。

13.2.8 [Measures 2009] 9.2.14: The passenger elevators in high-rise public buildings with the building height above 75m or with 25 or more floors are advised to stop in floors (odd-numbered and even-numbered floors) or run in sections (low-rise sections, medium-rise sections and high-rise sections). The passenger elevators in super high-rise buildings shall stop in floors and sections, See Fig. 13.2.8.

多台电梯宜采用群控，群控不宜超过 4 台。

For multiple elevators, group control shall be adopted, and each group is advised to be of not more than four elevators.

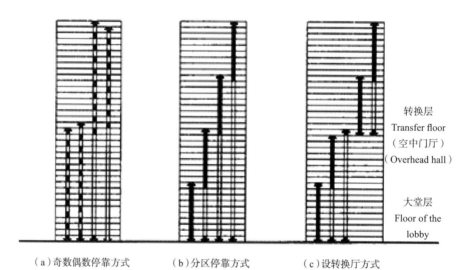

图 13.2.8 【09 措施】图 9.2.14：电梯按楼层分区服务的几种竖向布置
Fig. 13.2.8 [Measures 2009] Fig. 9.2.14: Several Vertical Arrangement Methods of Elevators in Floor Partition Service

1. 10 层以下采用全程服务（即一组电梯在建筑物的每层均开门）。10 层以上可采用分区服务，或在建筑物上部设置转换层以接力方式为上区服务。

1. Full service shall be adopted on the tenth floor and below (i.e., the doors of a group of elevators shall be opened on each floor of the buildings). Partition service can be adopted on the tenth floor and above, or the transfer floor shall be arranged at the upper part of the building to do service for the upper floors in the relay mode.

2. 分区时应考虑以乘客在轿厢内停留的时间为标准，一般采用 1min 较为理想，1.5~2.0min 为极限。

2. For partition service, the time of passengers remaining in the cars shall be used as the

criterion; generally, one minute is the ideal time, and 1.5 ~ 2.0 minutes is the limit time.

3. 分区标准应通过计算确定，一般上区层数应少些，下区层数应多些。

3. The partition criterion shall be determined through calculation; generally, the number of storeys in the upper partition shall be less, and the number of storeys in the lower partition shall be more.

4. 电梯分区宜以建筑高度 50m 或 10 ~ 12 个电梯停站为一个区。

4. For elevator partition, the building height of 50m or 10 ~ 12 elevator landings is advised to be a partition.

1）第 1 个 50m 采用 1.75m/s 的常规速度，然后每隔 50m 升一级，每升一级速度加 1.00 ~ 1.5m/s，即高度 50 ~ 100m 段的梯速用 2.50m/s，100 ~ 150m 段用 3.50m/s，150 ~ 200m 段用 4.50m/s，200 ~ 250m 段用 5.50m/s，以此类推。

1）At the first 50m, the normal speed of 1.75tm/s shall be adopted; and then one level shall be increased every 50m the speed will be increased by 1.00 ~ 1.5m/s per level, that is, the speed of the elevator shall be 2.50m/s for the floor height of the building 50 ~ 100m, 3.50m/s for the floor height of 100 ~ 150m, 4.50m/s for the floor height of the building 150 ~ 200m, 5.50m/s for the floor height of 200 ~ 250m, etc.

2）随着科技的发展，在超高层的办公建筑中，为提高电梯的运输效率，也可将电梯的速度每隔 50m 提高 1.50m/s。

2）With the development of science and technology, in super high-rise office buildings, the elevator speed can be improved by 1.50m/s every 50m to improve the transport efficiency of the elevators.

5. 通往最高层的观光层电梯应是直达的。

5. The sightseeing elevators to the highest floor shall be of the non-stop mode.

13.2.9 【09 措施】9.2.15 条：观光电梯的设置应满足下列要求。

13.2.9 [Measures 2009] 9.2.15: The arrangement of sightseeing elevators shall meet the following requirements.

1. 观光电梯具有垂直运输和观景双重功能，适用于高层旅馆、商业建筑、游乐场等公共建筑；它的井道壁和轿厢壁至少在同一侧透明。

1. Sightseeing elevators shall be provided with the double functions of vertical transport and viewing, and shall be suitable for public buildings of high-rise hotels, commercial buildings, amusement parks, etc.; the shaft wall and the car wall shall be transparent on the same side at least.

2. 观光电梯在建筑物的位置应选择使乘客获得视野广阔、景色优美的方位和景象。根据建筑的平面布局可露明在中庭，或嵌在主要的外墙面部位，或设于独立的玻璃井筒中。

2. For the positions of sightseeing elevators in the building, the places where passengers can have a wide view and enjoy the beautiful scenery shall be selected. The elevators can be arranged in the atriums in an exposed way, inlaid at the main exterior wall positions, or arranged in the independent glass shafts.

3. 观光电梯造型与平面形式多样，具体工程设计按电梯厂提供的技术参数和土建条件确定。

3. The shapes and plane forms of sightseeing elevators are multiple, and the specific engineering design shall be determined as per the technical parameters and civil engineering conditions provided by the elevator manufacturer.

13.2.10 【09措施】9.2.16条：无机房电梯的设置应满足下列要求。

13.2.10 [Measures 2009] 9.2.16: For the arrangement of elevators without machine rooms, the requirements in the following shall be conformed to.

1. 该电梯无须设置专用机房，其特点是将驱动主机安装在井道或轿厢上，控制柜放在维修人员能接近的位置。

1. Elevators do not require a dedicated machine room, and are characterized in that the driving hosts are installed on the shafts or cars, and the control cabinets are placed in a position accessible to the maintenance personnel.

2. 当电梯额定速度为1.0m/s时，最大提升高度为40m，最多楼层为16层；当电梯额定速度为1.60m/s和1.70m/s时，最大提升速度为80m，最多楼层数为24层。

2. When the rated speed of elevators is 1.0m/s, the maximum lifting height is 40m, and the maximum number of floors shall be 16 floors; when the rated speed of the elevators is 1.60m/s and 1.70m/s, the maximum lifting speed is 80m, and the maximum number of floors shall be 24 floors.

3. 多层住宅增设电梯时，宜配置无机房电梯。

3. When an elevator is added in a multi-storey residential building, the elevator without machine room is advised;

4. 无机房电梯的顶层高度根据电梯速度、载重量和轿厢高度确定，一般来说，载重量1t以下的电梯，顶层高度可按4.50m计；1t及以上的电梯，顶层高度可按4.80~5.00m计，施工图设计应以实际选用的电梯为准。

4. The height of the top floors of elevators without machine rooms shall be determined in accordance with the speeds, loading capacities and height of the elevators car, and, in general, for elevators with the loading capacity less than 1t, the height of the top floors can be calculated as 4.50m; for elevators with the loading capacity more than 1t, the height of the top floors can be calculated as 4.80~5.00m, and the construction drawing design shall be based on the elevators

actually used.

5. 无机房电梯主要技术参数参见表 13.2.10。

5. For main technical parameters of elevators without machine rooms, see Tab. 13.2.10.

【09 措施】表 9.2.16：无机房电梯主要技术参数　　　表 13.2.10

[Measures 2009] Tab. 9.2.16: Main Technical Parameters of Elevators Without Machine Rooms　　Tab. 13.2.10

额定载重量（kg） Rated loading capacities（kg）	乘客人数 Number of passengers	额定速度（m/s） Rated speed（m/s）	门宽（mm） Width（mm）	轿厢尺寸（mm） Car dimensions（mm）			井道尺寸（mm） Shaft dimensions（mm）	
				宽度 A Width A	宽度 B Width B	高度 Height	宽度 C Width C	深度 D Depth D
450	6	1.00	800	1100	1150	2200	1800	1650
630	8	1.00	800	1100	1400	2200	1800	1700
		1.65, 1.75		1100	1400	2280	1750	1850
800	10	1.00	800	1350	1400	2200	1900	1800
		1.65, 1.75	800	1400	1350	2280	2000	1850
			900	1350	1400	2280	1950	1900
1000	13	1.00	900	1600	1400	2200	2150	1900
				1100	2100	2200	2000	2400
		1.65, 1.75	900	1600	1400	2280	2200	1950
				1100	2100	2280	1950	2450

注：本表摘自通力电梯有限公司手册。其他电梯厂也有无机房电梯，具体工程设计时，应按供货电梯厂提供的技术参数。
Note: this table is extracted from the manual of KONE Elevators Co., Ltd. Other elevator manufacturers also have elevators without machine rooms, and the specific engineering design shall be completed as per the technical parameters provided by the elevator manufacturers.

13.2.11 【09 措施】9.2.17 条：液压电梯的设置应满足下列要求。

13.2.11 [Measures 2009] 9.2.17: The arrangement of hydraulic elevators shall meet the following requirements.

1. 液压电梯是以液压力传动的垂直运输设备，适用于行程高度小（一般小于等于 40m，货梯速度为 0.5m/s 为 20m）、机房不设在顶部的建筑物。货梯、客梯、住宅梯和病床梯可采用液压电梯。

1. Hydraulic elevators are vertical transport equipment driven by hydraulic pressure and suitable for buildings with small travel distances（generally less than or equal to 40m, and 20m at the freight elevator speed of 0.5m/s）and machine rooms not arranged at the tops. For the freight elevators, passenger elevators, residential elevators and sickbed elevators, hydraulic elevators can be adopted.

2. 电梯的额定载重量为 400 ~ 2000kg，额定速度为 0.10 ~ 1.00m/s（除非有附加要求，否则不应大于 1m/s）。

2. The rated loading capacity of the elevators is 400 ~ 2000kg, and the rated speed is 0.10 to 1.00m/s (not greater than 1m/s, unless there are additional requirements) .

3. 电梯每小时启动运行的次数不应大于 60 次。

3. The starting and operation times of the elevators per hour shall be not greater than 60.

4. 电梯的动力液压油缸应与驱动的轿厢处于同一井道，动力液压油缸可以伸到地下或其他空间。

4. The power hydraulic cylinder of the elevator shall be placed in the same shaft as the driven car, and the power hydraulic cylinder can extend to the underground or other spaces.

5. 电梯的液压站、电控柜及其附属设备必须安装在同一专用房间里，该房间应有独立的门、墙、地面和顶板。与电梯无关的物品不得置于其内。

5. The hydraulic pressure stations, electric control cabinet and accessory equipment of the elevator must be installed in the same special room, in which the independent door, walls, floor and ceiling shall be arranged. Any objects irrelevant to the elevator shall not be placed in it.

6. 机房宜靠近井道，有困难时，可布置在远离井道不大于 8m 的独立机房内。如果机房无法与井道毗连，则用于驱动电梯轿厢的液压管路和电气线路都必须从预埋的管道或专门砌筑的槽穿过。对于不毗邻的机房和轿厢之间应设置永久性的通信设备。

6. The machine room shall be located near the shaft, and if it is difficult to be located in this way, it can be arranged in an independent machine room with the distance to the shaft not greater than 8m. If the machine room can not be adjacent to the shaft, the hydraulic pipelines and electrical circuits for driving the elevator car must penetrate the embedded pipelines or specially built trunkings. Between the non-adjacent machine rooms and the cars, the permanent communication equipment shall be arranged.

7. 电梯机房尺寸不应小于 1900mm × 2100mm × 2000mm（宽 × 深 × 高），底坑深度应不小于 1.2m。

7. The dimensions of the machine room of the elevator shall be not less than 1900mm × 2100mm × 2000mm (width × depth × height) , and the depth of the pit shall be not less than 1.2m.

8. 机房内所安装的设备之间应留有足以操作和维修的人行通道和空间位置。

8. Between the equipment installed in the machine room, the pedestrian passage and space sufficient for operation and maintenance shall be reserved.

9. 标准液压电梯型式与参数范围见表 13.2.11，具体工程设计按电梯厂提供的技术参数和土建条件确定。

13 电梯、自动扶梯、自动人行道 Elevators, Escalators and Moving Walkways

9. For the types and parameter range of standard hydraulic elevators, see Tab. 13.2.11. The specific engineering design shall be determined as per the technical parameters and civil engineering conditions provided by the elevator manufacturer.

【09 措施】表 9.2.17：标准液压电梯型式与参数范围　　表 13.2.11
[Measures 2009] Tab. 9.2.17: Types and Parameter Range of Standard Hydraulic Elevators　　Tab. 13.2.11

序号 No.	型式 Type	额定重量（kg） Rated weight（kg）	额定速度（m/s） Rated speed（m/s）	轿厢最大行程（m） Maximum travels of cars（m）
1	单缸中心直顶式 Single-cylindercenter center-plunger straight-top type	600 ~ 5000	0.1 ~ 0.4	12
2	单缸侧置直顶式 Single-cylinder side-mounted straight-top type	400 ~ 630	0.1 ~ 0.63	7
3	双缸侧置直顶式 Double-cylinder side-mounted straight-top type	2000 ~ 5000	0.1 ~ 0.4	7
4	单缸侧置倍频式 Single-cylinder side-mounted frequency-multiplication type	400 ~ 1000	0.2 ~ 1.0	12
5	双缸侧置倍频式 Double-cylinder side-placed double-frequency type	2000 ~ 5000	0.2 ~ 0.4	12

注：本表摘自《液压电梯》JC 5071。
Note: This table is extracted from *Hydraulic Elevator* JC 5071.

13.3　自动扶梯

13.3　Escalators

13.3.1　【09 措施】9.3.1 条：自动扶梯应布置在建筑物入口处经合理安排的流线上。自动扶梯平面、立面、剖面见图 13.3.1，具体工程设计时应以供货厂家土建技术条件为准。

13.3.1　[Measures 2009] 9.3.1: Escalators shall be arranged on the streamlines reasonably arranged at the entrances of the buildings. For the plan, elevation and section of escalators, see Fig. 13.3.1. The specific design of the project shall be subject to the civil engineering technical conditions of the supplier.

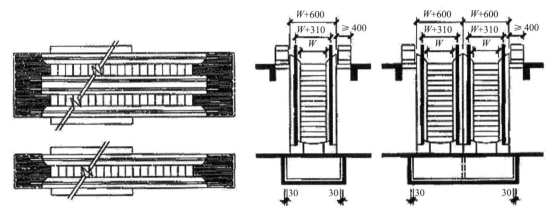

(a) 单台及双台并排平面
(a) Plan of Single Elevator and Double Elevators Side by Side

(b) 单台及双台并排立面
(b) Elevation View of Single Elevator and Double Elevators Side by Side

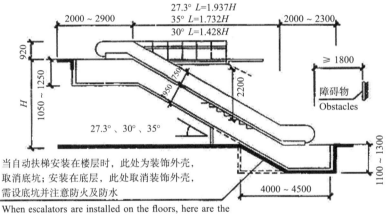

(c) 纵剖面
(c) Longitudinal profile

图 13.3.1【09 措施】图 9.3.1：自动扶梯平面、立面及剖面图
Fig. 13.3.1 [Measures 2009] Fig. 9.3.1: Plan, Elevation and Section Views of Escalators

13.3.2【09 措施】9.3.2 条：自动扶梯宜上下成对布置，宜采用使上行或下行者能连续到达各层，即在各层换梯时，不宜沿梯绕行，以方便使用者，并减少人流拥挤现象。自动扶梯的几种布置形式见图 13.3.2。

13.3.2 [Measures 2009] 9.3.2: Escalator are advised to be arranged up and down in pairs The mode of upward or downward movement is advised to be adopted to reach all the floors in succession, that is, when escalator change is made on all the floors, walking around the escalator shall be avoided to facilitate the users and reduce the phenomenon of crowds. For several arrangement forms of escalators, see Fig. 13.3.2.

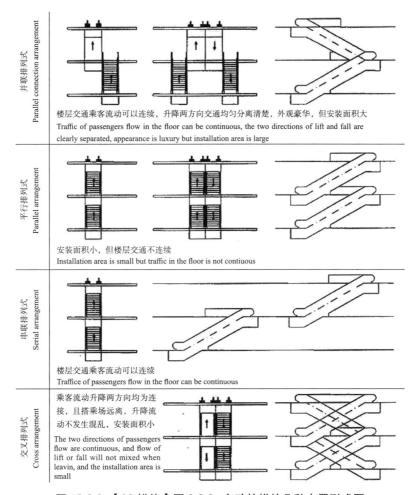

图 13.3.2 【09 措施】图 9.3.2：自动扶梯的几种布置形式图
Fig. 13.3.2 [Measures 2009] Fig. 9.3.2: Drawing of Several Arrangement Forms of Escalators

13.3.3 【通则】6.8.2.7 条：自动扶梯和层间相通的自动人行道单向设置时，应就近布置相匹配的楼梯。

13.3.3 [Code for design of civil buildings] 6.8.2.7: If escalators and moving walkways connecting between floors are arranged unidirectionally, the matching staircases shall be arranged nearby.

13.3.4 【商店建规】4.1.8 条：商店建筑内设置的自动扶梯，倾斜角度不应大于 30°；自动扶梯上下两端水平距离 3m 范围内应保持畅通，不得兼作他用；扶手带中心线与平行墙面或楼板开口边缘间的距离、相邻设置的自动扶梯或自动人行道的两梯（道）之间扶手带中心线的水平距离应大于 0.50m，否则应采取措施，以防对人员造成伤害。

13.3.4 [Code for design of civil buildings] 4.1.8: The inclination angles of escalators in business halls in malls shall be less than or equal to 30°; and the horizontal parts of the upper and lower ends shall conform to the requirement for safety operation length specified by the relevant departments; and the places within the arrange of 3 m shall not be used for other purposes.The horizontal distance between the center line of the handrail belt and the edge of the parallel wall or floor opening, the center line of the handrail strap between the escalator with the adjacent set or the two staircases (sidewalks) of the automatic sidewalk shall be greater than 0.50m, otherwise measures should be taken to prevent harm to personnel.

13.3.5 【09措施】9.3.5条：自动扶梯出入口畅通区的宽度至少等于扶手带中心线之间的距离，且不应小于 2.50m。如该区宽度增至扶手带中心距的两倍以上，则其纵深尺寸允许减少至 2.00m。应将该畅通区看作整个交通系统的组成部分。当畅通区有密集人流穿行时，其宽度应加大。

13.3.5 [Measures 2009] 9.3.5: The width of unobstructed areas at entrances and exits of escalators shall be at least equal to the distances between the center lines of the handrail belts, and shall be not less than 2.50m. If the width of the area is increased to more than twice the center distance of the handrails, the depth is allowed to be reduced to 2.00m. The unobstructed area shall be regarded as part of the entire transport system. When there are dense people flows passing through the unobstructed area, the width shall be increased.

13.3.6 【09措施】9.3.6条：在人员进出相对不大集中的场所宜配置带光电感应装置自动扶梯，或采用VVVF运行的自动扶梯，以节约能源。上述的自动扶梯，重载时全速运行，轻载时低速运行，无人时延时停梯。

13.3.6 [Measures 2009] 9.3.6: In places with the centralized access of moderate personnel flow, escalators with photoelectric sensors or VVVF escalators shall be equipped to save energy. The escalators above shall be able to run at full speed under heavy load, at low speed under light load, and suspend in a delay way under no load.

13.3.7 【09措施】9.3.7条：露天设置的自动扶梯，应选用室外型或半室外型自动扶梯。

13.3.7 [Measures 2009] 9.3.7: For escalators in the open air, the outdoor type or semi-outdoor type escalators shall be selected.

13.3.8 【通则】6.8.2.5条：自动扶梯的梯级、自动人行道的踏板或胶带上空，垂直净高不应小于 2.30m。

13.3.8 [Code for design of civil buildings] 6.8.2.5: The ladder of the escalator, the pedal of the automatic sidewalk or the tape, the vertical net height should not be less than 2.30m.

13.4 自动行人道

13.4 Moving Sidewalk

13.4.1 【09措施】9.4.1条：自动人行道最大倾斜角为小于等于12°，适于大型交通建筑。自动人行道平面、剖面图见图13.4.1。

13.4.1 [Measures 2009]9.4.1: Moving walkways with the maximum inclination angle less than or equal to 12° are suitable for large-scale transport buildings. For the plan and section views of moving walkways, see Fig. 13.4.1.

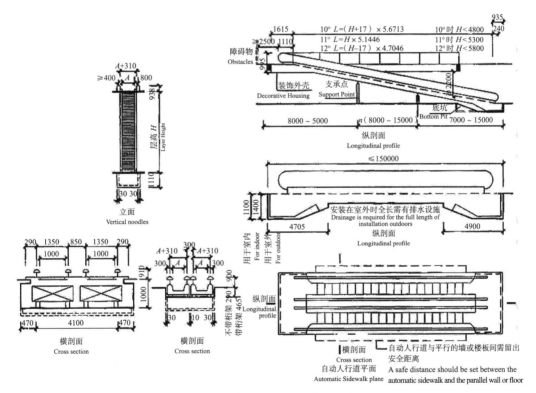

图 13.4.1 【09措施】图 9.4.1：自动人行道平面、剖面

Fig. 13.4.1 [Measures 2009] Fig. 9.4.1: Plan and Sections of Moving Walkways

13.4.2 【09措施】9.4.2条：自动人行道主要技术参数见表13.4.2，设计时应以供货厂土建技术条件为准。

13.4.2 [Measures 2009] 9.4.2: For the main technical parameters of moving walkways, see Tab. 13.4.2. The design shall be subject to the civil engineering technical conditions of suppliers.

【09 措施】表 9.4.2：自动人行道主要技术参数 　　　表 13.4.2
[Measures 2009] Tab. 9.4.2: Main Technical Parameters of Moving Walkways　Tab. 13.4.2

类型 Type	倾斜角 Inclination angle	踏板宽度 A（mm） Tread width A（mm）	额定速度（m/s） Rated speed（m/s）	理论运送能力（人/h） Theoretical transport capacity（person/h）	提升高度（m） Lifting height（m）	电源 Power supply
水平型 Horizontal type	0°~4°	800, 1000, 1200	0.50, 0.65, 0.75, 0.90	9000, 11250, 13500	2.2~6.0	动力三相交流 380V, 50Hz, 功率 3.7~15kW; 照明 220V, 50Hz Three-phase power AC 380 V, 50Hz, power 3.7~15kW; Lighting 220 V, 50 Hz
倾斜型 Inclination type	10°, 11°, 12°	800, 1000		6750, 9000		

注：1. 水平型自动人行道可全天候每天运行 24h，倾斜型室内每日可运行 16h。
Notes:1. Horizontal moving walkways shall be able to run 24 hours a day, and indoor inclination type walkways shall be able to run 16 hours a day.
2. 倾斜型自动人行道的倾斜角不应超过 12°。
2. The inclination angles of inclination-type moving walkways shall be not more than 12°.
3. 扶手带顶面与踏板面或胶带面间的垂直距离为 0.9~1.1m。
3. The vertical distances between the top of handrail belts and the tread surfaces or belt surfaces shall be 0.9~1.1 m.
4. 本表摘自《自动扶梯和自动人行道的制造与安装安全规范》GB 16899。
4. This table is extracted from *Safety code for manufacture and Installation of escalator and moving walkway GB 16899*.

13.4.3【09 措施】9.4.3 条：自动人行道出入口畅通区的宽度至少等于扶手带中心线之间的距离，且不应小于 2.50m。如该区宽度增至扶手带中心距的两倍以上，则其纵深尺寸允许减少至 2.00m。应将该畅通区看作整个交通系统的组成部分。当畅通区有密集人流穿行时，其宽度应加大。

13.4.3 [Measures 2009] 9.4.3: The width of unobstructed area at entrances and exits of moving walkways shall be at least equal to the distances between the center lines of handrail belts, and shall be not less than 2.50m. If the width of the area is increased to more than twice the center distances of the handrails, the depth is allowed to be reduced to 2.00m. The unobstructed area shall be regarded as part of the entire transport system. When there are dense people flows passing through the unobstructed area, the width shall be increased.

13.4.4【09 措施】9.4.4 条：在人员使用不太集中的场所宜采用配置带光电感应系统的自动人行道，或采用 VVVF 运行的自动扶梯，以节约能源。上述的自动人行道，重载时全速运行，轻载时低速运行，无人时延时停梯。

13.4.4 [Measures 2009] 9.4.4: In places with the centralized access of moderate personnel flow, moving walkways with photoelectric sensors or VVVF moving walkways shall be equipped to save energy. The moving walkways above shall be able to run at full speed under heavy load, at low speed under light load, and suspend in a delay way under no load.

13.5 防火设计要点

13.5 Key Points for Fire Protection Design

13.5.1 【防火规】5.5.3 条：自动扶梯和电梯不应计作安全疏散设施。

13.5.1 [Code for fire protection design of buildings] 5.5.3: Escalators and elevators shall not be used as safty evacuation facilities.

13.5.2 【防火规】7.3.1 条：下列建筑应设置消防电梯。

13.5.2 [Code for fire protection design of buildings] 7.3.1:fire elevators shall be arranged in the following buildings.

1. 建筑高度大于 33m 的住宅建筑；

1. Residential building with the building higher than 33m;

2. 一类高层公共建筑和建筑高度大于 32m 的二类高层公共建筑、5 层及以上且总建筑面积大于 3000m² （包括设置在其他建筑内五层及以上楼层）的老年人照料设施；

2. Class I high-rise public buildings and Class II high-rise public buildings with the building height greater than 32m, care facilities for the elderly at level 5 floors and above with a total floor area greater than 3000m² (including those set up on 5 floors and above in other buildings)；

3. 设置消防电梯的建筑的地下或半地下室、埋深大于 10m 且总建筑面积大于 3000m² 的其他地下或半地下建筑（室）。

3. Underground or semi-underground rooms in buildings provided with fire elevators and other underground or semi-underground (building) rooms with the burial depth greater than 10 m and the total building area greater than 3000m².

13.5.3 【防火规】7.3.2 条：消防电梯应分别设置在不同防火分区内，且每个防火分区不应少于 1 台。

13.5.3 [Code for fire protection design of buildings] 7.3.2: Fire elevators shall be arranged in different fire compartments respectively, and at least one set in each fire compartment.

13.5.4 【防火规】7.3.5 条：除设置在仓库连廊、冷库穿堂或谷物筒仓工作塔内的消防电梯外，消防电梯应设置前室，并应符合下列规定：

13.5.4 [Code for fire protection design of buildings] 7.3.5: Except for fire elevators arranged in warehouse sidewalks, refrigerated warehouse lobbies or grain silo work towers, anterooms shall be arranged for fire elevators, and the following provisions shall be met:

1. 前室宜靠外墙设置，并应在首层直通室外或经过长度不大于 30m 的通道通向室外。

1. Anterooms shall be arranged near the exterior walls, and shall directly lead to outdoors on the first floor or lead to outdoors through the passages with the length not greater than 30m.

2. 前室的使用面积不应小于 6.0m²；前室的短边不应小于 2.4m²；与防烟楼梯间合用的前室，应符合本书 12.3.6 条和 12.3.3 条的规定。

2. The usable floor area of the anteroom shall not be less than 6.0 m²; the short side of the anteroom should not be less than 2.4m²; the anteroom shared with the smoke-proof staircase shall conform to the provisions in clauses 12.3.6 and 12.3.3 of this book.

3. 除前室的出入口、前室内设置的正压送风口和《建筑设计防火规范》（2018 年版）GB 50016—2014 5.5.27 条（见本书表 12.3.5-2）规定的户门外，前室内不应开设其他门、窗、洞口。

3. Except for entrances and exits of anterooms, positive-pressure air supply inlets set in the anteroom, and doors specified in Clause 5.5.27 of *Code for Fire Protection Design of Buildings* GB 5001—2014（refer to Tab. 12.3.5-2 of this book）, any other doors, windows and openings shall not be arranged in the anteroom.

4. 前室或合用前室的门应采用乙级防火门，不应设置卷帘。

4. For the doors of the anteroom or shared anteroom, Grade B fire doors shall be adopted, and roller shutters shall not be arranged.

13.5.5 【防火规】7.3.6 条：消防电梯井、机房与相邻电梯井、机房之间应设置耐火极限不低于 2.00h 的防火隔墙，隔墙上的门应采用甲级防火门。

13.5.5 [Code for fire protection design of buildings] 7.3.6: The fireproof partition walls with fire resistance not less than 2.00 h shall be arranged between the fire elevator shaft and machine room and the adjacent elevator shafts and machine rooms; and Grade A fire doors shall be adopted for the doors in the partition walls.

13.5.6 【防火规】7.3.7 条：消防电梯的井底应设置排水设施，排水井的容量不应小于 2m³，排水泵的排水量不应小于 10L/s。消防电梯间前室的门口宜设置挡水设施。

13.5.6 [Code for fire protection design of buildings] 7.3.7: The drainage facilities shall be arranged at the bottom of the fire elevator shaft; the capacity of the drainage wells shall be not less than 2m³; and the water displacement of the drainage well shall be not less than 10 L/s.

Water retaining facilities shall be advised to be arranged at the doors of the anteroom of the fire elevator room.

13.5.7 【防火规】7.3.8 条：消防电梯应符合下列规定。

13.5.7 [Code for fire protection design of buildings] 7.3.8: Fire elevators shall conform to the following provisions.

1. 应能每层停靠；

1. Fire elevators can be stopped at each floor;

2. 电梯的载重量不应小于 800kg；

2. The load capacity of the elevator shall be not less than 800kg;

3. 电梯从首层至顶层的运行时间不宜大于 60s；

3. The running time of the elevator from the first floor to the top floor shall be advised to be not greater than 60 seconds;

4. 电梯的动力与控制电缆、电线、控制面板应采取防水措施；

4. Waterproof measures shall be adopted for the power supply, control cables, wires and control panel of the elevator;

5. 在首层的消防电梯入口处应设置供消防队员专用的操作按钮；

5. The operation button dedicated to firefighters shall be arranged at the entrance of the fire elevator on the first floor;

6. 电梯轿厢的内部装修应采用不燃材料；

6. The incombustible materials shall be used in the interior decoration of the elevator car;

7. 电梯轿厢内部应设置专用消防对讲电话。

7. The dedicated fire intercom shall be arranged in the elevator car.

13.5.8 【防火规】6.2.9 条：建筑内的电梯井等竖井应符合下列规定。

13.5.8 [Code for fire protection design of buildings] 6.2.9: Shafts such as elevator shafts in buildings shall conform to the following provisions.

1. 电梯井应独立设置，井内严禁敷设可燃气体和甲、乙、丙类液体管道，不应敷设与电梯无关的电缆、电线等。电梯井的井壁除设置电梯门、安全逃生门和通气孔洞外，不应设置其他开口。

1. The elevator shafts shall be independently arranged.It is prohibited to lay pipelines for combustible gases and liquids of Classes A, B and C. Cables, wires and others, which are irrelevant with the elevator, shall not be laid. On the walls of the elevator shaft, except that elevator doors, safety escape doors and ventilation holes are arranged, other openings shall not be arranged.

2. 电缆井、管道井、排烟道、排气道、垃圾道等竖向井道，应分别独立设置。井壁的耐火极限不应低于1.00h，井壁上的检查门应采用丙级防火门。

2. Vertical wells, such as cable wells, pipeline shafts, smoke exhaust passage, air exhaust passage, refuse chutes, shall be independently arranged. The fire endurance of shaft walls shall not be less than 1.00h , and for the inspection doors on the shaft walls, Grade C fire doors shall be adopted.

3. 建筑内的电缆井、管道井应在每层楼板处采用不低于楼板耐火极限的不燃材料或防火封堵材料封堵。建筑内的电缆井、管道井与房间、走道等相连通的孔隙应采用防火封堵材料封堵。

3. Cable and pipeline shafts in the building shall be sealed at each floor slab with incombustible materials or fireproof blocking materials with the fire endurance not less than that for floor slabs at the floor slabs on each floor. The pores of cable wells, pipeline shafts and rooms, sidewalks and so on in the building shall be blocked by fireproof blocking materials.

4. 建筑内的垃圾道宜靠外墙设置，垃圾道的排气口应直接开向室外，垃圾斗应采用不燃材料制作，并应能自行关闭。

4. The refuse chutes in buildings shall be advised to be arranged close to exterior walls; the exhaust vents of the refuse chutes shall be opened toward the outside directly. The refuse buckets shall be made of incombustible materials and can be automatically closed.

5. 电梯层门的耐火极限不应低于1.00h，并应符合现行国家标准《电梯层门耐火试验完整性、隔热性和热通量测定法》GB /T 27903 规定的完整性和隔热性要求。

5. The fire endurance of the elevator landing doors shall be not less than 1.00 hour and shall conform to the integrity and heat shielding requirements specified in the current national standard *Fire Resistance Test for Lift Landing Doors - Methods of Measuring Integrity, Thermal Insulation and Heat Flux* GB /T 27903.

13.5.9 【住设规】9.4.4 条：当住宅建筑中的楼梯、电梯直通住宅楼层下部的汽车库时，楼梯、电梯在汽车库出入口部位应采取防火分隔措施。

13.5.9 [Design code of residential buildings] 9.4.4: When staircases and elevators in residential buildings are directly connected to the garages under the residential buildings, for staircases and elevators, fire separation measures shall be taken at the entrances and exits of the garages.

13.5.10 【09措施】9.5.9 条：自动扶梯和开敞式楼梯一样，上下层应视为一个防火分区，应符合防火规范所规定的有关防火分区等要求。若分属两区时应有防火卷帘等隔绝措施，如图13.5.10所示。

13.5.10 [Measures 2009] 9.5.9: For escalators, it shall be the same to consider the upper and lower floors as one fire compartment as for open type staircases, and the requirements for fire compartments, etc. as specified in the fire protection specifications shall be conformed to. If the upper and lower floors are divided into two compartments, isolation measures such as fireproof roller shutters, etc., shall be adopted, see Fig. 13.5.10.

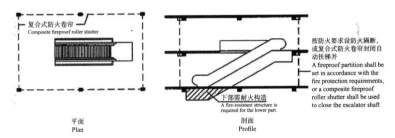

图 13.5.10 【09 措施】图 9.5.9：自动扶梯防火分隔
Fig. 13.5.10　[Measures 2009] Fig. 9.5.9: Fire Separation for Escalators

13.5.11　【09 措施】9.5.10 条：设置自动扶梯的开敞空间应按防火规范要求加强防火措施。机房、楼板底和机械传动部分除留设检修孔和通风口外，均应以不燃烧材料包覆。

13.5.11　[Measures 2009] 9.5.10: For open spaces with escalators, the fire prevention measures shall be strengthened in accordance with the requirements in the code for fire protection. The machine rooms, floor bottoms and mechanical transmission parts shall be covered with noncombustible materials except vents and holes for inspection and maintenance.

13.6　局部布置及构造

13.6　Local Arrangement and Construction

13.6.1　【09 措施】9.6.1 条：电梯井道、底坑和顶板应坚固，选用具有足够强度和不产生粉尘的材料，耐火极限不应低于 1.00h 的不燃烧体。井道厚度，钢筋混凝土墙不应小于 200mm，或承重砌体墙时不应小于 240mm，或根据结构计算确定。当井道采用砌体墙时，应设框架柱和水平圈梁与框架梁，以满足固定轿厢和配重导轨之用。水平圈梁宜设在各层预留门洞上方，高度不宜小于 350mm，垂直中距宜为 2.5m 左右。框架梁高不宜小于 500mm。

13.6.1　[Measures 2009] 9.6.1: The shaft, pit and ceiling of an elevator shall be firm, and the material selected shall be with enough strength and do not generate dust and the incombustible material shall be with the fire endurance not less than 1.00 hour. For the thickness of elevator shafts, the reinforced concrete walls shall be not less than 200mm, and the masonry load-bearing walls shall be not less than 240mm, or it shall be determined on the basis of structural calculation. When the masonry walls are adopted for shafts, the frame columns and horizontal ring beams and frame beams shall be arranged for fixation of the car and counterweight rail. The horizontal ring beam shall be arranged above the reserved door opening on each floor, the height shall be advised to be not less than 350mm, and the vertical center-to-center distance shall be about 2.5m. The height of the frame beam shall be advised to be not less than 500mm.

13.6.2 【09措施】9.6.3条：电梯井道不宜设置在能够到达的空间上部。如确有人们能到达的空间存在，底坑地面最小应按支承5000Pa荷载设计，或将对重缓冲器安装在一直延伸到坚固地面上的实心柱墩上或由厂家附加对重安全钳。上述做法应得到电梯供货厂的书面文件确认其安全。

13.6.2 [Measures 2009] 9.6.3: Elevator shafts shall not be arranged above the accessible spaces. If the accessible space exists, the pit ground shall be designed to bear at least 5000Pa load, or the counterweight buffer shall be installed on the solid column pier extending to the solid ground, or the counterweight safety tongs shall be added by the manufacturer. The above construction method shall be confirmed by written documents from the elevator supplier.

13.6.3 【09措施】9.6.4条：电梯井道除层门开口、通风孔、排烟口、安装门、检修门和检修人孔外，不得有其他与电梯无关的开口。

13.6.3 [Measures 2009] 9.6.4: No other openings irrelevant to elevators on elevator shafts shall be arranged except for landing door openings, ventilation holes, smoke exhaust ports, installation doors, doors for inspection and maintenance and inspection holes.

13.6.4 【09措施】9.6.6条：当相邻两层门地坎间距离超过11m时，其间应设安全门，其高度不得小于1.8m，宽度不得小于0.35m。安全门和检修门应具有和层门一样的机械强度和耐久性能，且均不得向井道里开启，门本身应是无孔的。

13.6.4 [Measures 2009] 9.6.6: When the distance between the sills of the two adjacent landing doors exceeds 11m, safety doors shall be arranged between the sills. The height of the door shall not be less than 1.8m, and the width shall not be less than 0.35m. The safety doors and access doors for inspection and maintenance shall be provided with the same mechanical strength and durability as the landing doors, which shall not be opened toward the shafts. No holes shall be drilled in the doors.

13.6.5 【09措施】9.6.7条：高速直流乘客电梯的井道上部应做隔声层，隔声层应做800mm×800mm的进出口。

13.6.5 [Measures 2009] 9.6.7: A sound insulation layer shall be constructed at the upper part of the shaft of high-speed passenger elevators, and a 800mm×800mm outlet shall be made in the insulation layer.

13.6.6 【09措施】9.6.8条：多台并列成排电梯井道内部尺寸应符合下列规定。

13.6.6 [Measures 2009] 9.6.8: The internal size of multiple parallel elevators shall conform to the following provisions.

1. 共用井道总宽度：单梯井道宽度之和＋单梯井道之间的分界宽度之和。每个分界宽度最小按100～200mm计。当两轿厢相对一面设有安全门时，位于该两台电梯之间的井道壁不应为实体墙，应设钢或钢筋混凝土梁，分界宽度大于等于100mm。

1. Total width of the shared shafts: sum of the width of the single elevator shafts and sum of the division width between single elevator shafts. Each division width is 100～200mm at least. When safety doors are arranged on the opposite sides of two cars, the shaft walls between the two elevators shall not be solid walls, but steel or reinforced concrete beams shall be arranged, and the division width shall be greater than or equal to 100mm.

2. 共用井道各组成部分深度与这些电梯单独安装时井道的深度相同。

2. The depth of the parts of the shared shafts shall be the same as the shaft depth when these elevators are separately installed.

3. 底坑深度按群梯中速度最快的电梯确定。

3. The depth of the pit shall be determined as per the fastest elevator in the elevator group.

4. 顶层高度按群梯中速度最快的电梯确定。

4. The height of the top floor shall be determined as per the fastest elevator in the elevator group.

5. 多台电梯中，电梯厅门间的墙宜为填充墙，不宜为钢筋混凝土抗震墙。

5. For multiple elevators, the walls between elevator hall doors shall be advised to be infill walls and not reinforced concrete earthquake resistant walls.

13.6.7 【09措施】9.6.9条：多台并列成排电梯共用机房内部尺寸应符合下列规定。

13.6.7 [Measures 2009] 9.6.9: The internal sizes of shared machine rooms of multiple parallel elevators shall conform to the following provisions.

1. 多台电梯共用机房的最小宽度，应等于共用井道的总宽度加上最大的1台电梯单独安装时所侧向延伸长度之和。

1. The minimum width of a shared machine room for multiple elevators shall be equal to the sum of the total width of the shared shaft and the lateral extending length required during the separate installation of the largest elevator.

2. 多台电梯共用机房的最大深度，应等于电梯单独安装所需最深井道加上2100mm。

2. The maximum depth of a shared machine room for multiple elevators shall be equal to the sum of the deepest shaft required for separate installation of the elevators plus 2100mm.

3. 多台电梯共用机房最小高度，应等于其中最高机房的高度。

3. The minimum height of a shared machine room for multiple elevators shall be equal to the height of the highest machine room.

13.6.8 【09措施】9.6.10条：机房的剖面位置和工作环境。

13.6.8 [Measures 2009] 9.6.10: The sectional locations and working environment of machine rooms.

1. 机房的剖面位置：

1. Sectional positions of machine rooms：

1）乘客电梯、住宅电梯、病床电梯、载货电梯的机房位于顶站上部；

1）Machine rooms of passenger elevators, residential elevators, sickbed elevators and freight elevators shall be located on the top of the highest station;

2）杂物电梯的机房位于顶站上部或位于本层；

2）The machine room of the sundries elevator shall be located on the highest floor or on the top of the highest station;

3）液压电梯的机房位于底层或地下。

3）The machine room of the hydraulic elevator shall be located on the ground floor or underground.

2. 机房的工作环境：

2. Working conditions of the machine room：

1）机房应为专用的房间，围护结构应保温隔热，室内应有良好通风、防尘，宜有自然采光。环境温度应保持在5~40℃之间，相对湿度不大于85%。

1）The machine room shall be dedicated, and the enclosure structures shall be provided with the insulation and heat shielding devices; good ventilation and dustproofing shall be provided indoor, and natural lighting shall be adopted. The ambient temperature shall be kept between 5 and 40℃, and the relative humidity shall be not greater than 85%.

2）介质中无爆炸危险、无足以腐蚀金属和破坏绝缘的气体及导电尘埃。

2）In the medium, there shall be no explosive hazard and no gas and conductive dust able to corrode metals and damage insulation.

3）供电电压波动在±7%范围以内。

3）The voltage fluctuation of power supply shall be within the range of ±7%.

13.6.9 【09措施】9.6.11条：通向机房的通道、楼梯和门的宽度不应小于1200mm，门的高度不应小于2000mm。楼梯的坡度小于等于45°。上电梯机房应通过楼梯到达，也可经过一段屋顶到达，但不应经过垂直爬梯。机房门的位置还应考虑电梯更新时机组吊装与进出方便。

13.6.9 [Measures 2009] 9.6.11: The width of passages, staircases and doors to machine rooms shall be not less than 1200mm, and the height of the doors shall be not less than 2000 mm. The gradients of the staircases shall be less than or equal to 45°. The machine room of elevators shall be arrived through the staircases or a section of roof, rather than the vertical ladders. Convenient lifting and entrance and exit of the unit during the replacement of the elevator shall be considered for the determination of the door position of the machine room.

13.6.10 【09措施】9.6.12条：机房地面应平整、坚固、防滑和不起尘。机房地面允许有不同高度，当高差大于0.5m时，应设防护栏杆和钢梯。

13.6.10 [Measures 2009] 9.6.12: The ground of machine room shall be flat, solid and anti-skid, and dust emission is not allowed. Different heights are allowed for the floor of the machine room, and when the height difference is greater than 0.5m, protective railings and steel ladders shall be arranged.

13.6.11 【09措施】9.6.13条：机房顶板上部不宜设置水箱，如不得不设置时，不得利用机房顶板作为水箱底板，且水箱间地面应有可靠的防水措施。也不应在机房内直接穿越水管和蒸汽管。

13.6.11 [Measures 2009] 9.6.13: Water tanks shall not be arranged on the top boards of machine room. If water tanks have to be arranged, the top boards of machine room shall not be used as the bottom boards of water tanks and reliable waterproof measures shall be taken for the floors of water tank rooms. And the water pipes and steam pipes shall not pass through the machine room directly.

13.6.12 【09措施】9.6.14条：机房可向井道两个相邻侧面延伸，液压电梯机房宜靠近井道。

13.6.12 [Measures 2009] 9.6.14: The machine room can extend to two adjacent sides of the shaft, and the machine room for hydraulic elevators shall be arranged near the shaft.

13.6.13 【09措施】9.6.15条：机房顶部应设起吊钢梁或吊钩，其中心位置宜与电梯井纵横轴的交点对中。吊钩承受的荷载对于额定载重量3000kg以下的电梯不应小于2000kg；对于额定载重量大于3000kg电梯，应不少于3000kg。或根据生产厂的要求。

13.6.13 [Measures 2009] 9.6.15: The lifting steel beams or hooks shall be arranged on the top of the machine room, the central position of which shall match the center of the vertical and cross axles of the elevator shaft. The hook load shall be not less than 2000kg for the elevators with the rated loading capacity less than 3000kg; and not less than 3000kg for the elevators with the rated loading capacity greater than 3000kg. Or the requirements of the manufacturers shall be followed.

13.6.14 【09措施】9.6.16条：设置曳引机承重梁和有关预埋铁件，必须埋入承重墙内或直接传力至承重梁的支墩上。承重梁的支撑长度应超过墙中心20mm且不应少于

75mm。

13.6.14 [Measures 2009] 9.6.16: The bearing beams of the traction machine and relevant embedded iron components must be embedded into the bearing walls or directly transmitted to the supporting piers of the bearing beams. The support length of the bearing beams shall be 20mm greater than the centers of the walls and shall be not less than 75mm.

13.6.15【09措施】9.6.17条：相邻两层站间的距离，当层门入口高度为2000mm时，应不小于2450mm；层门入口高度为2100mm时，应不小于2550mm。

13.6.15 [Measures 2009] 9.6.17: When the height of entrances of landing doors is 2000mm, the distance between two adjacent stations shall be not less than 2450mm; when the height of entrances of landing doors is 2100mm, the distance shall be not less than 2550mm.

13.6.16【09措施】9.6.18条：层门尺寸指门套装修后的净尺寸，土建层门的洞口尺寸应大于层门尺寸，留出装修的余量，一般宽度为层门两边各加100mm，高度为层门加70~100mm。

13.6.16 [Measures 2009] 9.6.18: The sizes of landing doors refer to the net sizes of door pockets after decoration, and the dimensions of door openings in civil engineering shall be greater than the sizes of the landing doors. The allowance for decoration shall be reserved, and generally, the width shall be the width of landing doors plus 100mm on both sides respectively and the height shall be landing doors plus 70~100mm.

13.6.17【09措施】9.6.20条：底坑深度超过900mm时，需根据要求设置固定金属梯或金属爬梯。金属梯或金属爬梯不得凸入电梯运行空间，且不应影响电梯运行部件的运行。当生产厂自带该梯时，设计不必考虑。

13.6.17 [Measures 2009] 9.6.20: When the pit depth exceeds 900mm, a fixed metal ladder or metal climbing ladder shall be arranged in accordance with the requirements. The metal ladder or metal climbing ladder shall not extrude into the running space of the elevator, and shall not impact the operation of the elevator running components. When the ladder is provided by the manufacturer, this can be ignored in design.

13.6.18【09措施】9.6.21条：底坑深度超过2500mm时，应设带锁的检修门，检修门高度大于1400mm，宽度大于600mm，检修门不得向井道内开启。

13.6.18 [Measures 2009] 9.6.21: When the pit depth exceeds 2500mm, the access door with locks shall be arranged, the height of the access door shall be greater than 1400mm, the width shall be greater than 600mm, and the access door shall not be opened toward the inside of the shaft.

13.6.19【09措施】9.6.22条：同一井道安装有多台电梯时，相邻电梯井道之间可为钢筋混凝土隔墙或钢梁（每层设置），用以安装导轨支架，墙厚200mm，梁的宽度为100mm。在井道下部不同的电梯运行部件之间应设置护栏，高度为底坑底面以上2.5m。

13.6.19 [Measures 2009] 9.6.22: When several elevators are installed in the same shaft, reinforced concrete partition walls or steel beams can be arranged between the adjacent elevator shafts (arranged on each floor) , to facilitate the installation of the track brackets; the thickness of the wall shall be 200mm; and the width of the beam shall be 100mm. Between the elevator running components of the elevators in the lower part of the shaft, railings shall be arranged, with the height of 2.5m above the pit bottom.

13.6.20 【09措施】9.6.23条：电梯详图中应按电梯生产厂要求，在井道和机房详图中表示导轨预埋件、厅门牛腿、厅门门套、机房工字钢梁（或混凝土梁）和顶部检修吊钩的位置、规格等，层数指示灯及按钮留洞位置。为电梯检修，必须满足吊钩底的净空高度要求，当不能满足时，可通过增加层高或吊钩梁为反梁解决。

13.6.20 [Measures 2009] 9.6.23: In the detailed drawings of elevators, the positions, specifications, etc. of the rail embedded components, corbels and door pockets of the hall doors, steel I-shaped beams (or concrete beams) in the machine rooms, and maintenance hooks at the top, as well as the reserved hole positions for floor number indicator lights and buttons, shall be shown in the detailed drawings of the shafts and machine rooms as per the requirements of the elevator manufacturers. For elevator inspection and maintenance, the net height requirements for the bottom of the hook shall be met, and when the requirements can not be met, the floor height is advised to be raised, or the hoisting beam can be changed to be the reversed beam.

13.6.21 【09措施】9.6.24条：自动扶梯和自动人行道起止平行墙面深度除满足设备安装尺寸外，应根据梯长和使用场所的人流留有足够的等候及缓冲面积；当畅通区宽度至少等于扶手带中心线之间距离时，扶手带转向端距前面障碍物应大于等于2.5m；当该区宽度增至扶手带中心距2倍以上时，其纵深尺寸允许减至2.0m。

13.6.21 [Measures 2009] 9.6.24: The depth of the parallel starting and ending walls of escalators and moving walkways shall meet the installation dimensions of equipment, and sufficient waiting and buffer areas shall be reserved in accordance with the length of the escalators and the people flow in the use places; when the width of unobstructed areas is at least equal to the distance between the center lines of the handrail belts, the distances between the steering ends of the handrail belts and the obstacles in front shall be greater than or equal to 2.5m; when the width of the areas is increased to more than twice the center distance of the handrails, the sizes in depth are allowed to be reduced to 2.0m.

13.6.22 【09措施】9.6.25条：自动扶梯和自动人行道与平行墙面间、扶手与楼板开口边缘及相邻平行梯的扶手带的水平距离不应小于0.5m。当既有建筑不能满足上述距离时，特别是在楼板交叉处及各交叉设置的自动扶梯或自动人行道之间，应采取措施防止障碍物引起人员伤害，可在外盖板上方设置一个无锐利边缘的垂直防碰挡板，其高度不应小于 0.3m，例如一个无孔三角板。

13.6.22 [Measures 2009] 9.6.25: The horizontal distances between escalators and moving walkways and parallel walls, and between the handrails and the opening edge of the floor and the handrails of adjacent parallel escalators shall be not less than 0.5m. When the above distance can not be met in existing buildings, especially between crossings of floor slabs and escalators or moving walkways crosswise set up, measures shall be taken to prevent obstacles from causing personal injury. A vertical anti-collision baffle without sharp edge can be arranged above the outer cover plates, and the height shall be not less than 0.3m, such as a triangular plate without holes.

13.6.23 【09 措施】9.6.27 条：倾斜式自动人行道距楼板开洞处净高应大于等于 2.0m。出口处扶手带转向端距前面障碍物水平距离大于等于 2.5m。

13.6.23 [Measures 2009] 9.6.27: The net height between inclination-type moving walkways and openings of floor slabs shall be greater than or equal to 2.0m. The horizontal distance between the turning ends of handrail belts at exits and obstacles in front shall be greater than or equal to 2.5m.

13.6.24 【09 措施】9.6.28 条：自动扶梯扶手带外缘与墙壁或其他障碍物之间的水平距离不得小于 80mm。相互邻近平行或交错设置的自动扶梯，扶手带的外缘间的距离不得小于 120mm。

13.6.24 [Measures 2009] 9.6.28: The horizontal distances between the outer edges of escalators and walls or other obstacles shall be not less than 80mm. For escalators arranged in the mutually parellel or staggered way, the distances between the outer edges of the handrails shall be not less than 120mm.

13.6.25 【09 措施】9.6.29 条：自动人行道地沟排水应符合下列规定。

13.6.25 [Measures 2009] 9.6.29: The drainage of trenches for moving walkways shall conform to the following provisions.

1. 室内自动人行道按有无集水可能而设置。

1. For indoor moving walkways, such arrangement shall be made as per the possibility of collecting water.

2. 室外自动扶梯无论全露天或在雨篷下，其地沟均需全长设置下水排放系统。

2. For any outdoor escalator, whether it is completely in the open air or under a canopy, such arrangement shall be provided with a water discharge system for the full length of the trenches.

13.6.26 【09 措施】9.6.30 条：自动扶梯或自动人行道在露天运行时，宜加顶棚和围护。

13.6.26 [Measures 2009] 9.6.30: When escalators or moving walkways are running in the open air, it is advisable to add a ceiling and enclosure.

14

门窗
Doors and Windows

14.1 一般规定

14.1 General Provisions

14.1.1 【09措施】10.1.1条：建筑外门、外窗物理性能应根据其所在地区的气候、周围环境以及住宅建筑的高度、体形系数等因素进行确定，并符合设计要求。

14.1.1 [Measures 2009] 10.1.1: The physical properties of exterior doors and exterior windows of buildings shall be determined in accordance with the climatic factors of the areas where the buildings are located, surroundings, height and shape coefficients of the residential buildings, etc., and shall conform to the design requirements.

14.1.2 【09措施】10.1.2条：建筑外门、外窗的立面形式、构造节点以及材料，应视住宅建筑中客厅、卧室、起居室、厨房、卫生间的不同使用功能进行设计，力求美观、安全、易于清洁和使用方便。

14.1.2 [Measures 2009] 10.1.2: The facade forms, structure nodes and materials of the exterior doors and exterior windows of buildings shall be designed on the basis of the different functions of guest rooms, bedrooms, living rooms, kitchens and toilets in residential buildings, and aesthetics, safety, easy cleaning and convenient use shall be strived for.

14.1.3 【09措施】10.1.3条：建筑外窗（包括阳台门）的保温性能应符合建筑节能设计标准。

14.1.3 [Measures 2009] 10.1.3: Thermal insulation performance of exterior windows (including balcony doors) of a building shall conform to the energy-saving design standard for buildings.

14.1.4 【09措施】10.1.4条：在快速路、主干路、次干路和支路道路红线两侧50m范围内，新建住宅建筑临街一侧应设计、采用具有隔声性能的建筑外窗（包括阳台门）。

14.1.4 [Measures 2009] 10.1.4: In the range of 50m to both sides of the boundary lines of the express ways, trunk roads, secondary trunk roads and branch roads, the external windows of buildings with the performance of sound insulation (including balcony doors) shall be designed and adopted on the street-facing side of the new residential building.

14.1.5 【09措施】10.1.5条：建筑外窗上宜设置可以调节的换气装置。

14.1.5 [Measures 2009] 10.1.5: On the exterior windows in buildings, adjustable ventilation devices shall be advised to be arranged.

14.1.6 【09措施】10.1.6条：高层塔式住宅建筑和主体朝向为东西向住宅建筑主要居住空间的东、西向建筑外窗，宜设置活动外遮阳设施。

14.1.6 [Measures 2009] 10.1.6: For the east and west exterior windows of the main living spaces of the high-rise residential tower building and east-west oriented residential building, movable external sun-shading facilities shall be advised to be arranged.

14.1.7 【09措施】10.1.7条：面临走廊或凹口的建筑外窗应避免视线干扰，采取遮挡措施，朝向走廊开启的建筑外窗不应妨碍交通。

14.1.7 [Measures 2009] 10.1.7: For exterior windows facing corridors or recesses in the building, interference of sightline shall be avoided; shielding measures can be adopted; and the exterior windows opened toward corridors of buildings shall not interfere with traffic.

14.1.8 【09措施】10.1.8条：建筑外窗可开启部位必须设计配置纱窗，纱窗的安装方式及结构应易于拆装、清洗及更换。

14.1.8 [Measures 2009] 10.1.8: At the openable positions of the exterior windows in buildings, window screens must be designed, and the installation methods and structures of the window screens shall facilitate disassembly and assembly, cleaning and replacement.

14.1.9 【09措施】10.1.9条：建筑外门、外窗用玻璃必须采用中空玻璃（不包括封闭阳台的外窗），其空气层厚度（两层玻璃间距）不小于9mm，严禁使用单层玻璃及简易双层玻璃。

14.1.9 [Measures 2009] 10.1.9: Insulating glass must be adopted as the glass for exterior doors and exterior windows of buildings (excluding the exterior windows of the enclosed balcony), the thickness of the air layer (spacing between two layers of glass) shall be not less than 9mm, and the single-layer glass and simple double-layer glass shall not be used.

14.1.10 建筑外窗宜为内平开下悬开启形式，中高层、高层及超过100m高度的住宅建筑严禁设计、采用外平开窗。采用推拉门窗时，窗扇必须有防脱落措施。

14.1.10 The inward casement with lower hinging opening mode shall be adopted for the exterior windows of buildings, and the outward casement windows shall not be designed or used in the medium- and high-rise residential buildings, high-rise residential buildings and residential buildings with the height above 100m. When the sliding windows and doors are used, the shedding preventive measures must be provided for window sashs.

14.2 门、窗设计

14.2 Design of Door and Window

14.2.1 【建筑外门窗气密、水密、抗风压性能分级及检测方法】4.1.1 条:抗风压性能。分级指标值 P_3 见表 14.2.1 规定。

14.2.1 [Graduations and test methods of air permeability, watertightness, wind load resistance performance for building external windows and doors] 4.1.1: Wind Load Resistance Performance. For the grading index values, see the provisions in Tab. 14.2.1.

建筑外门窗抗风压性能分级表（kPa） 表 14.2.1

Grading Table of Wind Load Resistance Performance of Exterior Doors and Windows in Buildings (kPa) Tab. 14.2.1

分级 Grading	1	2	3	4	5
指标值 P_3 Index value P_3	$1.0 \leq P_3 < 1.5$	$1.5 \leq P_3 < 2.0$	$2.0 \leq P_3 < 2.5$	$2.5 \leq P_3 < 3.0$	$3.0 \leq P_3 < 3.5$
分级 Grading	6	7	8	9	—
指标值 P_3 Index value P_3	$3.5 \leq P_3 < 4.0$	$4.0 \leq P_3 < 4.5$	$4.5 \leq P_3 < 5.0$	$P_3 \geq 5.0$	—

注:1. 本表摘自《建筑外门窗气密、水密、抗风压性能分级及检测方法》GB/T 7106—2008。设计时根据当地规定选定等级。
Notes:1. This table is extracted from *Graduations and Test Methods of Air Permeability, Watertightness, Wind Load Resistance Performance for Building External Windows and Doors* GB/T 7106—2008. In the design, the grades shall be selected in accordance with the local regulations.
2. 第 9 级应在分级后同时注明具体检测压力值。
2. For Grade 9, the specific testing pressure value shall be marked at the same time after grading.
3. 在各分级指标值中，窗（门）主要受力构件（面板）相对挠度：单层、夹层玻璃小于等于 $L/120$；中空玻璃挠度小于等于 $L/180$。
3. In all the grading index values, the relative deflections of the main stressed components (panels) of the windows (doors): less than or equal to $L/120$ for single-layer glass and laminated glass; less than or equal to $L/180$ for insulating glass.
4. 抗风压性能。窗的强度应能满足所在地区的最大正、负风压作用时的要求。尤其是风力较大的地区（如沿海地区等）及高层建筑。高层建筑或位于大风压的建筑设计应提出窗的具体强度指标或其抗风压性能等级。
4. Wind load resistance performance. The strength of the windows shall be able to meet the requirements for the maximum positive and negative wind pressure in the area where the windows are adopted, especially in the areas with big wind (coastal regions, etc.) and high-rise buildings. In the design of high-rise buildings or buildings in high wind pressure, the specific strength indicator or wind pressure resistance performance grade of the windows shall be proposed.

14.2.2 【建筑外门窗气密、水密、抗风压性能分级及检测方法】4.2.2 条：建筑外门窗水密分级指标值 ΔP 见表 14.2.2 规定。

14.2.2 [Graduations and test methods of air permeability, watertightness, wind load resistance performance for building external windows and doors] 4.2.2: For the grading index value ΔP of watertightness of exterior doors and windows in buildings, see the provisions in Tab. 14.2.2.

【建筑外门窗气密、水密、抗风压性能分级及检测方法】4.2.2 表 2：建筑外门窗水密分级表　　表 14.2.2

[Graduations and test methods of air permeability, watertightness, wind load resistance performance for building external windows and doors] Tab. 2 in 4.2.2: Grading Table of Watertightness of Exterior Doors and Windows in Buildings　　Tab. 14.2.2

分级 Grading	1	2	3	4	5	6
分级指标 ΔP Grading index ΔP	$100 \leq \Delta P < 150$	$150 \leq \Delta P < 250$	$250 \leq \Delta P < 350$	$350 \leq \Delta P < 500$	$500 \leq \Delta P < 700$	$\Delta P \geq 700$

注：第 6 级应在分级后同时注明具体测试压力差值。
Note: for Grade 6, the difference of the specific testing pressure shall be simultaneously indicated after grading.

14.2.3 【通则】7.1.1 条：采光性能。分级指标值 T_r 见表 14.2.3-1 ~表 14.2.3-5 规定。

14.2.3 [Code for design of civil buildings] 7.1.1: Lighting performance. The rating indicator value T_r is shown in Tab. 14.2.3-1 to Tab. 14.2.3-5.

【通则】7.1.1-1：居住建筑的采光系数标准值　　表 14.2.3-1

[Code for design of civil buildings] 7.1.1-1: The standard value of the lighting coefficient of residential buildings　　Tab. 14.2.3-1

采光等级 Lighting grade	房间名称 Room name	侧面采光 Side Lighting	
		采光系数最低值 C_{min}（%） Minimum lighting coefficient C_{min}（%）	室内天然临界照度（lx） Indoor natural critical illumination（lx）
IV	起居室（厅）、卧式、书房、厨房 Living room (Hall), horizontal, study, kitchen	1	50
V	卫生间、过厅、楼梯间、餐厅 Toilet, hall, stairwell, dining room	0.5	25

【通则】7.1.1-2：办公建筑的采光系数标准值　　　　　　　　　　表 14.2.3-2
[Code for design of civil buildings] 7.1.1-2:
The Standard Value of the Lighting Coefficient of Office Buildings　　Tab. 14.2.3-2

采光等级 Lighting grade	房间名称 Room name	侧面采光 Side lighting	
		采光系数最低值 C_{min}（%） Minimum lighting coefficient C_{min}（%）	室内天然临界照度（lx） Indoor natural critical illumination（lx）
Ⅱ	设计室、绘图室 Design room, drawing room	3	150
Ⅲ	办公室、视屏工作室、会议室 Office, video Studio, meeting room	2	100
Ⅳ	复印室、档案室 Copy Room, Archives	1	50
Ⅴ	走道、楼梯间、卫生间 Walkway, stairwell, toilet	0.5	25

【通则】7.1.1-3：学校建筑的采光系数标准值　　　　　　　　　　表 14.2.3-3
[Code for design of civil buildings] 7.1.1-3:
The Standard Value of the Lighting Coefficient of School Buildings　　Tab. 14.2.3-3

采光等级 Lighting grade	房间名称 Room name	侧面采光 Side lighting	
		采光系数最低值 C_{min}（%） Minimum lighting coefficient C_{min}（%）	室内天然临界照度（lx） Indoor natural critical illumination（lx）
Ⅲ	教室、阶梯教室、实验室、报告厅 Classrooms, ladder classrooms, laboratories, reporting halls	2	100
Ⅴ	走道、楼梯间、卫生间 Walkway, stairwell, toilet	0.5	25

【通则】7.1.1-4：图书馆建筑的采光系数标准值　　　　　　　　　表 14.2.3-4
[Code for design of civil buildings] 7.1.1-4:
The Standard Value of the Lighting Coefficient of Libraryl Buildings　　Tab. 14.2.3-4

采光等级 Lighting grade	房间名称 Room name	侧面采光 Side lighting		顶部采光 Top lighting	
		采光系数最低值 C_{min}（%） Minimum lighting coefficient C_{min}（%）	室内天然临界照度（lx） Indoor natural critical illumination（lx）	采光系数最低值 C_{min}（%） Minimum iighting coefficient C_{min}（%）	室内天然临界照度（lx）Indoor natural critical illumination（lx）
Ⅲ	阅览室、开架书库 Reading room, open-shelf library	2	100	—	—
Ⅳ	目录室 Catalog room	1	50	1.5	75
Ⅴ	书库、走道、楼梯间、卫生间 Library, walkway, stairwell, toilet	0.5	25	—	—

【通则】7.1.1-5：医院建筑的采光系数标准值　　　表 14.2.3-5
[Code for design of civil buildings] 7.1.1-5:
The Standard Value of the Lighting Coefficient of Hospital Buildings　　Tab. 14.2.3-5

采光等级 Lighting grade	房间名称 Room name	侧面采光 Side Lighting		顶部采光 Top Lighting	
		采光系数最低值 C_{min} (%) Minimum lighting coefficient C_{min} (%)	室内天然临界照度 (lx) Indoor natural critical illumination (lx)	采光系数最低值 C_{min} (%) Minimum lighting coefficient C_{min} (%)	室内天然临界照度 (lx) Indoor natural critical illumination (lx)
Ⅲ	诊室、药房、治疗室、化验室 Clinics, pharmacies, treatment rooms, laboratories	2	100	—	—
Ⅳ	候诊室、挂号处、综合大厅、病房、医生办公室（护士室）Waiting room, registration office, general hall, inpatient ward, doctor's office (nurse's room)	1	50	1.5	75
Ⅴ	走道、楼梯间、卫生间 Walkway, stairwell, toilet	0.5	25	—	—

注：表 14.2.3-1～表 14.2.3-5 所列采光系数标准值适用于Ⅲ类光气候区。其他地区采光系数标准值应乘以光气候系数。
Note: The standard values for lighting coefficients listed in Tab. 14.2.3-1 to Tab.14.2.3-5 are applicable to the Ⅲ light climate zone. The standard value of lighting coefficient in other areas should be multiplied by the light climate coefficient.

14.2.4 【隔声规】4.2.5～4.2.6 条：建筑门窗的空气声隔声性能见表 14.2.4-1、表 14.2.4-2。

14.2.4 [Code for design of sound insulation of civil buildings] 4.2.5-4.2.6: Air sound insulation performance of building doors and windows see Tab.14.2.4-1, Tab.14.2.4-2.

【隔声规】4.2.5：外窗的空气声隔声性能　　　表 14.2.4-1
[Code for design of sound insulation of civil buildings] 4.2.5:
External Window Air Sound Insulation Performance　　Tab. 14.2.4-1

名称 Name	空气隔声单值（dB）Air sound insulation single value (dB)	
交通干线两侧卧室、起居室（厅）的窗 Windows on both bedrooms and living rooms (halls) of the main traffic trunk	计权隔声量 + 交通噪声频谱修正量 R_w+C_{tr} weighted sound reduction index + traffic noise spectrum correction amount R_w+C_{tr}	≥ 45
其他窗 Other windows	计权隔声量 + 交通噪声频谱修正量 R_w+C_{tr} weighted sound reduction index + traffic noise spectrum correction amount R_w+C_{tr}	≥ 25

【隔声规】4.2.6：户门的空气声隔声性能　　　　　　　　表 14.2.4-2
[Code for design of sound in sulation of civil buildings] 4.2.6:
Portal Air Sound Insulation Performance　　　　　　　　　Tab. 14.2.4-2

名称 Name	空气隔声单值 （dB）Air sound insulation single value（dB）	
户门 Portal	计权隔声量 + 噪声频谱修正量 R_w+C Weighted sound reduction index + noise spectrum correction R_w+C	≥ 25

14.2.5 【防火规】6.4.11 条：门的开启方式及选用要点。

14.2.5 [Code for fire protection design of buildings] 6.4.11: The opening mode and selection key points of doors.

1.【防火规】6.4.11 条：门的开启方式常见的有：固定门、平开门、推拉门、弹簧门、提升推拉门、推拉下悬门、内平开下悬门、转门、折叠门、折叠平开门、折叠推拉门、卷门等多种形式。

1. [Code for fire protection design of buildings] 6.4.11: The common door opening way include fixed door, flat door, sliding door, spring doors, lifting sliding doors, push and pull down the suspension door, internal horizontal-openning suspended doors, turnstiles, folding doors, folding flat doors, folding sliding doors, rolling doors and other forms.

2.【防火规】6.4.11 条：建筑内的疏散门应符合下列规定。

2. [Code for fire protection design of buildings] 6.4.11: The evacuation doors in the building shall comply with the following requirements.

（1）民用建筑和厂房的疏散门，应采用向疏散方向开启的平开门，不应采用推拉门、卷帘门、吊门、转门和折叠门。除甲、乙类生产车间外，人数不超过 60 人且每樘门的平均疏散人数不超过 30 人的房间，其疏散门的开启方向不限。

（1）The evacuation doors of civil buildings and factories shall be opened in a flat door open to the direction of evacuation, and no sliding doors, shutter doors, hanging doors, turnstiles and folding doors shall be used. In addition to the production workshop in categories A and B, the number of people without more than 60 and the average number of evacuees per combination several door is not more than 30, and the opening direction of the evacuation door is not limited.

（2）仓库的疏散门应采用向疏散方向开启的平开门，但丙、丁、戊类仓库首层靠墙的外侧可采用推拉门或卷帘门。

（2）The evacuation door of the warehouse should be opened in the direction of evacuation of the flat door, but the first floor of the categories C, D and E warehouse against the outside of the wall can be used sliding door or shutter door.

（3）开向疏散楼梯或疏散楼梯间的门，当其完全开启时，不应减少楼梯平台的有效宽度。

（3）For the door opening to the evacuation staircase or evacuate the stairwell, and when it is fully opened, the effective width of the stair platform should not be reduced.

（4）人员密集场所内平时需要控制人员随意出入的疏散门和设置门禁系统的住宅、宿舍、公寓建筑的外门，应保证火灾时不需使用钥匙等任何工具即能从内部易于打开，并应在显著位置设置具有使用提示的标识。

（4）In densely populated places, it is usually necessary to control the arbitrary access of personnel to the evacuation door and set up the access control system of the residential, dormitory, apartment building outside the door, should ensure that the fire without the use of keys and other tools can be easily opened from the inside, and should be in a prominent location with the use of tips for the identification.

（5）【住宅门窗技术】4.1.10 条：建筑外窗宜为内平开下悬开启形式，中高层、高层及超过 100m 高度的住宅建筑严禁设计、采用外平开窗。采用推拉门窗时，窗扇必须有防脱落措施。

（5）[Technology of residential buildings' door and window] 4.1.10: Exterior window should be internal horizontal-openning suspension form, middle and high level, high-rise and more than 100m height of residential buildings strictly prohibited design, the use of external flat window. When using push and pull doors and windows, the window sash must have anti-shedding measures.

15

建筑物无障碍设计
Accessibility Design for Buildings

15.1 无障碍设计范围

15.1 Accessibility Design Scope

建筑无障碍设计要求见表 15.1。

Building accessibility design requirements see Tab.15.1.

15.2 无障碍设施设计要求

15.2 Design Requirements for Accessibility Facilities

15.2.1 缘石坡道设计要求（图 15.2.1）。

15.2.1 Kerbstone Ramp design requirements (Fig. 15.2.1) .

1.【无规】3.1.1 条：缘石坡道应符合下列规定。

1. [Code for accessibility design] 3.1.1: Kerbstone ramp shall comply with the following provisions.

1）缘石坡道的坡面应平整、防滑。

1）The slope of the kerbstone ramp should be flat and non-slip.

2）缘石坡道的坡口与车行道之间宜没有高差；当有高差时，高出车行道的地面不应大于 10mm。

2）There should be no high difference between the groove of the kerbstone ramp and the roadway, and when there is a high difference, the ground above the roadway should not be greater than 10mm.

3）宜优先选用全宽式单面坡缘石坡道。

3）It is advisable to choose the full-width single-sided slope kerbstone ramp.

2.【无规】3.1.2 条：缘石坡道的坡度应符合下列规定。

2. [Code for accessibility design] 3.1.2: The slope of the kerbstone ramp shall conform to the following provisions.

1）全宽式单面坡缘石坡道的坡度不应大于 1：20。

1）The slope of the full-width type single-sided slope kerbstone ramp should not be greater

15 建筑物无障碍设计　Accessibility Design for Buildings

建筑无障碍设计要求［无障碍设计 12J926—7］
building Accessibility design requirements [Design of accessibility facilities 12J926—7]

表 15.1　Tab. 15.1

无障碍设施 Barrier-free facilities 建筑类型 Type of building		室外道路 Outdoor roads	建筑出入口 Building entrance and exit	无障碍通道 Barrier-free access	无障碍楼梯 Barrier-free stairs	无障碍电梯 Barrier-free elevators	无障碍厕所 Barrier-free toilets	无障碍厕位 Barrier-free toilet position	轮椅席位 Wheelchair seats	低位服务设施 Low service facilities	无障碍停车位 Barrier-free parking spaces	休息区 Lounge area	无障碍浴室 Barrier-free bathroom	盲道 Blind road	标识 Identity	信息系统 Information system
居住建筑 Residential buildings	住宅及公寓 Residences and apartments	○	○													
	宿舍建筑 Dormitory building	○	○	○		○	○								○	
办公、科研、司法 Office, scientific research, justice	为公众办理业务与信访接待的办公建筑 Office buildings for the public to handle business and petition reception	○	○	○	○	○	○	○	○	○	○	○			○	○
	其他办公建筑 Other office buildings	○	○			○		○	○	○	○				○	○
教育建筑 Educational architecture	普通教育建筑 General education buildings	○	○		○	○	○	○								
	残疾生源教育建筑 Educational architecture for students with disabilities	○	○		○	○	○	○	○						○	

续表

无障碍设施 Barrier-free facilities 建筑类型 Type of building	室外道路 Outdoor roads	建筑出入口 Building entrance and exit	无障碍通道 Barrier-free access	无障碍楼梯 Barrier-free stairs	无障碍电梯 Barrier-free elevators	无障碍厕所 Barrier-free toilets	无障碍厕位 Barrier-free toilet position	轮椅席位 Wheelchair seats	低位服务设施 Low service facilities	无障碍停车位 Barrier-free parking spaces	休息区 Lounge area	无障碍浴室 Barrier-free bathroom	盲道 Blind road	标识 Identity	信息系统 Information system
医疗康复建筑 Medical rehabilitation buildings	○	○	○	○	○	○	○			○	○	○		○	○
福利及特殊服务建筑 Welfare and special service buildings	○	○	○	○	○	○	○		○	○	○	○		○	○
体育建筑 Sports architecture	○	○	○	○	○	○	○	○		○	○			○	○
文化建筑 Cultural architecture	○	○	○	○	○	○	○	○	○	○	○		○*	○	○
商业服务建筑 Commercial service building	○	○	○	○	○	○	○		○	○				○	○
汽车客运站 Automobile passenger terminal	○	○	○	○	○	○	○		○	○				○	○

注：1. 表中"○"为各类建筑中应设置无障碍设施的主要内容，设计中还应结合《无障碍设计规范》的具体要求进行设计；

2. 表中"*"表示仅在盲人专用图书室（角）时设置。

Notes: 1. the Table "○" for all types of buildings should be set up accessibility facilities of the main content, the design should also be combined with the specific requirements of the Barrier-free design Code design;

2. The "*" in the Table indicates that it is set only in the dedicated blind library (corner).

than 1∶20.

2）三面坡缘石坡道正面及侧面的坡度不应大于1∶12。

2）The slope on the front and side of the three-sided slope kerbstone ramp should not be greater than 1∶12.

3）其他形式的缘石坡道的坡度均不应大于1∶12。

3）The slope of other forms of kerbstone ramp should not be greater than 1∶12.

3.【无规】3.1.3：缘石坡道的宽度应符合下列规定。

3. [Code for accessibility design] 3.1.3: The width of the kerbstone ramp shall conform to the following provisions.

1）全宽式单面坡缘石坡道的宽度应与人行道宽度相同。

1）The width of the full-type type single-sided slope kerbstone ramp should be the same as the width of the sidewalk.

2）三面坡缘石坡道的正面坡道宽度不应小于1.20m。

2）The width of the front ramp of the three-sided slope kerbstone ramp should not be less than 1.20m.

3）其他形式的缘石坡道的坡口宽度均不应小于1.50m。

3）The groove width of other forms of kerbstone ramp should not be less than 1.50m.

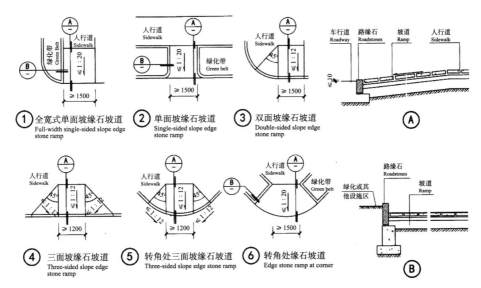

图 15.2.1 【无障碍设计 12J926—A4】缘石坡道的做法

Fig. 15.2.1 [accessibility design 12J926—A4] The Practice of Kerbstone Ramp

15.2.2 盲道设计要求（图 15.2.2-1~图 15.2.2-3）。

15.2.2 Blind road design requirements（see Fig. 15.2.2-1 ~ Fig. 15.2.2-3）.

【无规】3.2.2、3.2.3 条：盲道的设计要求：行进盲道的宽度宜为 250～500mm；行进盲道宜在距围墙、花台、绿化带 250～500mm 处设置；行进盲道在起点、终点、转弯处及其他有需要处应设提示盲道，当盲道的宽度不大于 300mm 时，提示盲道的宽度应大于行进盲道的宽度。

[Code for accessibility design] 3.2.2 and 3.2.3: Blind road design requirements: The width of the travel blind Road should be 250 ~ 500mm; travel Blind Road should be set in the distance from the wall, flower Table, green belt 250 ~ 500mm; the marching blind road should be prompted blind road at the beginning, the end, the turn and other needs. When the width of the blind road is not greater than 300mm, the width of the prompt blind road should be larger than the width of the Marching blind Road.

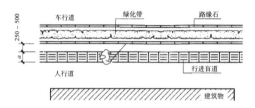

图 15.2.2-1 沿绿化带行进盲道

Fig. 15.2.2-1　Marching Blind Road Along the Green Belt

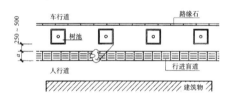

图 15.2.2-2 沿树池行进盲道

Fig. 15.2.2-2　Marching Blind Road Along the Tree Pool

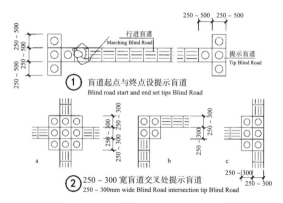

图 15.2.2-3 提示盲道做法示意

Fig. 15.2.2-3　Tip Blind Road Practice Signal

15.2.3 无障碍出入口设计要求。

15.2.3 Accessibility access design requirements.

1.【无规】3.3.1 条：无障碍出入口包括以下几种类别。

1. [Code for accessibility design] 3.3.1: accessibility entrances and inlets include the following categories.

1）平坡出入口；

1）Flat slope entrance and exit;

2）同时设置台阶和轮椅坡道的出入口；

2）set up the entrance and exit of steps and wheelchair ramps at the same time;

3）同时设置台阶和升降平台的出入口。

3）set up the entrance and exit of the steps and lifting platform at the same time.

2.【无规】3.3.2 条：无障碍出入口应符合下列规定。

2. [Code for accessibility design] 3.3.2: accessibility entrances and inlets shall comply with the following require ments.

1）出入口的地面应平整、防滑；

1）The ground of the entrance and exit shall be flat and non-slip;

2）室外地面滤水箅子的孔洞宽度不应大于 15mm；

2）The hole width of the outdoor ground filter grate should not be greater than 15mm;

3）同时设置台阶和升降平台的出入口宜只应用于受场地限制无法改造坡道的工程。

3）At the same time set up the entrance and exit of the steps and lifting platform should only be used for site restrictions can not be modified ramp works.

4）除平坡出入口外，在门完全开启的状态下，建筑物无障碍出入口的平台的净深度不应小于 1.50m；

4）In addition to the flat slope entrance and exit, in the state of the door is fully open, the net depth of the building accessibility entrance platform should not be less than 1.50m;

5）建筑物无障碍出入口的门厅、过厅如设置两道门，门扇同时开启时两道门的间距不应小于 1.50m；

5）Building accessibility entrance and exit of the foyer, Hall if two doors are set, when the door is opened at the same time, the spacing between the two doors should not be less than 1.50m;

6）建筑物无障碍出入口的上方应设置雨篷。

6）A rain shed should be set up above the accessibility entrance of the building.

3.【无规】3.3.3 条：无障碍出入口的轮椅坡道及平坡出入口的坡度应符合下列规定。

3. [Code for accessibility design] 3.3.3: The slope of the wheelchair ramp and the flat slope entrance and exit of the accessibility entrance shall comply with the following requirements.

1）平坡出入口的地面坡度不应大于1∶20，当场地条件比较好时，不宜大于1∶30；

1) The ground slope of the flat slope entrance and exit should not be greater than 1∶20, when the site conditions are better, it should not be greater than 1∶30;

2）同时设置台阶和轮椅坡道的出入口，轮椅坡道的最大高度和水平长度应符合表15.2.3。

2) At the same time set the entrance and exit of the steps and wheelchair ramps, the maximum height and horizontal length of the wheelchair ramp should conform to Tab. 15.2.3.

【无规】表3.4.4：轮椅坡道的最大高度和水平长度　　　　表 15.2.3

[Code for accessibility facilities] Tab. 3.4.4:
Maximum Height and Horizontal Length of Wheelchair Ramp　　　Tab. 15.2.3

坡度 Slope	1∶20	1∶16	1∶12	1∶10	1∶8
最大高度（m） Maximum height (m)	1.20	0.90	0.75	0.60	0.30
水平长度（m） Horizontal length (m)	24.00	14.40	9.00	6.00	2.40

注：其他坡度可用插入法进行计算。

Note: Other slopes can be calculated by inserting method.

4.【无规】3.4条：轮椅坡道宜设计成直线形、直角形或折返形；轮椅坡道的净宽度不应小于1.00m，无障碍出入口的轮椅坡道净宽度不应小于1.20m；轮椅坡道的高度超过300mm且坡度大于1∶20时，应在两侧设置扶手，坡道与休息平台的扶手应保持连贯；轮椅坡道的坡面应平整、防滑、无反光；轮椅坡道起点、终点和中间休息平台的水平长度不应小于1.50m；轮椅坡道临空侧应设置安全阻挡措施；轮椅坡道应设置无障碍标志。

4. [Code for accessibility design] 3.4: Wheelchair ramps should be designed to be linear, rectangular or reentrant; the net width of the wheelchair ramp should not be less than 1. 00m, the net width of the wheelchair ramp without barrier access should not be less than 1.20m; when the height of the wheelchair ramp exceeds 300mm and the slope is greater than 1∶20, handrails should be arranged on both sides, and the handrails of the ramp and rest platform should be consistent; the slope of the wheelchair ramp should be flat, non-slip and non-reflective; wheelchair ramp starting point The horizontal length of the end and middle rest platforms should not be less than 1.50m; The wheelchair ramp on the air side should be set up safety blocking measures; Wheelchair ramps should be provided with accessibility signs.

5.【无障碍设计12J926】D2：同时设置台阶及轮椅坡道的出入口形式，见图15.2.3。

5. [Design of accessibility facilities12J926] D2: Set up steps and wheelchair ramp access form at the same time, see Fig. 15.2.3.

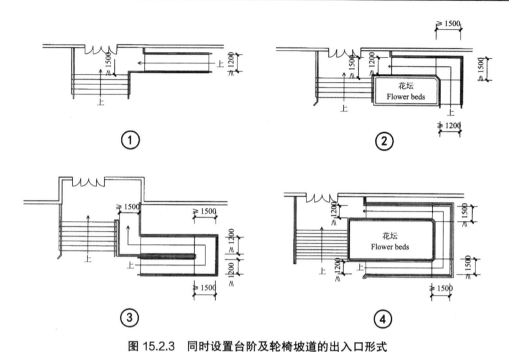

图 15.2.3 同时设置台阶及轮椅坡道的出入口形式
Fig. 15.2.3 Set Up Steps and Wheelchair Ramp Access Form at the Same Time

15.2.4 无障碍通道设计要求。

15.2.4 Accessibility sidewalk design requirements.

1.【无规】3.5.1 条：无障碍通道的宽度应符合表 15.2.4 要求。

1. [Code for accessibility design] 3.5.1: The width of the accessibility sidewalk shall conform to Tab. 15.2.4.

无障碍通道宽度要求　　　　　　　　　　　表 15.2.4
Accessibility Sidewalk Width Requirements　　　Tab. 15.2.4

通道类型 Channel type	宽度要求 Width requirements
室内通道 Indoor channels	≥ 1.20m
人流较多或较集中的大型公共建筑的室内走道 Indoor walkways for large public buildings with more or more concentrated crowds	≥ 1.80m
室外通道 Outdoor channels	≥ 1.50m
检票口、结算口轮椅通道 Ticket gate, clearing mouth wheelchair access	≥ 0.90m

2.【无规】3.5.2条：无障碍通道应符合下列规定。

2. [Code for accessibility design] 3.5.2: accessibility sideways shall comply with the following provisions.

1）无障碍通道应连续，其地面应平整、防滑、反光小或无反光，并不宜设置厚地毯；

1）Accessibility sidewalk should be continuous, its ground should be flat, anti-slip, reflective small or non-reflective, and should not set thick carpet;

2）无障碍通道上有高差时，应设置轮椅坡道；

2）Wheelchair ramps should be set up when there is a high deviation on the accessibility sidewalk;

3）室外通道上的雨水箅子的孔洞宽度不应大于15mm；

3）The hole width of the rainwater grate on the outdoor sidewalk should not be greater than 15mm;

4）固定在无障碍通道的墙、立柱上的物体或标牌距地面的高度不应小于2.00m；如小于2.00m时，探出部分的宽度不应大于100mm；如突出部分大于100mm，则其距地面的高度应小于600mm；

4）The height of a wall, column, or sign fixed to a accessibility sidewalk shall not be less than 2.00m from the ground; if less than 2.00m, the width of the probing part should not be greater than 100mm, and if the protruding part is greater than 100mm, the height from the ground should be less than 600mm;

5）斜向的自动扶梯、楼梯等下部空间可以进入时，应设置安全挡牌。

5）The inclined escalator, staircase and other lower space can be entered when the safety card should be set.

15.2.5【无规】3.5.3条：门的无障碍设计要求。

15.2.5 [Code for accessibility design] 3.5.3: Accessibility door design.

不应采用力度大的弹簧门，并不宜采用弹簧门、玻璃门；当采用玻璃门时，应有醒目的提示标志；自动门开启后通行净宽度不应小于1.00m，其他门开启后的通行净宽度不应小于800mm（图15.2.5-1），有条件时，不宜小于900mm；在门扇内外应留有直径不小于1.50m的轮椅回转空间（图15.2.5-2）；门槛高度及门内外地面高差不应大于15mm，并以斜面过渡。

Suggest those spring doors inappropriate to large forces not be applied with spring doors and glass ones. When using glass door, some eye-catching warning signs shall be pasted thereon. After unlocking automatic door, net passage width shall not be less than 1.00m. And after opening other doors, net passage width shall not be lower than 800mm (Fig. 15.2.5-1). When available, such net passage width shall not be less than 900mm. There shall keep not-less-than-1.50m wheelchair swiveling space inside and outside door leaf (Fig. 15.2.5-2). And both sill height and floor height difference between inside and outside of door shall not surpass 15mm and required to be transitioned with slopes.

15 建筑物无障碍设计　Accessibility Design for Buildings

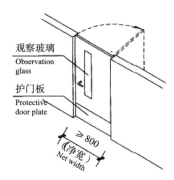

图 15.2.5-1　无障碍平开门
Fig. 15.2.5-1　Accessibility Flat Door

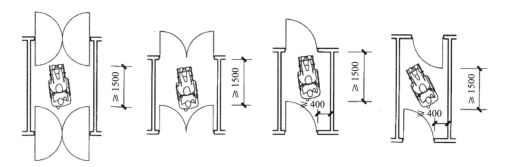

图 15.2.5-2　两道门间距要求
Fig. 15.2.5-2　Two Door Spacing Requirement

15.2.6　无障碍楼梯、台阶设计要求。

15.2.6　Accessibility staircase, step design.

1.【无规】3.5.6 条：无障碍楼梯的设计要求：宜采用直线形楼梯；公共建筑楼梯的踏步宽度不应小于 280mm，踏步高度不应大于 160mm；不应采用无踢面和直角形突缘的踏步；宜在两侧均做扶手；栏杆式楼梯在栏杆下方宜设置安全阻挡措施。参见图 15.2.6。

1. [Code for accessibility design] 3.5.6: The design requirements of the accessibility staircase: it is advisable to adopt a straight staircase; the step width of the staircase of the public building should not be less than 280mm, the step height should not be greater than 160mm; the tread without the kicking surface and the right angle flange should not be used; it is advisable to make handrails on both sides. See Fig. 15.2.6.

2.【无规】3.6.2 条：台阶的无障碍设计应符合下列规定。

2. [Code for accessibility design] 3.6.2: The accessibility design of the steps shall be in accordance with the following provisions.

1）公共建筑的室内外台阶踏步宽度不宜小于 300mm，踏步高度不宜大于 150mm，并不应小于 100mm；

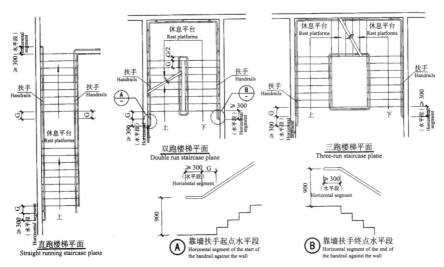

图 15.2.6 【无障碍设计 12J926】F1：无障碍楼梯的设计要求
Fig. 15.2.6 [Design of accessibility facilities 12J926] F1: Design Requirements for Accessibility Stairs

1）The indoor and outdoor step width of the public building should not be less than 300mm, the tread height should not be greater than 150mm, should not be less than 100mm;

2）踏步应防滑；

2）Tread should be anti-skid;

3）三级及三级以上的台阶应在两侧设置扶手；

3）Three steps and the above should be set on both sides of the handrail;

4）台阶上行及下行的第一阶宜在颜色或材质上与其他阶有明显区别。

4）The first order of step uplink and downlink should be significantly different from other orders in color or material.

15.2.7 无障碍电梯的设计要求。

15.2.7 Design requirements for accessibility elevators.

1.【无规】8.1.4 条：建筑内设有电梯时，至少应设置 1 部无障碍电梯。

1. [Code for accessibility design] 8.1.4: When there is an elevator in the building, at least 1 accessibility elevators should be set up.

2.【无规】3.7.1 条：无障碍电梯厅的设计要求：呼叫按钮高度为 0.90～1.10m；电梯门洞的净宽度不宜小于 900mm；电梯出入口处宜设提示盲道；候梯厅应设电梯运行显示装置和抵达音响。参见表 15.2.7。

2. [Code for accessibility design] 3.7.1: The design requirements of the accessibility elevator hall: The call button height is 0.90～1.10m; elevator door hole net width should not be less than 900mm; elevator out of the entrance should be set to prompt blind road; The waiting hall should be set up elevator operation display device and arrival sound. See Tab. 15.2.7.

【无障碍设计12J926】G1：无障碍电梯的候梯厅深度要求　　　　表15.2.7

[Code for accessibility facilities 12J926] G1:
The Depth Requirements of the Waiting Room for Accessibility Elevators　　Tab. 15.2.7

电梯类别 Elevator categories	布置方式 Layout method	候梯厅深度 Hou staircase hall depth
住宅电梯 Residential elevator	单台 Single-Station	≥1.5m 且 ≥ b ≥1.5m and ≥ b
	多台单侧排列 Multiple single-sided permutations	≥1.5m 且 ≥ B ≥1.5m and ≥ B
	多台双侧排列 Multiple two-sided arrangement	≥相对电梯 B 之和且 <3.5m ≥ Relative to the sum of elevator B and <3.5m
公共建筑电梯 Public building elevator	单台 Single-Station	≥1.8m 且 ≥1.5b ≥1.8m and ≥1.5b
	多台单侧排列 Multiple single-sided permutations	且 ≥1.5B，当电梯群为4台时应 ≥2.4m And ≥1.5B, when the elevator group is 4 units should be ≥2.4m
	多台双侧排列 Multiple two-sided arrangement	≥相对电梯 B 之和且 <4.5m ≥ Relative to the sum of elevator B and <4.5m
病床电梯 Bed elevator	单台 Single-station	≥1.8m 且 ≥1.5b ≥1.8m and ≥1.5b
	多台单侧排列 Multiple single-sided permutations	≥1.5B
	多台双侧排列 Multiple two-sided arrangement	≥相对电梯 B 之和 ≥ Relative to the sum of elevator B

注：b 为轿厢深度，B 为电梯群中最大轿厢深度。
Note: b is the car depth, and B is the maximum car depth in lift groups.

3.【无规】3.7.2条：无障碍电梯的轿厢设计要求：轿厢门开启的净宽度不应小于800mm；轿厢的三面壁上应设高850~900mm扶手，轿厢的最小规格为深度不应小于1.40m，宽度不应小于1.10m；中型规格为深度不应小于1.60m，宽度不应小于1.40m。

3. [Code for accessibility design] 3.7.2: Taccessibility elevator car design requirements: the net width of the car door opening should not be less than 800mm; the three wall of the car should be set with a high 850-900mm handrail, the minimum specification for the car should not be less than 1. 40m, the width should not be less than 1.10m; medium specification for depth should not be less than 1. 60m, the width should not be less than 1.40m.

15.2.8【无规】3.8.1、3.8.2条：无障碍单层扶手的高度应为850~900mm，无障碍双层扶手的上层扶手高度应为850~900mm，下层扶手高度应为650~700mm。扶手应保持连贯，靠墙面的扶手的起点和终点处应水平延伸不小于300mm的长度。

15.2.8 [Code for accessibility design] 3.8.1 and 3.8.2: Accessibility single-layer handrail height should be 850-900mm, accessibility double-decker handrail upper handrail height should be 850-900mm, the height of the lower handrail should be 650-700mm. The handrail shall remain coherent and shall extend horizontally to a length not less than 300mm at the beginning and end of the handrail against the wall.

15.2.9 公共厕所及无障碍厕所。

15.2.9 Public toilets and accessible toilets.

1.【无规】3.9.1 条：公共厕所的无障碍设计要求：女厕所的无障碍设施包括至少 1 个无障碍厕位和 1 个无障碍洗手盆；男厕所的无障碍设施包括至少 1 个无障碍厕位、1 个无障碍小便器和 1 个无障碍洗手盆；厕所的入口和通道回转直径不小于 1.50m；门净宽不应小于 800mm。参见图 15.2.9-1。

1. [Codes for accessibility design] 3.9.1: Accessibility design requirements for public toilets: accessibility facilities for the ladies' room include at least 1 accessibility toilets and 1 accessibility washbasins; accessible facilities in men's toilets include at least 1 accessibility toilets, 1 accessibility urinals and 1 accessibility washbasins; the inlet and sidewalk rotation diameter of the toilet is not less than 1.50m; the net width of the door should not be less than 800mm. See Fig. 15.2.9-1.

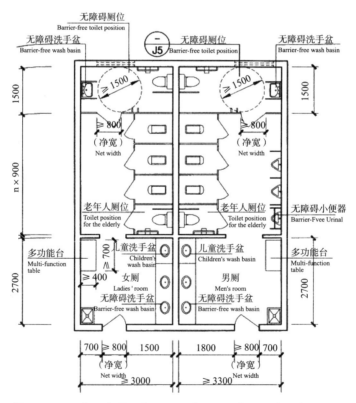

图 15.2.9-1 【无障碍设计 12J926】J3：公共厕所的无障碍设计

Fig. 15.2.9-1 [Design of accessibility facilities 12J926] J3: Accessibility Design of Public Toilets

2.【无规】3.9.2 条：无障碍厕位设计要求：厕位宜做到 2.00m×1.50m，不应小于 1.80m×1.00m；门宜向外开启，如内开需在开启后厕位内留有直径不小于 1.50m 的轮椅回转空间；门净宽不应小于 800mm；厕位两侧距地面 700mm 处应设长度不小于 700mm 的水平安全抓杆，另一侧应设高 1.40m 的垂直安全抓杆。参见图 15.2.9-2。

2. [Codes for accessibility design] 3.9.2: Accessibility toilet design requirements: toilet position should be 2.00m × 1.50m, should not be less than 1. 80m × 1.00m; the door should be opened outward, if the inner opening needs to be left in the toilet position in the diameter of not less than 1.50m wheelchair rotation space; the net width of the door should not be less than 800mm; the toilet bit on both sides of the ground 700mm should be set a length of not less than 700mm horizontal safety grip, the other side should be set high 1. 40m vertical safety grip.See Fig. 15.2.9-2.

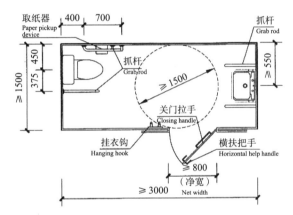

图 15.2.9-2 【无障碍设计 12J926】J5：无障碍厕位示意图

Fig. 15.2.9-2 [Design of accessibility facilities 12J926] J5: Diagram of Accessibility Toilet Position

3.【无规】3.9.3 条：无障碍厕所的无障碍设计要求：位置宜靠近公共厕所，回转直径不小于 1.50m；面积不应小于 4.00m²；门净宽不应小于 800mm；内部应设坐便器、洗手盆、多功能台、挂衣钩和呼叫按钮；多功能台长度不宜小于 700mm，宽度不宜小于 400mm，高度宜为 600mm；挂衣钩距地高度不应大于 1.20m；在坐便器旁的墙面上应设高 400~500mm 的救助呼叫按钮。

3. [Code for accessibility design] 3.9.3: Accessibility toilet accessibility design requirements: the location should be close to the public toilet, the rotation diameter is not less than 1. 50m; the area should not be less than 4.00m²; the net width of the door should not be less than 800mm; the inside should be a toilet, wash basin, multi-function table, hanging hook and call button; the length of the multifunction table should not be less than 700mm, the width should not be less than 400mm, the height should be 600mm; the height of the hook should not be greater than

1.20m; a high 400 ~ 500mm rescue call button should be set on the wall next to the toilet.

15.2.10 【无规】3.10.2 条：无障碍淋浴间设计要求：无障碍淋浴间的短边宽度不应小于 1.50m；浴间坐台高度宜为 450mm，深度不宜小于 450mm；淋浴间应设距地面高 700mm 的水平抓杆和高 1.40 ~ 1.60m 的垂直抓杆；淋浴间内的淋浴喷头的控制开关的高度距地面不应大于 1.20m；毛巾架的高度不应大于 1.20m。参见图 15.2.10。

15.2.10 [Codes for accessibility design] 3.10.2: Accessibility shower design requirements: the short edge width of the accessibility shower should not be less than 1.50m; bath sit on height should be 450mm, depth should not be less than 450mm; shower should be set from the ground high 700mm horizontal grip and 1 higher 1.40 ~ 1.60m vertical grip; the height of the control switch for the shower nozzle in the shower should not be greater than 1.20m from the ground; the height of the towel rack should not be greater than 1.20m. See Fig. 15.2.10.

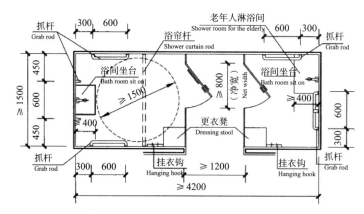

图 15.2.10 【无障碍设计 12J926】K2：无障碍淋浴间示意图
Fig. 15.2.10 [Design of accessibility facilities 12J926] K2: Accessibility Shower Diagram

15.2.11 【无规】3.11 条：无障碍客房的设计要求：无障碍客房应设在便于到达、进出和疏散的位置；房间及卫生间内均应有空间能保证轮椅进行回转，回转直径不小于 1.50m；床间距离不应小于 1.20m；床的使用高度为 450mm；客房及卫生间应设高 400 ~ 500mm 的救助呼叫按钮；客房应设置为听力障碍者服务的闪光提示门铃。见图 15.2.11。

15.2.11 [Code for accessibility design] 3.11: Design requirements for accessible rooms: accessible rooms should be located in accessible, inbound and outgoing locations; rooms and toilets should have space to ensure wheelchair rotation, rotation diameter is not less than 1.50m; the distance between the beds should not be less than 1.20m; the height of the bed is 450mm; rooms and toilets should be provided with a high 400 ~ 500mm rescue call button; rooms should be set up for the hearing impaired service flash prompt doorbell. See Fig. 15.2.11.

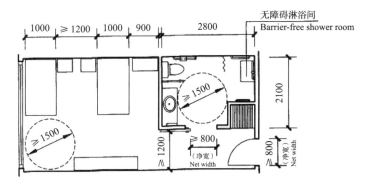

图 15.2.11 【无障碍设计 12J926】L1：无障碍客房示意图
Fig. 15.2.11　[Design of accessibility facilities 12J926] L1: Accessibility Rooms Diagram

15.2.12　【无规】3.12 条：无障碍住房及宿舍的设计要求：通往卧室、起居室（厅）、厨房、卫生间、储藏室及阳台的通道应为无障碍通道，并在一侧或两侧设置扶手；单人卧室面积不应小于 7.00m²，双人卧室面积不应小于 10.50m²，兼起居室的卧室面积不应小于 16.00m²，起居室面积不应小于 14.00m²，厨房面积不应小于 6.00m²；设坐便器、洗浴器（浴盆或淋浴）、洗面盆三件卫生洁具的卫生间面积不应小于 4.00m²；设坐便器、洗浴器二件卫生洁具的卫生间面积不应小于 3.00m²；设坐便器、洗面盆二件卫生洁具的卫生间面积不应小于 2.50m²；单设坐便器的卫生间面积不应小于 2.00m²；厨房操作台下方净宽和高度都不应小于 650mm，深度不应小于 250mm；居室和卫生间内应设求助呼叫按钮；家具和电器控制开关的位置和高度应方便乘轮椅者靠近和使用；供听力障碍者使用的住宅和公寓应安装闪光提示门铃。参见图 15.2.12。

15.2.12　[Code for accessibility design] 3.12: Design requirements for accessible housing and dormitories: access to bedrooms, living rooms (halls) , kitchens, toilets, storage rooms and balconies should be accessibility and with handrails on one or both sides; single bedroom area should not be less than 7.00m², double bedroom area should not be less than 10.50m², and the bedroom area of the living room should not be less than 16.00m², the living room area should not be less than 14.00m², the kitchen area should not be less than 6.00m²; toilet, bath (bath or shower), washbasin three pieces of sanitary ware bathroom area should not be less than 4.00m²; toilet, bath two sanitary ware bathroom area should not be less than 3.00m²; toilet and washbasin two pieces of sanitary ware bathroom area should not be less than 2.50m²; the bathroom area of a single toilet should not be less than 2.00m²; the net width and height below the kitchen operator should not be less than 650mm, the depth should not be less than 250mm; the room and toilet should be equipped with a help call button; the location and height of the furniture and electrical control switch should be easily accessible and used by wheelchair users; a flash prompt doorbell should be installed in homes and apartments for persons with hearing impairment. See Fig. 15.2.12.

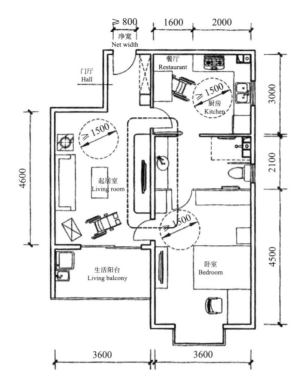

图 15.2.12 【无障碍设计 12J926】L4：无障碍住房及宿舍示意图
Fig. 15.2.12 [Design of accessibility facilities 12J926] L4: Map of Accessibility Housing and Hostels

15.2.13 【无规】3.13 条：轮椅席位的设计要求：应设在便于到达疏散口及通道的附近，不得设在公共通道范围内；通道宽度不应小于1.20m；每个轮椅席位的占地面积不应小于1.10m×0.80m；在轮椅席位旁或在邻近的观众席内宜设置1∶1的陪护席位。

15.2.13 [Code for accessibility design] 3.13: Wheelchair seat design requirements: should be located in the vicinity of the evacuation port and access, should not be located in the range of public access, sidewalk width should not be less than 1.20m; the floor area of each wheelchair seat should not be less than 1.10m×0.80m; 1∶1 escort seats should be set up next to a wheelchair seat or in a neighbouring auditorium.

15.2.14 【无规】3.14 条：无障碍机动车停车位的设计要求：应将通行方便、行走距离路线最短的停车位设为无障碍机动车停车位；停车位地面坡度不应大于1∶50；一侧应设宽度不小于1.20m 的轮椅通道；地面应涂有停车线、轮椅通道线和无障碍标志。参见图15.2.14。

15.2.14 [Code for accessibility design] 3.14: Design requirements for accessibility motor vehicle parking spaces: parking spaces with convenient access and the shortest walking distance routes should be set up as accessible motor vehicle parking spaces; the ground slope of parking spaces should not be greater than 1∶50; and the side should be set at a width not less than 1.20m wheelchair access; the ground should be coated with parking lines, wheelchair access lines and accessibility signs.See Fig. 15.2.14.

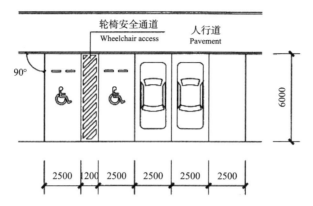

图 15.2.14 【无障碍设计 12J926】N1：无障碍机动车停车位示意图
Fig. 15.2.14 [Design of accessibility facilities 12J926] N1: Diagram of Accessibility Motor Vehicle Parking Spaces

16

厨房
Kitchen

16.1 一般规定

16.1 General Provisions

16.1.1 【09措施】12.1.1条：厨房设计时，应对厨房家具、设备布置及各专业管线设施等综合考虑，并为厨房家具和设备的更新发展留有余地。

16.1.1 [Measures 2009] 12.1.1: During the design for kitchens, the arrangement of furniture and equipments in kitchen and pipeline facilities shall be put into comprehensive consideration, and appropriate margin shall be reserved for renewal and development of furniture and equipment in kitchen.

16.1.2 【09措施】12.1.2条：厨房、特别是公用厨房的平面布置，应重视按照食品加工的工艺流程及人员炊事行为特征进行设计，做到流程合理、使用方便，并妥善安排排气道、烟囱的位置。

16.1.2 [Measures 2009] 12.1.2: In plan layout of kitchens, especially public kitchens, the design shall be carried out as per food processing procedures and the cooking behavior characteristics of the personnel, to achieve the effect of reasonable process and easy use. Positions of the exhaust passage and the chimneys shall be properly arranged.

16.1.3 【09措施】12.1.3条：厨房、特别是公用厨房应重视原料及成品有方便、有效地水平和垂直的运输方式，并做到生、熟分开，洁、污分区。

16.1.3 [Measures 2009] 12.1.3: In kitchens, especially public kitchens, attention shall be paid to the convenient and efficient horizontal and vertical transport for raw materials and finished products, to achieve the effect of separate transport of raw food and cooked food, and storage of clean items and dirty items in different areas.

16.1.4 【09措施】12.1.4条：厨房作为食品加工场所，设计时应注意。

16.1.4 [Measures 2009] 12.1.4: As food processing places, kitchens shall be designed with attention paid to.

1. 防水、防火及防止有害气体的泄漏；

1. Waterproof, fire prevention and harmful gas leakage prevention;

2. 防止气、水、声、有机垃圾等对环境的污染；

2. Prevention of pollution of air, water, sound, organic waste, etc to the environment;

3. 节能、节水、节电、节约燃气等。

3. Energy saving, water saving, electricity saving, gas saving, etc.

16.1.5 【09措施】12.1.5条：当厨房内使用液化石油气瓶作为燃料时,不得设置在地下室、半地下室或通风不良的场所。商业用房使用的气瓶组严禁与燃气燃烧器具布置在同一房间内。

16.1.5 [Measures 2009] 12.1.5: When liquefied petroleum gas cylinders are used in kitchens, the kitchens shall not be arranged in basements, semi-basements or poorly ventilated places. In commercial buildings, gas cylinder groups and gas-burning appliances must not be arranged in the same room.

16.2 住宅厨房

16.2 Kitchen of the Residential Buildings

16.2.1 【09措施】12.2.1条：厨房的净宽、净长要求和平面布置示意见图16.2.1。厨房的净高：安装燃气灶的不宜小于2.2m；安装燃气热水器和燃气壁挂炉的不宜小于2.4m。厨房门洞口净宽度不宜小于0.8m。

16.2.1 [Measures 2009] 12.2.1: The net width and the net length requirement and the layout schematic diagram are shown in Fig. 16.2.1. Net height of kitchens: kitchens with gas stoves shall be appropriately not less than 2.2m; and kitchens with gas water heaters and gas hanging stoves shall be appropriately not less than 2.4m. The net width of kitchen door openings shall be appropriately not less than 0.8m.

16.2.2 【09措施】12.2.2条、【住设规】5.3.3条：厨房设备的布置应方便操作,符合洗、切、烧的炊事流程,操作面最小净长2.1m。厨房内应设置炉灶、洗涤池、操作台、吸油烟机等设备和家具或预留位置。燃气灶尽量避免布置在贴邻外窗口处。

16.2.2 [Measures 2009] 12.2.2. [Design code of residential buildings]5.3.3: The arrangement of the kitchen equipment shall facilitate operation and conform to the cooking process of washing, cutting and cooling, and the minimum net length of the operation tables shall be 2.1m. The equipment and furniture of cooking ranges, sinks, operation platforms, range hoods and others shall be arranged, or positions shall be reserved in the kitchen. It shall be avoided to arrange gas stoves near exterior windows.

16.2.3 【09措施】12.2.3条：厨房应有直接采光、自然通风,或通过住宅的阳台通风采光。当厨房外为封闭阳台时,应确保阳台窗有足够的自然通风和采光。当高层住宅厨房布置确有困难时,其外窗可开向公共外走廊（如窗上部分亮子做成固定通风百叶）,但

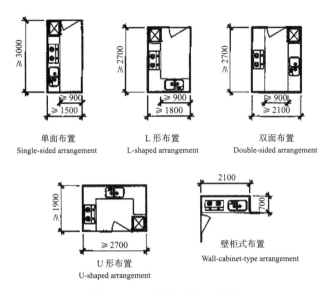

图 16.2.1. 厨房净宽、净长
Fig. 16.2.1 Net width and Length of Kitchens

注：壁柜型厨房宜用于家庭人口少且在家做饭概率少的家庭，灶具为电气或电磁灶
Note: wall-cabinet type kitchens shall be used for households with a small family size and low cooking rate at home, and for cooking utensils, electric or electromagnetic ranges shall be adopted

该公共走廊必须有良好的自然通风和采光。

16.2.3 [Measures 2009] 12.2.3: Kitchens shall have direct daylighting and natural ventilation, or the ventilation and daylighting can be achieved through the balcony of the residential building. When there is a closed balcony outside a kitchen, the balcony shall be ensured to have enough natural ventilation and daylighting. When there are difficulties in arrangement of kitchens in high-rise residential buildings, the exterior windows can be opened toward the external public corridors (such as fixed ventilation louvers can be constructed into the transoms of the window), but the public corridors must have good natural ventilation and daylighting.

16.2.4 【09 措施】12.2.4 条：厨房天然采光标准，其侧面采光，窗洞口面积不应小于地面面积的 1/7。其自然通风的通风开口面积不应小于地面面积的 1/10，且不得小于 $0.6m^2$。如通过阳台采光通风时，应以有效的最小采光通风面积计算（距地 800mm 以上为有效采光面积）。按阳台窗计算时，地面面积应包括阳台面积；按阳台内门窗计算时，应乘以 0.7 的折减系数。

16.2.4 [Measures 2009] 12.2.4: As per the standard for natural daylighting in kitchens, the window opening area for daylighting of the side shall not be less than 1/7 of the ground area. The area of the ventilation opening for natural ventilation shall be not less than 1/10 of the ground area, and shall not be less than $0.6m^2$. For daylighting and ventilation through the balcony, the calculation shall be done as the minimum effective lighting and ventilation areas (the effective daylighting area shall be more than 800 mm above the ground). For the calculation on the basis of the balcony window, the balcony area shall be included in the ground area; and it shall be multiplied by a reduction factor of 0.7 in the calculation on the basis of the interior doors and windows of balconies.

16.2.5【09措施】12.2.5条：厨房门下方应设进风固定百叶，有效截面积不小于$0.02m^2$；或在门扇下方与地面之间留15～20mm进风缝隙。

16.2.5 [Measures 2009] 12.2.5: The fixed air inlet louver shall be arranged on the lower part of the kitchen door, and the effective cross-sectional area shall be not less than $0.02m^2$; or 15-20mm air inlet gap shall be reserved between the lower part of the door leaf and ground.

16.2.6【09措施】12.2.6条：厨房不应布置在地下室。当布置在半地下室时，必须满足采光、通风的要求，并采取防水、防潮、排水及安全防护措施。

16.2.6 [Measures 2009] 12.2.6: Kitchens shall not be arranged in basements. When kitchens are arranged in semi-basements, the requirements for daylighting and ventilation must be met, and waterproof, moisture proofing, drainage and safety protective measures must be taken.

16.2.7【09措施】12.2.7条：低、多层住宅的厨房如采用煤为燃料的炉灶，必须设置直接通向室外的符合国家安全要求的烟道。烟道应采用不燃性材料制作。每层次烟道下端及主烟道的最下端应设掏灰口。

16.2.7 [Measures 2009] 12.2.7: For kitchens in low-rise and multi-storey buildings, if coal is used as the fuel, flues must be arranged and they are directly connected to the outside and meet the national safety requirements. The flues shall be made of incombustible materials. And dust clearing ports shall be arranged on the lower ends of the secondary flues and the lowest ends of the main flues on each floor.

16.2.8【09措施】12.2.8条：当厨房设有吸油烟机或燃气热水器时，应设专用排气管道排至室外。吸油烟机与燃气热水器的排气管道严禁合用。

16.2.8 [Measures 2009] 12.2.8: When range hoods or gas water heaters are arranged in the kitchen, the dedicated exhaust pipelines to the outdoor shall be arranged. The exhaust ducts of the range hoods and gas water heaters must not be shared mutually.

16.2.9【09措施】12.2.9条：厨房当上下层或毗连房间合用烟囱或排气管道时，应有防止串烟、串气的设施。高层住宅厨房采用垂直排油烟系统时，该系统应有分层的防火隔离措施。

16.2.9 [Measures 2009] 12.2.9: When the chimneys or exhaust pipelines are shared between upper and lower floors or between adjacent rooms and the kitchen, facilities shall be arranged to prevent smokes and gases from running into other chimneys or exhaust pipelines. For kitchens in high-rise residential buildings, when the vertical fume exhaust systems are adopted, the layered fire preventive and isolation measures shall be provided for the systems.

16.2.10 【09措施】12.2.10条：厨房如采用管道燃气，与燃气引入管贴邻或相邻，以及下部有管道通过的房间，其地面以下空间应采取防止燃气积聚的措施。如在地面至室内地坪面的墙身，采用密实性钢筋混凝土浇筑，或将室内地面以下空间与室外空气流通等措施。

16.2.10 [Measures 2009] 12.2.10: If pipeline gas is used in the kitchen, measures shall be taken to prevent accumulation of gas in the space below the ground of the rooms adjacent to gas introduction pipes and with pipelines going through the lower parts of rooms. For example, the walls between the ground and the indoor floor shall be poured with dense reinforced concrete, or the space below the indoor floor shall be ventilated.

16.2.11 【09措施】12.2.11条：烟囱或排气管道应伸出屋面。伸出高度应根据屋面形式、排出口周围遮挡物的高度、距离及寒冷地区积雪深度等因素确定，但不应小于0.6m。顶部应有防雨水、防倒灌、防强风的措施。

16.2.11 [Measures 2009] 12.2.11: Chimneys or exhaust pipelines shall extend from the roof. The extension height shall be determined in accordance with the factors of the roof forms, height and distances of the obstructions around the exhaust port, snow depth in cold regions, etc, and shall not be less than 0.6 m. The measures to prevent rain water, backflow and strong wind shall be adopted on the top.

16.2.12 【09措施】12.2.12条：厨房内灶具、吸油烟机及洗涤池等易产生噪声的设备，不宜安装在与卧室相邻的隔墙上。吊柜应挂装在有承重能力的墙上，如安装在轻质墙上应有安全可靠的固定措施。不同墙体材料上的安装构造可参见国标图集墙体类中的相关图集。

16.2.12 [Measures 2009] 12.2.12: The noise-prone equipment such as cookers, range hoods and sinks in kitchens shall not be installed on the partition walls adjacent to bedrooms. Wall cupboards shall be mounted on the walls with the bearing capacity, and if they are mounted on the light walls, the safe and reliable fixing measures shall be adopted. For the installation structures of different wall materials, refer to the relevant drawing collection of walls in the national standard drawing collection.

16.2.13 【09措施】12.2.13条：厨房内各种设备及管线设施等应进行综合设计，合理安排。管线宜隐蔽，并注意与厨房家具的配合。

16.2.13 [Measures 2009] 12.2.13: The various devices, pipeline and other facilities in kitchens shall be comprehensively designed and reasonably arranged. Pipelines shall be appropriately concealed, and have good coordination with the furniture in kitchens.

16.2.14 【09措施】12.2.14条：厨房内装修应易于清洁、防火、防潮，地面应防滑。

楼面不宜设地漏，当设有地漏时，楼地面应考虑防水、排水坡度和地漏返味措施。

16.2.14 [Measures 2009] 12.2.14: The decoration in kitchens shall be easy to clean, fireproof, and dampproof, and the floors shall be anti-slip. Floor drains shall not be advised to be set in the kitechen floor. When floor drains are arranged, consideration shall be given to waterproof, drainage gradients and preventive measures against the backflow of odor.

16.3 公用厨房

16.3 Communal Kitchen

16.3.1 【09措施】12.3.1条：公用厨房应按原料验收、储藏、冷冻等处理、主、副食加工（包括粗加工、切配、烹调等）、备餐、食具洗涤、消毒、存放等工艺流程合理布置（图16.3.1），严格做到原料与成品分开，生食与熟食分隔加工和存放。并应符合下列规定：

16.3.1 [Measures 2009] 12.3.1: Public kitchen should be in accordance with raw material acceptance, storage, freezing and other treatment, main, side food processing (including roughing, cutting, cooking, etc.）, prepared meals, utensils washing, disinfection, storage and other processes reasonable layout (see Fig. 16.3.1), strictly to separate raw materials and finished products, raw food and cooked food separated processing and storage, And shall comply with the following provisions:

1. 副食粗加工宜分设动物性食品、植物性食品、水产的工作台和清洗池，粗加工后的原料送入细加工间避免反流。遗留的废弃物应妥善处理；

1. Side food roughing should be divided into animal food, plant food, aquatic Table and cleaning pool, after roughing raw materials into the fine processing room to avoid reflux. The waste left behind should be properly disposed;

2. 冷荤成品应在单间内进行拼配，在其入口处应设有洗手设施的前室；

2. Cold meat The finished product shall be blending in a single room and the front room of the hand-washing facility shall be provided at its entrance;

3. 冷食制作间的入口处应设有通过式消毒设施；

3. The entrance to the cold food production room should be equipped with through-type disinfection facilities;

4. 垂直运输的食梯应生、熟分设。

4. Vertical transport of the food ladder should be raw, cooked and divided.

图 16.3.1　公用厨房组成与流程示意图
Fig. 16.3.1　Common kitchen composition and process diagram

16.3.2　【09措施】12.3.2条：公用厨房应附设厨房工作人员专用的更衣室、淋浴室、卫生间等。设在公共建筑内的公用厨房应有单独的人流、货流路线，不应干扰其他部分的使用。公用厨房如设置在地下室，宜避免紧邻锅炉房、变电间等易燃、易爆及忌水、汽的房间。

16.3.2　[Measures 2009] 12.3.2: The public kitchen should be equipped with a dressing room for kitchen staff, shower, toilet, etc. A communal kitchen located in a public building should have a separate flow of traffic, cargo routes and should not interfere with the use of other parts. Public kitchen if set in the basement, it is advisable to avoid close to the boiler room, variable room and other flammable, explosive and water, steam room.

16.3.3　【防火规】6.2.3条：宿舍、公寓建筑中的公共厨房和其他建筑（公共建筑）内的厨房，应采用耐火极限不低于2.00h的防火隔墙与其他部位分隔，墙上的门、窗应采用乙级防火门、窗，确有困难时，可采用防火卷帘。

16.3.3　[Code for fire protection design of buildings] 6.2.3: Kitchen in dormitories, public kitchens in apartment buildings and other buildings (public buildings) shall be based on fire resistance limits of not less than 2.00h of fireproof partitions separated from other parts, wall doors, Windows should use Class B fire doors, windows, there is a real difficulty, you can use fireproof shutter.

16.3.4　【09措施】12.3.7条：厨房设计必须注意对废水、废气、噪声、隔油的处理，并应符合有关部门的规定。

16.3.4　[Measures 2009] 12.3.7: Kitchen design must pay attention to the treatment of waste water, exhaust gas, noise, oil insulation, and should comply with the regulations of the relevant departments.

16.3.5　【09措施】12.3.8条：具备自然进风条件的厨房，应采用自然补风、机械排风系统。不具备自然进风条件的厨房，应采用机械送、排风系统。在进行烹饪作业时，厨房内应保持负压。

16.3.5　[Measures 2009] 12.3.8: Kitchen with natural air intake conditions shall be made of natural ventilation and mechanical exhaust system. Kitchens that do not have natural air intake conditions should be equipped with mechanical delivery and exhaust systems. Negative pressure should be maintained in the kitchen while cooking is in operation.

16.3.6 【09措施】12.3.9条：厨房的通风排气应符合下列规定。

16.3.6 [Measures 2009] 12.3.9: The ventilation and exhaust of the kitchen shall conform to the following provisions.

1. 各加工间均应处理好通风排气，并应防止厨房油烟气味污染餐厅；

1. Each processing room should be handled well ventilation and exhaust, and should prevent the smell of kitchen fumes pollution restaurant;

2. 热加工间应采用机械排风，也可设置出屋面的排风竖井或设有挡风板的天窗等有效自然通风设施；

2. Thermal processing room should use mechanical exhaust, can also be set out of the roof of the exhaust shaft or with the windshield of the skylight and other effective natural ventilation facilities;

3. 产生油烟的设备上部，应加设附有机械排风及油烟过滤器的排气装置，过滤器应便于清洗和更换；

3. The upper part of the equipment that produces the oil fume shall be equipped with an exhaust device with mechanical exhaust and fume filters, and the filter shall be easy to clean and replace;

4. 产生大量蒸汽的设备除应加设机械排风外，尚宜分隔成小间，防止结露并做好凝结水的引泄；

4. Equipment producing a large amount of steam, in addition to the mechanical exhaust should be added, it is advisable to separate into small rooms, to prevent condensation and do a good job of condensation water leakage;

5. 公共建筑中营业面积大于1000m² 的餐饮场所，其厨房烹饪操作间的排油烟罩及烹饪部位应设置自动灭火装置，且应在燃气或燃油管道上设置紧急事故自动切断装置。

5. Public buildings in the business area of more than 1000m² dining places, its kitchen cooking operation between the hood and cooking parts should be set up automatic fire extinguishing device, and should be in the gas or fuel pipeline set up an emergency automatic cutting device.

16.3.7 【09措施】12.3.10条：厨房使用燃煤炉灶时，烟囱应单独设置，烟道与排气道不得共用一个管道系统，烟囱材料应符合相应的耐火极限及出屋面的规定。

16.3.7 [Measures 2009] 12.3.10: When using a coal-fired stove in the kitchen, the chimney shall be set separately, the flue and the exhaust sidewalk shall not share a piping system, and the chimney material shall conform to the corresponding fire resistance limit and the requirements of the roof.

16.3.8【09措施】12.3.10条：厨房使用燃煤炉灶时，烟囱应单独设置，烟道与排气道不得共用一个管道系统，烟囱材料应符合相应的耐火极限及出屋面的规定。

16.3.8 [Measures 2009] 12.3.10: When using a coal-fired stove in the kitchen, the chimney shall be set separately, the flue and the exhaust sidewalk shall not share a piping system, and the chimney material shall conform to the corresponding fire resistance limit and the requirements of the roof.

16.3.9【09措施】12.3.11条：粗加工、切配、餐具清洗消毒、烹调等需经常冲洗场所的地面应设置排水沟，地面设计时应考虑下层空间的净空高度。

16.3.9 [Measures 2009] 12.3.11: Roughing, cutting, Tableware cleaning and disinfection, cooking and so on need to often rinse the site of the ground should be set up drains, ground design should take into account the clearance height of the lower space.

1. 当为新建厨房或有条件时，优选采用结构降板方式留出排水沟空间。

1. When the new kitchen or conditional, preferably the use of structural drop plate way to set aside the drain space.

2. 无条件或为改造工程时，采用架高地面方式解决排水沟所需的高度要求，此时应注意解决好厨房架高空间与其他空间入口的高差问题。

2. Unconditional is for the renovation of the project, the use of a high ground way to solve the drainage requirements of the height required, at this time should pay attention to solve the kitchen frame high space and other space inlet high difference problem.

3. 排水沟净空高度根据厨房工艺要求、排水量、排水沟坡度及长度等因素确定，但至少不小于200mm。每段排水沟的最低处宜设沉渣池，排水口设于池侧壁，且至少高出池底100mm。

3. Drain clearance height is determined according to kitchen process requirements, displacement, drain slope and length, but at least not less than 200mm. The lowest part of each drain should be a slag pool, the drain is located on the side wall of the pool, and at least 100mm above the bottom of the pool.

4. 沟壁宜选择光滑、不易挂油污的材料；沟侧壁与底面宜采用弧角交接。

4. Trench wall should choose smooth, not suitable for oil pollution materials, trench side wall and bottom surface should be used arc angle transfer.

5. 凉菜间、裱花间、备餐、集体用餐分装等对清洁要求高的专间内不得设置明沟。

5. Cold room, framed flower room, meal preparation, group dining and other high cleaning requirements of the special room must not be set up in the Minggou.

6. 与排水沟构造有关的国标图集有：《建筑防腐蚀构造》08J333、《窗井、设备吊装口、排水沟、集水坑》07J306。

6. The national standard Atlas related to the drainage structure is: *building anti-corrosion structure* 08J333, *window well, equipment hoisting port, drain, set puddle* 07J306.

16.3.10 【09措施】12.3.12条：厨房的含油废水应与其他排水分流设计。含油废水应经隔油设施处理，存油部分应便于清运和管理。经隔油处理后达到规定排放标准的含油废水方可排出。

16.3.10 [Measures 2009] 12.3.12: The oily wastewater in the kitchen should be designed with other drainage diversions. Oily wastewater should be treated by oil isolation facilities, and the oil storage part should be convenient for clearance and management. Oil-bearing wastewater that meets the prescribed emission standards after oil isolation can be discharged.

16.3.11 【09措施】12.3.13条：隔油池不应设在厨房、饮食制作间内，但应便于清运。

16.3.11 [Measures 2009] 12.3.13: Oil trap should not be located in the kitchen, food production room, but should be easy to clearance.

16.3.12 【09措施】12.3.14条：厨房内的主、副食品库应考虑防水、防虫及防鼠害。

16.3.12 [Measures 2009] 12.3.14: The main and paid food bank in the kitchen should consider waterproofing, insect protection and rodent protection.

16.3.13 【09措施】12.3.15条：固体废弃物堆放地不应设在公共场所，应采取妥善的存放方式，防止二次污染。

16.3.13 [Measures 2009] 12.3.15: Solid waste dumps should not be located in public places and should be properly stored in such a way as to prevent two of pollution.

16.3.14 【09措施】12.3.16条：厨房内装修应易于清洁、防火、防潮，地面应防滑。

16.3.14 [Measures 2009] 12.3.16: Kitchen decoration should be easy to clean, fireproof, moisture-proof, the ground should be anti-skid.

16.3.15 【09措施】12.3.17条：各加工间室内构造应符合下列规定。

16.3.15 [Measures 2009] 12.3.17: The interior structure of each processing room shall conform to the following provisions.

1. 地面应采用耐磨、不渗水、耐腐蚀、防滑、易清洗的材料，并应处理好地面排水。

1. The ground should use wear-resistant, non-seepage, corrosion-resistant, anti-skid, easy to clean materials, and should handle the ground drainage.

2. 墙面、隔断及工作台、水池等设施均应采用无毒、光滑易洁的材料，各阴角宜做成弧形。粗加工、切配、就餐用具清洗消毒和烹调等需经常冲洗的场所、易潮湿场所，应设不低于1.5m高宜清洗、光滑的墙裙，各类专间的墙裙应铺设到墙顶（或吊顶）。

2. Wall, partition and workbench, pool and other facilities should be non-toxic, smooth and easy to clean materials, each yin angle should be made into an arc. Roughing, cutting, dining

utensils cleaning and disinfection and cooking need to often rinse the place, easy to wet places, should be set not less than 1.5m high appropriate cleaning, smooth wall skirt, all kinds of special wall skirt should be laid to the top of the wall（or ceiling）.

3. 顶棚应选用无毒、无异味、不吸水、表面光洁、耐腐蚀、耐温、浅色材料。顶棚与横梁或墙壁结合处，宜有一定弧度（曲率半径在3cm以上）；水蒸气较多场所的顶棚应有适当坡度，在结构上减少凝结水滴落。专间、备餐、烹调、干净餐具存放及其他半成品、成品暴露等场所的顶棚若为不平整的结构或有管道通过时，应设平整易于清洁的吊顶。

3. Roof should be non-toxic, non-odor, do not absorb water, smooth surface, corrosion resistance, temperature resistance, light color materials. When the ceiling is combined with a beam or wall, it is advisable to have a certain arc（curvature radius above 3cm）, and the ceiling of more water vapour should have the appropriate slope, which reduces the condensation water drip in structure. For special rooms, meal preparation, cooking, clean Tableware storage and other semi-finished products, finished product exposure and other places of the roof if it is uneven structure or pipe through, should be set up easy to clean ceiling.

4. 与外界相通的门窗应选用易于拆下清洗且不生锈的、防蝇纱网的门窗，或设置空气幕。

4. Doors and windows connected to the outside world should be easy to remove clean and not rusty, fly-proof yarn mesh doors and windows, or set up air curtain.

16.3.16 【09措施】12.3.18条：加工间直接采光时，其侧面采光窗洞口面积不宜小于地面面积的1/6。自然通风时，通风开口面积不应小于地面面积的1/10。

16.3.16 [Measures 2009] 12.3.18: When the direct lighting between the machining room, its side lighting window hole area should not be less than 1/6 of the ground area.When naturally ventilated, the ventilation opening area should not be less than 1/10 of the ground area.

16.3.17 【09措施】12.3.19条：厨房的电源进线应留有一定余量，配电箱应留有一定数量的备用回路插座。电气设备、灯具、管路应有防潮措施并采用漏电保护器。

16.3.17 [Measures 2009] 12.3.19: Kitchen power inlet should be left with a certain margin, distribution box should leave a certain number of spare loop socket. Electrical equipment, lamps, piping should be moisture-proof measures and the use of leakage protector.

17

卫生间
Toilets

17.1　一般规定

17.1　General Provisions

17.1.1　【09措施】13.1.1条：卫生间设计中应注意合理布置卫生器具，使管道集中、隐蔽，重视平面及空间的充分利用。

17.1.1　[Measures 2009] 13.1.1: In the design for toilets, attention shall be paid to reasonable arrangement of sanitary ware. Pipelines shall be arranged in a concentrated and concealed way significance shall be paid to full utilization of the plane and space.

17.1.2　【09措施】13.1.2条：卫生间特别是公用卫生间应注意保持良好的通风换气和采光。无自然通风的卫生间应采取有效的机械通风换气措施。

17.1.2　[Measures 2009] 13.1.2: In toilets, especially public toilets, appropriate ventilation and lighting shall be provided. For toilets without natural ventilation, effective mechanical ventilation measures shall be adopted.

17.1.3　【09措施】13.1.3条：卫生间应有良好的防水、防潮、排水、防滑及隔声功能。

17.1.3　[Measures 2009] 13.1.3: Toilets shall be provided with good waterproof, moisture proofing, drainage, skid resistance and sound insulation.

17.1.4　【09措施】13.1.4条：公共卫生间的位置选择应注意使用方便、位置隐蔽，并注意气味、潮气、噪声等对其他房间的影响和干扰。

17.1.4　[Measures 2009] 13.1.4: For position selection for public toilets, the effect of easy use and concealed position shall be achieved, and attention shall be paid to the influence and interference of odor, moisture, noise, etc. on other rooms.

17.1.5　【通则】6.5.1条：厕所、盥洗室、浴室应符合下列规定。

17.1.5　[Code for design of civil buildings] 6.5.1: Toilets, washrooms and bathrooms shall conform to the following provisions.

1. 建筑物的厕所、盥洗室、浴室不应直接布置在餐厅、食品加工、食品贮存、医药、医疗、变配电等有严格卫生要求或防水、防潮要求用房的上层；除本套住宅外，住宅卫生间不应直接布置在下层的卧室、起居室、厨房和餐厅的上层。

1. Toilets, washrooms and bathrooms in buildings shall not be directly arranged above the floors of dining halls, food processing places, food storage rooms, pharmaceutical rooms, medical rooms, electric transformer and distribution rooms, and rooms with strict hygiene

requirements or waterproof and moisture proofing requirements. Except for this residence, toilets in residential buildings shall not be directly arranged above bedrooms, living rooms, kitchens and dining halls on the lower floor.

2. 卫生设备配置的数量应符合专用建筑设计规范的规定，在公用厕所男女厕位的比例中，应适当加大女厕位比例。

2. The quantities of the sanitary equipment arranged shall conform to the provisions of the special specifications for building design. In the proportion of toilet positions for men and women in public toilets, the proportion of female toilet positions shall be appropriately increased.

3. 卫生用房宜有天然采光和不向邻室对流的自然通风，无直接自然通风和严寒及寒冷地区用房宜设自然通风道；当自然通风不能满足通风换气要求时，应采用机械通风。

3. Sanitary rooms shall be appropriately provided with natural daylighting and natural ventilation without convection current to adjacent rooms; for rooms without direct natural ventilation and in cold and severely cold regions, natural ventilation ducts shall be appropriately arranged; when the natural ventilation shall not meet the ventilation requirements, mechanical ventilation shall be adopted.

4. 楼地面、楼地面沟槽、管道穿楼板及楼板接墙面处应严密防水、防渗漏。

4. Strict waterproof and leakage prevention shall be carried out for floors, grooves, positions where pipelines penetrate the floor and junctions between floors and walls.

5. 楼地面、墙面或墙裙的面层应采用不吸水、不吸污、耐腐蚀、易清洗的材料。

5. The non-water-absorbent, non-sewage-absorbent, corrosion resistant and easy-to-clean materials shall be adopted for the surface layers of the floor ground, walls or wall skirting.

6. 楼地面应防滑，楼地面标高宜略低于走道标高，并应有坡度坡向地漏或水沟。

6. The floor ground shall be anti-slip; the floor elevation shall be slightly less than that of the corridor; and the floor shall slope to floor drains or ditches.

7. 室内上下水管和浴室顶棚应防冷凝水下滴，浴室热水管应防止烫人。

7. The indoor water supply and drainage pipes and bathroom ceilings shall be equipped with preventive devices for condensate drip, and scald prevention shall be carried out for the hot water pipes in bathrooms.

8. 公用男女厕所宜分设前室，或有遮挡措施。

8. For public toilets for men and women, anterooms shall be appropriately arranged, or shielding measures shall be taken.

9. 公用厕所宜设置独立的清洁间。

9. Public toilets shall be with separate broom closet.

17.2 公共卫生间

17.2 Public Toilets

17.2.1 【公厕设计标准】3.0.6 条：附属式公共厕所应按场所和建筑设计要求分为一类和二类。附属式公共厕所类别的设置应符合表 17.2.1 的规定。

17.2.1 [Standards for design of urban public toilets] 3.0.6: Accessory public toilets are classified into class I and class II in accordance with the design requirements of places and buildings. The classification of accessory public toilets shall conform to the provisions in Tab. 17.2.1.

附属式公共厕所类别　　　　　　　　　　　　　　　表 17.2.1
Categories of Accessory Public Toilets　　　　　　　Tab. 17.2.1

设置场所 Arrangement site	类别 Category
大型商场、宾馆、饭店、展览馆、机场、车站、影剧院、大型体育场馆、综合性商业大楼和二、三级医院等公共建筑 Public buildings like large malls, hotels, restaurants, exhibition halls, airports, bus stations, movie theaters, large stadiums, comprehensive commercial buildings, grades II and III hospitals, etc	一类 Class I
一般商场（含超市）、专业性服务机关单位、体育场馆和一级医院等公共建筑 Public buildings such as general shopping malls (including supermarkets) , governmental institutes for pecial services, stadiums and class I hospitals, etc	二类 Class II

注：附属式公共厕所二类为设置场所的最低标准。
Note: For accessory public toilets, class II is the minimum standard in site arrangement.

17.2.2 【公厕设计标准】4.1.1 条及 4.1.2 条：在人流集中的场所，女厕位与男厕位（含小便站位，下同）的比例不应小于 2∶1。在其他场所，男女厕位比例可按下式计算：$R=1.5w/m$。

17.2.2 [Standards for design of urban public Toilets] 4.1.1 and 4.1.2: In places of high population, the proportion of female and male toilet positions (including urinal positions, similarly hereinafter) shall not be less than 2∶1. In other places, proportion of the male and female toilet positions shall be calculated as per the following formula: $R=1.5w/m$.

式中　　R——女厕位数与男厕位数的比值；

Where　R——Ratio between the female toilet positions and the male toilet positions;

　　　　1.5——女性与男性如厕占用时间比值；

　　　　1.5——Ratio of toilet time for males and females;

w——女性如厕测算人数；

w——Estimated number of females in toilet;

m——男性如厕测算人数。

m——Estimated number of males in toilet.

17.2.3 【公厕设计标准】4.1.4 条：公共厕所男女厕位（坐位、蹲位和站位）与其数量宜符合表 17.2.3-1 和表 17.2.3-2 的规定。

17.2.3 [Standards for design of urban public toilets] 4.1.4: The male and female toilet positions (sitting positions, squatting positions and standing positions) in public toilets and the quantities shall conform to the provisions in Tab. 17.2.3-1 and Tab. 17.2.3-2.

男厕位及数量（个） 表 17.2.3-1
Male Toilet Positions and Quantities (Piece) Tab. 17.2.3-1

男厕位总数 Total number of toilet positions for men	坐位 Sitting positions	蹲位 Squatting positions	站位 Standing positions
1	0	1	0
2	0	1	1
3	1	1	1
4	1	1	2
5~10	1	2~4	2~5
11~20	2	4~9	5~9
21~30	3	9~13	9~14

注：表中厕位不包含无障碍厕位。

Note: the toilet positions in the table does not include the barrier-free toilet positions.

女厕位及数量（个） 表 17.2.3-2
Female Toilet Positions and Quantity (Piece) Tab. 17.2.3-2

女厕位总数 Total number of toilet positions female	坐位 Sitting positions	蹲位 Squatting positions
1	0	1
2	1	1
3~6	1	2~5
7~10	2	5~8
11~20	3	8~17
21~30	4	17~26

注：表中厕位不包含无障碍厕位。

Note: The toilet positions in the table does not include the barrier-free toilet positions.

17.2.4 【公厕设计标准】4.2.1 条：公共场所公共厕所厕位服务人数应符合表 17.2.4 的规定。

17.2.4 [Standards for design of urban public toilets] 4.2.1: The number of people served in the toilet positions of the public toilets in the public places shall conform to the provisions in Tab. 17.2.4.

公共场所公共厕所厕位服务人数　　　　表 17.2.4
Number of People Served in Toilet Positions of Public Toilets in Public Places　Tab. 17.2.4

公共场所 Public places	服务人员 [人/(厕位·d)] People Served [person/(toilet position · day)]	
	男 Male	女 Female
广场、街道 Squares and streets	500	350
车站、码头 Stations and docks	150	100
公园 Parks	200	130
体育场所 Sports places	150	100
海滨活动场所 Seashore activity places	60	40

17.2.5 【公厕设计标准】4.2.2 条：商场、超市和商业街公共厕所厕位数应符合表 17.2.5 的规定。

17.2.5 [Standard for design of urban public toilets] 4.2.2: The number of toilet positions in malls, supermarkets and commercial streets shall conform to the provisions in Tab. 17.2.5.

商场、超市和商业街公共厕所厕位数　　　　表 17.2.5
Number of Toilet Positions in Malls, Supermarkets and Commercial Streets　Tab. 17.2.5

购物面积（m²） Shopping area (m²)	男厕位（个） Number of male toilet positions (toilet position)	女厕位（个） Number of female toilet positions (toilet position)
500 以下 Below 500	1	2
501~1000	2	4
1001~2000	3	6
2001~4000	5	10
≥4000	每增加 2000 m² 男厕位增加 2 个，女厕位增加 4 个 Two male toilet positions and four female toilet positions shall be added for every increase of 2000m² in area	

注：1. 按男女如厕人数相当时考虑；
Notes: 1. The above arrangement is made on the basis of the equivalent quantities of males and females in toilet;
2. 商业街应按各商店的面积合并计算后，按上表比例配置。
2. For the commercial street, the arrangement of toilet positions in accordance with the proportions in the above table shall be made as per the total area of all the stores.

17.2.6 【公厕设计标准】4.2.3 条：饭馆、咖啡店、小吃店和快餐店等餐饮场所公共厕所厕位数应符合表 17.2.6 的规定。

17.2.6 [Standard for design of urban public toilets] 4.2.3: The number of public toilet positions in catering places such as restaurants, coffee shops, snack bars, fast food restaurants, etc. shall conform to the provisions in Tab. 17.2.6.

饭馆、咖啡店等餐饮场所公共厕所厕位数　　　　　表 17.2.6
Quantity of Public Toilet Positions in Catering Places of Restaurants, Coffee Shops, etc　　Tab. 17.2.6

设施 Facilities	男 Male	女 Female
厕位 Toilet positions	50 座位以下至少设 1 个；100 座位以下设 2 个；超过 100 座位每增加 100 座位增设 1 个 At least one position shall be arranged when the seat quantity is below 50; two positions shall be arranged when the seat quantity is below 100; one position shall be increased for every 100 seats when the seat quantity exceeds 100	50 座位以下至少设 2 个；100 座位以下设 3 个；超过 100 座位每增加 65 座位增设 1 个 At least two positions shall be arranged when the seat quantity is below 50; three positions shall be arranged when the seat quantity is below 100; one position shall be increased for every 65 seats when the seat quantity exceeds 100

注：按男女如厕人数相当时考虑。
Note: The above arrangement is made on the basis of the equivalent quantities of males and females in toilet.

17.2.7 【公厕设计标准】4.2.4 条：体育场馆、展览馆、影剧院、音乐厅等公共文体娱乐场所公共厕所厕位数应符合表 17.2.7 的规定。

17.2.7 [Standard for design of urban public toilets] 4.2.4: The quantities of public toilet positions in cultural entertainment places of stadiums, exhibition halls, movie theaters, concert halls, etc, shall conform to the provisions in Tab. 17.2.7.

体育场馆、展览馆等公共文体娱乐场所公共厕所厕位数　　　　　表 17.2.7
Quantities of Public Toilet Positions in Cultural Entertainment Places of Stadiums, Exhibition Halls, etc.　　Tab. 17.2.7

设施 Facilities	男 Male	女 Female
坐位、蹲位 Sitting positions and squatting positions	250 座以下设 1 个，每增加 1~500 座增设 1 个 One toilet position shall be arranged when the seat quantity is below 250, and one position shall be increased every time when 1 to 500 seats are increased	不超过 40 座的设 1 个；41~70 座设 3 个；71~100 座设 4 个；每增加 1~40 座增设 1 个 One toilet position shall be arranged when the seat quantity is below 40; three toilet positions shall be arranged when the seat quantity is 41-70; four toilet positions shall be arranged when the seat quantity is 71-100; one additional toilet position shall be arranged when 1-40 seats are added

续表

设施 Facilities	男 Male	女 Female
站位 Standing positions	100 座以下设 2 个，每增加 1~80 座增设 1 个 Two toilet positions shall be arranged when the seat quantity is below 100, and one additional toilet position shall be arranged when 1~80 seats are added	无 No

注：1. 若附有其他服务设施内容（如餐饮等），应按相应内容增加配置；

Notes: 1. If other service facilities (such as catering, etc) are affiliated, the toilet positions shall be added as per the corresponding content;

2. 有人员聚集场所的广场内，应增建馆外人员使用的附属或独立厕所。

2. Accessory or separate public toilets for outsiders shall be arranged in squares at which people gather.

17.2.8 【公厕设计标准】4.2.5 条：机场、火车站、公共汽（电）车和长途汽车始末站、地下铁道的车站、城市轻轨车站、交通枢纽站、高速路休息区、综合性服务楼和服务性单位公共厕所厕位数应符合表 17.2.8 的规定。

17.2.8 [Standard for design of urban public toilets] 4.2.5: The number of toilet positions in public toilets in airports, railway stations, terminal stations of buses (trams) and coaches, underground railway stations, urban light rail stations, transportation hub stations, highway service areas, comprehensive service buildings and service institutions shall conform to the provisions in Tab. 17.2.8.

机场、火车站、综合性服务楼和服务性单位公共厕所厕位数　　　表 17.2.8

Number of Toilet Positions of Public Toilets in Airports, Railway Stations, Comprehensive Service Buildings and Service Institutions　　　Tab. 17.2.8

设施 Facilities	男（人/h） Male (person/h)	女（人/h） Female (person/h)
厕位 Toilet positions	100 人以下设 2 个；每增加 60 人增设 1 个 Two positions shall be provided when the number is below 100; one additional position is set for every 60 additional people	100 人以下设 4 个；每增加 30 人增设 1 个 Four toilet positions shall be arranged for people below 100; one additional toilet position is set for every 30 additional people

17.2.9 【公厕设计标准】4.2.6 条：公共厕所的男女厕所间应至少各设一个无障碍厕位。

17.2.9 [Standard for design of urban public toilets] 4.2.6: At least one barrier-free toilet position shall be arranged in the male and female public toilet compartments separately.

17.2.10 【公厕设计标准】4.2.8 条：洗手盆应按厕位数设置，洗手盆数量设置要求应符合表 17.2.10 的规定。

17.2.10 [Standard for design of urban public toilets] 4.2.8: Wash basins shall be arranged on the basis of the number of toilet positions, and the number of wash basins to be arranged shall conform to the provisions in Tab. 17.2.10.

洗手盆数量设置要求　　　　　　　　　　　　表 17.2.10
Requirements for the Number of Wash Basins to be Arranged　　Tab. 17.2.10

厕位数（个） Number of toilet positions (piece)	洗手盆数（个） Number of wash basins (piece)	备注 Remarks
4 以下 Below 4	1	1）男女厕所宜分别计算，分别设置； 1) Male and female toilets shall be calculated separately and set respectively; 2）当女厕所洗手盆数 $n \geq 5$ 时，实际设置数 N 应按下式计算：$N=0.8n$ 2) When the number of wash basins in the female toilet "n" is greater than or equal to five, the actual number of wash basins arranged "N" shall be calculated as per the following formula: $N=0.8n$
5～8	2	
9～21	每增 4 厕位增设 1 个 One additional wash basin is set for every four additional toilet positions	
22 以上 Above 22	每增 5 厕位增设 1 个 One additional wash basin is set for every five additional toilet positions	

注：洗手盆为 1 个时可不设儿童洗手盆。
Note: When there is one wash basin, children's wash basin may not be set.

17.2.11 【公厕设计标准】4.2.9 条：公共厕所应至少设置一个清洁池。

17.2.11 [Standard for design of urban public toilets] 4.2.9: At least one broom sink shall be arranged in a public toilet.

17.2.12 【公厕设计标准】4.2.10 条：公共厕所第三卫生间应在下列各类厕所中设置。

17.2.12 [Standard for design of urban public toilets] 4.2.10: Unisex toilets shall be arranged in following public toilets.

1. 一类固定式公共厕所；

1. Class I fixed-type public toilets;

2. 二级及以上医院的公共厕所；

2. Public toilets in hospitals of grade II or above;

3. 商业区、重要公共设施及重要交通客运设施区域的活动式公共厕所。

3. Movable public toilets in commercial areas, and areas with important public facilities and important traffic passenger transport facilities.

17.2.13 【公厕设计标准】4.3.1 条：公共厕所的平面设计应符合下列规定。

17.2.13 [Standard for design of urban public toilets] 4.3.1: The plan design for public toilets shall conform to the following provisions.

1. 大门应能双向开启；

1. The door can be bidirectionally opened;

2. 宜将大便间、小便间、洗手间分区设置；

2. The closet pans, urinals and washrooms shall be arranged in separate partitions;

3. 厕所内应分设男、女通道，在男、女进门处应设视线屏蔽；

3. Male and female passages shall be separately arranged in toilets, and sight shielding objects shall be arranged at the male and female entrances respectively;

4. 当男、女厕所厕位分别超过20个时，应设双出入口；

4. When the number of toilet positions for male and female respectively exceeds 20, two-way entrances and exits shall be arranged;

5. 每个大便器应有一个独立的厕位间。

5. Each closet pan shall be provided with an independent toilet cubicle.

17.2.14 【公厕设计标准】4.3.3 条：第三卫生间（图 17.2.14）的设置应符合下列规定。

17.2.14 [Standard for design of urban public toilets] 4.3.3: The arrangement of unisex toilets（Fig.17.2.14）shall conform to the following provisions.

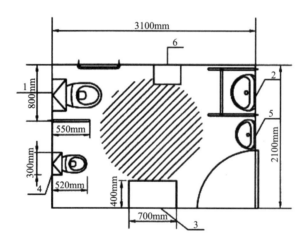

图 17.2.14 第三卫生间平面布置图
Fig. 17.2.14 Plan Layout of Third Toilet

1- 成人坐便器；2- 成人洗手盆；3- 可折叠的多功能台；4- 儿童坐便器；5- 儿童洗手盆；6- 可折叠的儿童安全座椅
1 - Adult pedestal pan; 2 - Adult wash basin; 3 - Foldable multi-functional counter; 4 - Children's pedestal pan; 5-Children's wash basin; 6 - Foldable children's safety seat

1. 位置宜靠近公共厕所入口，应方便行动不便者进入，轮椅回转直径不应小于 1.50m；

1. The position shall be close to the entrance of the public toilet, to facilitate the entry of the people with mobility problems, and the turn-around diameter for wheelchairs shall be not less than 1.50m;

2. 内部设施宜包括成人坐便器、成人洗手盆、多功能台、安全抓杆、挂衣钩和呼叫器、儿童坐便器、儿童洗手盆、儿童安全座椅；

2. The interior facilities shall appropriately include adult pedestal pans, adult wash basins, multi-functional tables, safety grab bars, clothes hooks, beepers, children's pedestal pans, children's pedestal pans and children's safety seats;

3. 使用面积不应小于 6.5m²；

3. The usable floor area shall be not less than 6.5m²;

4. 地面应防滑、不积水；

4. The floor shall be anti-slip and free of accumulated water;

5. 成人坐便器、洗手盆、多功能台、安全抓杆、挂衣钩、呼叫按钮的设置应符合现行国家标准《无障碍设计规范》GB 50763 的有关规定；

5. The arrangement of adult toilet bowls, washbasins, multi-functional counters, safety grab bars, clothes hooks and call buttons shall conform to the relevant provisions in the current national standard *Code for Accessibility design* GB 50763;

6. 多功能台和儿童安全座椅应可折叠并设有安全带，儿童安全座椅长度宜为 280mm，宽度宜为 260mm，高度宜为 500mm，离地高度宜为 400mm。

6. Multi-functional counters and children's safety seats shall be foldable and provided with safety belts, the length of children's safety seats shall be 280 mm, the width 250 mm, the height 500 mm, and the ground clearance 400 mm.

17.2.15【公厕设计标准】4.4.2 条：公共厕所卫生洁具的使用空间应符合表 17.2.15 的规定。

17.2.15 [Standard for design of urban public toilets] 4.4.2: The use space of sanitary ware in public toilets shall conform to the provisions in Tab. 17.2.15.

常用卫生洁具平面尺寸和使用空间　　　　表 17.2.15
Plane Dimensions and Use Spaces of Common Sanitary Ware　　Tab. 17.2.15

洁具 Sanitary ware	平面尺寸（mm×mm） Plane dimensions (mm × mm)	使用空间（mm）（宽×进深） Use space (mm)(width × depth)
洗手盆 Wash basin	500 × 400	800 × 600
坐便器（低位、整体水箱） Pedestal pan (low-level and integral water tank)	700 × 500	800 × 600
蹲便器 Squatting pan	800 × 500	800 × 600
卫生间便盆（靠墙式或悬挂式） Pedestal pans in toilets (wall-mounted or suspended-type)	600 × 400	800 × 600
碗形小便器 Bowl-shaped urinals	400 × 400	700 × 500

续表

洁具 Sanitary ware	平面尺寸（mm×mm） Plane dimensions（mm×mm）	使用空间（mm）（宽 × 进深） Use space（mm）（width × depth）
水槽（桶/清洁工用） Sink（barrels/for cleaners）	500 × 400	800 × 800
烘手器 Hand dryer	400 × 300	650 × 600

注：使用空间是指除了洁具占用的空间，使用者在使用时所需空间及日常清洁和维护所需空间。使用空间与洁具尺寸是相互联系的。洁具的尺寸将决定使用空间的位置。

Notes: Use space means the space required by the user during use and required for daily cleaning and maintenance apart from the space occupied by the sanitary ware. The use space and the sizes of the sanitary ware are interrelated. The dimensions of the sanitary ware will determine the position of use space.

17.2.16【公厕设计标准】4.4.3 条：公共厕所坐便器、蹲便器、小便器、烘手器和洗手盆需要的人体使用空间最小尺寸应满足图示要求（图 17.2.16-1 ~ 图 17.2.16-5）。

17.2.16 [Standard for design of urban public toilets] 4.4.3: The minimum dimensions of required use spaces for people of pedestal pans, squatting pans, urinals, hand dryers and wash basins in public toilets shall meet the requirements in Fig. 17.2.16-1 ~ Fig.17.2.16-5.

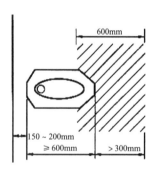

图 17.2.16-1 蹲便器人体使用空间
Fig. 17.2.16-1 Use Space for People of Squatting Pans

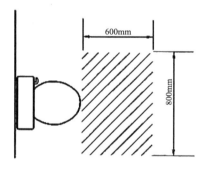

图 17.2.16-2 坐便器人体使用空间
Fig.17.2.16-2 Use Space for people of Pedestal Pans

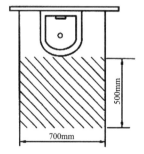

图 17.2.16-3 小便器人体使用空间
Fig. 17.2.16-3 Use Space for People of Urinals

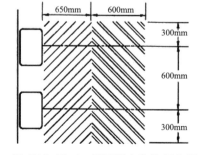

图 17.2.16-4 烘手器人体使用空间
Fig.17.2.16-4 Use Space for people of Hand Dryers

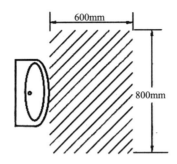

图 17.2.16-5 洗手盆人体使用空间
Fig. 17.2.16-5　Use Spaces for People of Wash Basins

17.2.17　【公厕设计标准】4.4.8 条：在洁具可能出现的每种组合形式中，一个洁具占用另一相邻洁具使用空间重叠最大部分可以增加到 100mm。平面组合可根据这一规定的数据设置（图 17.2.17）。

17.2.17　[Standard for design of urban public toilets] 4.4.8: Among all the possible existing combination forms of the sanitary ware, the maximum overlapping part of the use space of one sanitary ware occupied with another adjacent sanitary ware can be increased to 100mm. The plane combination can be arranged in accordance with the data in this Standard (Fig. 17.2.17) .

17.2.18　【公厕设计标准】4.4.9 条：在有坐便器的厕所间内设置洗手洁具时，厕所间的尺寸应由洁具的安装、门的宽度和开启方向来决定。450mm 的无障碍圆形空间不应被重叠使用空间占据。洁具的轴线间和临近墙面的距离不应小于 400mm。在有厕位隔间的地方应为坐便器和水箱设置宽 800mm，深 600mm 的使用空间（图 17.2.17），并应预备安装厕纸架，衣物挂钩和废物处理箱的空间。

17.2.18　[Standard for design of urban public toilets] 4.4.9: When hand-washing sanitary ware is arranged in the toilets with pedestal pans, the dimensions of the toilet rooms shall be determined through the installation situation of fixtures of the sanitary ware, widths and opening directions of doors. The 450mm barrier-free circular space shall not be occupied with the overlapped use space. The distance between the axis of the sanitary ware and the nearby wall shall be not less than 400mm. In the places with toilet cubicles, the 800mm (width) × 600mm (depth) use space shall be arranged for pedestal pans and water tanks, and the space for installation of toilet paper holders, clothes hooks and waste treatment boxes shall be reserved see Fig.17.2-17.

17.2.19　【通则】6.5.2 条：厕所和浴室隔间的平面尺寸不应小于表 17.2.19 的规定。

17.2.19　[Code for design of civil buildings] 6.5.2: The plane dimensions of cubicles in toilets and bathrooms shall be not less than the provisions in Tab. 17.2.19.

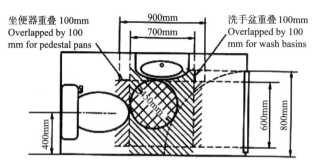

图 17.2.17 洁具平面组合使用空间重叠

Fig. 17.2.17 Overlapping of Plane Combined Use Space of Sanitary Ware

厕所和浴室隔间平面尺寸（图 17.2.19） 表 17.2.19

Plane Dimensions of Cubicles in Toilets and Bathrooms (see Fig. 17.2.19) Tab. 17.2.19

类别 Category	平面尺寸（m）（宽度 × 深度） Plane dimensions (m) (width × depth)
外开门的厕所隔间 Toilet cubicles with the outward opening door	0.90 × 1.20
内开门的厕所隔间 Toilet cubicles with doors to be opened inward	0.90 × 1.40
医院患者专用厕所隔间 Dedicated toilet cubicles for patients in hospitals	1.10 × 1.40
无障碍厕所隔间 Barrier-free toilet position cubicles	1.40 × 1.80（改建用 1.00 × 2.00） 1.40 × 1.80 (1.00 × 2.00 in reconstruction)
外开门淋浴隔间 Shower cubicles with outward opening doors	1.00 × 1.20
内设更衣凳的淋浴隔间 Shower cubicles with dressing stools	1.00 ×（1.00+0.60）
无障碍专用浴室隔间 Special barrier-free bathroom stalls	盆浴（门扇向外开启）2.00 × 2.25 Bathtubs (with the door leaves to be opened outward) 2.00 × 2.25 淋浴（门扇向外开启）1.50 × 2.35 Showers (with the door leaves to be opened outward) 1.50 × 2.35

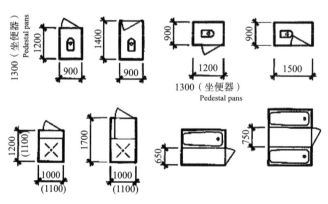

图 17.2.19 厕所、浴室隔间最小尺寸

Fig. 17.2.19 Minimum Dimensions of Cubicles of Toilets and Bathrooms

17.2.20 【通则】6.5.3 条：卫生设备间距应符合下列规定（图 17.2.20）。

17.2.20 [Code for design of civil buildings] 6.5.3: The spacing of sanitary equipment shall conform to the following provisions (Fig. 17.2.20).

1. 洗脸盆或盥洗槽水嘴中心与侧墙面净距不宜小于 0.55m。

1. The clear distance between centers of water nozzles of wash basins or sinks and side walls shall be not less than 0.55m.

2. 并列洗脸盆或盥洗槽水嘴中心间距不应小于 0.70m。

2. The center spacing between the faucets of paratactic washbasins or washing sinks shall be not less than 0.70m.

3. 单侧并列洗脸盆或盥洗槽外沿至对面墙的净距不应小于 1.25m。

3. The clear distance from the outer edges of paratactic washbasins on one side to the opposite walls shall be not less than 1.25m.

4. 双侧并列洗脸盆或盥洗槽外沿之间的净距不应小于 1.80m。

4. The clear distance between the outer edges of paratactic washbasins or sinks on both sides shall be not less than 1.80m.

5. 浴盆长边至对面墙面的净距不应小于 0.65m，无障碍盆浴间短边净宽度不应小于 2m。

5. The clear distance from the long edges of bath tubs to the opposite walls shall be not less than 0.65m; net width of the short sides of barrier-free bathtub rooms shall be not less than 2m.

6. 并列小便器的中心距离不应小于 0.65m。

6. The center distance between paratactic urinals shall be not less than 0.65m.

7. 单侧厕所隔间至对面墙面的净距：当采用内开门时，不应小于 1.10m；当采用外开门时不应小于 1.30m；双侧厕所隔间之间的净距：当采用内开门时，不应小于 1.10m；当采用外开门时不应小于 1.30m。

7. The clear distance from the one-side toilet cubicles to the opposite wall: when the inward opening door is adopted, the clear distance shall be not less than 1.10m; when the outward opening door is adopted, the clear distance shall be not less than 1.30m; the clear distance between the two-sided toilet cubicles: when the inward opening door is adopted, the clear distance shall be not less than 1.10m; when the outward opening door is adopted, the clear distance shall be not less than 1.30m.

8. 单侧厕所隔间至对面小便器或小便槽外沿的净距：当采用内开门时，不应小于 1.10m；当采用外开门时，不应小于 1.30m。

8. The clear distance from the one-sided toilet cubicles to the opposite urinal or the outer edge of the urinal trough: when the inward opening door is adopted, the clear distance shall be

not less than 1.10m; when the outward opening door is adopted, the clear distance shall be not less than 1.30m.

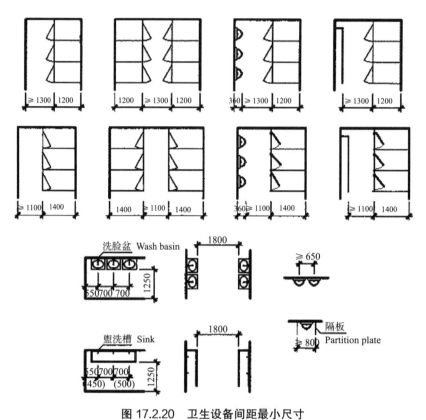

图 17.2.20 卫生设备间距最小尺寸
Fig. 17.2.20 Minimum Spacing of Sanitary Equipment

17.3 住宅卫生间

17.3 Toilets of Residential Buildings

17.3.1 【住设规】5.4.1 条及 5.4.2 条：每套住宅应设卫生间，应至少配置便器、洗浴器、洗面器三件卫生设备或为其预留设置位置及条件。三件卫生设备集中配置的卫生间的使用面积不应小于 2.50 m²。卫生间可根据使用功能要求组合不同的设备。不同组合的空间使用面积应符合下列规定（卫生间布置最小尺寸平面，见图 17.3.1-1～图 17.3.1-4）。

17.3.1 [Design code of residential buildings] 5.4.1 and 5.4.2: Toilets shall be arranged in each apartment, and the three sanitary devices of pedestal pans, bathing devices and washing basins shall be provided at least, or the positions and conditions shall be provided for the sanitary devices. The usable floor area of the toilet with three sanitary equipment in a concentrated way shall be not less than 2.50m². Different equipment can be combined in the toilet in accordance with the requirements for use functions. The areas of use spaces in different combinations shall conform to the following provisions (for the minimum-size plane arrangement for the toilet, see Fig. 17.3.1-1 ~ Fig. 17.3.1-4).

1. 设便器、洗面器时不应小于1.80m²;

1. The area shall be not less than 1.80m² when pedestal pans and washing basins are arranged;

2. 设便器、洗浴器时不应小于2.00m²;

2. The area shall be not less than 2.00m² when pedestal pans and bathing devices are arranged;

3. 设洗面器、洗浴器时不应小于2.00m²;

3. The area shall be not less than 2.00m² when washing basins and bathing devices are arranged;

4. 设洗面器、洗衣机时不应小于1.80m²;

4. The area shall be not less than 1.80m² when washing basins and washing machines are arranged;

5. 单设便器时不应小于1.10m²。

5. The areas shall be not less than 1.10m² when only pedestal pans is set.

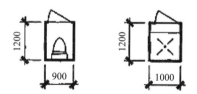

图 17.3.1-1 【09 措施】13.2.5（a）：单件布置
Fig. 17.3.1-1　[Measures 2009] 13.2.5（a）: Single-piece Arrangement

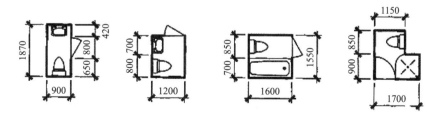

图 17.3.1-2 【09 措施】13.2.5（b）：两件布置
Fig. 17.3.1-2　[Measures 2009] 13.2.5（b）: Two-piece Arrangement

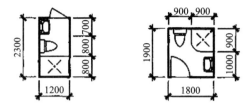

图 17.3.1-3 【09措施】13.2.5（c）：两件及淋浴布置
Fig. 17.3.1-3 [Measures 2009] 13.2.5（c）: Two-piece and Shower Arrangement

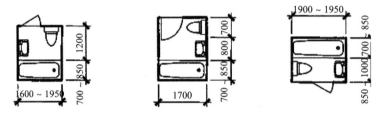

图 17.3.1-4 【09措施】13.2.5（d）：三件合设布置
Fig. 17.3.1-4 [Measures 2009] 13.2.5（d）: Arrangement for Joint Use of Three Pieces

17.3.2 【住设规】5.4.3 条：无前室的卫生间的门不应直接开向起居室（厅）或厨房。

17.3.2 [Design code of residential buildings] 5.4.3: The door of toilets without anterooms shall not be directly opened to living rooms（halls）or kitchens.

17.3.3 【住设规】5.4.4 条：卫生间不应直接布置在下层住户的卧室、起居室（厅）、厨房和餐厅的上层。

17.3.3 [Design code of residential buildings] 5.4.4: Toilets shall not be directly arranged on the floor above bedrooms, living rooms（halls）, kitchens and dining rooms of the household on the lower floor.

17.3.4 【住设规】5.4.5 条：当卫生间布置在本套内的卧室、起居室（厅）、厨房和餐厅的上层时，均应有防水和便于检修的措施。

17.3.4 [Design code of residential buildings] 5.4.5: When toilets are arranged on the upper floor of bedrooms, living rooms（halls）, kitchens and restaurants in the same residential house, waterproof measures and convenient maintenance measures shall be adopted.

17.3.5 【住设规】5.4.6 条：每套住宅应设置洗衣机的位置及条件。

17.3.5 [Design code of residential buildings] 5.4.6: In each residential house, the location and conditions for the washing machine shall be arranged.

17.3.6 【住设规】6.8.1 条：无外窗的卫生间应设共用排气道。

17.3.6 [Design code of residential buildings] 6.8.1: For toilets without exterior windows, shared exhaust passages shall be arranged.

17.3.7 【住设规】6.8.2 条：卫生间的共用排气道应采用能够防止各层回流的定型产品，

并应符合国家有关标准。排气道断面尺寸应根据层数确定，排气道接口部位应安装支管接口配件，卫生间排气道接口直径应大于80mm。

17.3.7 [Design code of residential buildings] 6.8.2: For the shared exhaust passages in toilets, the finalized products which can prevent backflow of all floors shall be adopted, and shall conform to the relevant national standards. The sectional dimensions of the exhaust passage shall be determined as per the number of the storeys. At the interface positions for the exhaust passage, interface fittings for branch tubes shall be installed, and the interface diameter for the exhaust passage in toilets shall be greater than 80mm.

17.3.8 【住设规】7.2.4.1条：明卫生间的直接自然通风开口面积不应小于该房间地板面积的1/20。

17.3.8 [Design code of residential buildings] 7.2.4.1: For toilets with ventiliation windows, the area of direct natural ventilation openings shall be not less than 1/20 of the area of the room floor.

17.3.9 【住设规】5.8.6条：卫生间的门应在下部设置有效截面积不小于$0.02m^2$的固定百叶，也可距地面留出不小于30mm的缝隙。

17.3.9 [Design code of residential buildings] 5.8.6: For doors in toilets, fixed louvers with the effective cross-sectional area not less than $0.02m^2$ shall be arranged at the lower part, and the gap with the distance to the ground not less than 30 mm can be reserved.

17.3.10 【住设规】5.8.7条：住宅卫生间门洞口宽度不应小于0.7m，门洞口高度不应小于2m。

17.3.10 [Design code of residential buildings] 5.8.7: The width of the door openings of the bathrooms in the residential house shall be not less than 0.7 m, and the height of the door opening shall be not less than 2m.

17.3.11 【住设规】6.8.4条：厨房的共用排气道与卫生间的共用排气道应分别设置。

17.3.11 [Design code of residential buildings] 6.8.4: The shared exhaust passages in kitchens and shared exhaust passages in toilets shall be separately arranged.

17.4 卫生间防水

17.4 Waterproofing in Toilets

17.4.1 【室内防水规】3.1.3条：室内需进行防水设防的区域，不应跨越变形缝、抗震缝等部位。

17.4.1 [Technical specification for interior waterproof of residential buildings] 3.1.3: The areas with indoor waterproof fortification requirements shall not traverse the positions of deformation joint ,aseismatic joint, etc.

17.4.2 【室内防水规】3.1.4 条：自身无防护功能的柔性防水层应设置保护层，保护层或饰面层应符合下列规定。

17.4.2 [Technical specification for Interior waterproof of residential buildings] 3.1.4: The protective layers shall be arranged on the flexible waterproof layers without protective functions, and the protective layer or finish layer shall conform to the following provisions.

1. 地面饰面层为石材、厚质地砖时，防水层上应用不小于 20mm 厚的 1∶3 水泥砂浆做保护层；

1. When the finish layer of floors is made of thick stone tiles, the protective layer of cement mortar with the thickness of at least 20mm shall be carried out on the waterproof layer;

2. 地面饰面为瓷砖、水泥砂浆时，防水层上应浇筑不小于 30mm 厚的细石混凝土做保护层；

2. When the ceramic tile and cement mortar are used as floor finish, the protective layer of fine-aggregated concrete with the the thickness of 30 mm at least shall be poured on the waterproof layer;

3. 墙面防水高度高于 250mm 时，防水层上应采取防止饰面层起壳剥落的措施。

3. When the height of the waterproof of wall surfaces is greater than 250mm, the measures for the prevention against flaking of the finish layers shall be adopted for the waterproof layers.

17.4.3 【室内防水规】3.1.5 条：楼地面向地漏处的排水坡度不宜小于 1%，地面不得有积水现象。

17.4.3 [Technical specification for interior waterproof of residential buildings] 3.1.5: The gradient of the drainage slopes from the floor grounds to the floor drains shall be appropriately not less than 1%, and no water can be accumulated on the grounds.

17.4.4 【室内防水规】3.1.6 条：地漏应设在人员不经常走动且便于维修和便于组织排水的部位。

17.4.4 [Technical specification for interior waterproof of residential buildings] 3.1.6: The floor drains shall be arranged at places without frequent activities of people and convenient for maintenance and drainage.

17.4.5 【室内防水规】3.2.1 条：厕浴间、厨房的墙体，宜设置高出楼地面 150mm 以上的现浇混凝土泛水。

17.4.5 [Technical specification for interior waterproof of residential buildings] 3.2.1: For the walls of toilets, bathrooms and kitchens, the cast-in-situ concrete flashing 150 mm higher than the floor grounds shall be arranged.

17.4.6 【室内防水规】3.2.2 条：主体为装配式房屋结构的厕所、厨房等部位的楼板应采用现浇混凝土结构。

17.4.6 [Technical specification for interior waterproof of residential buildings] 3.2.2: For the main body of assembly house structures, the floor slabs at the positions of toilets, kitchens, and others shall be of cast-in-situ concrete structures.

17.4.7 【室内防水规】3.2.3 条：厕浴间、厨房四周墙根防水层泛水高度不应小于 250mm，其他墙面防水以可能溅到水的范围为基准向外延伸不应小于 250mm。浴室花洒喷淋的临墙面防水高度不得低于 2m（图 17.4.7）。

17.4.7 [Technical specification for interior waterproof of residential buildings] 3.2.3: The flashing height of waterproof layers on the surrounding wall feet of toilets, bathrooms or kitchens shall be not less than 250mm, and for waterproofing of other wall surfaces, the outward expansion shall be not less than 250mm from the benchmark of the possible water splashing range. The waterproof height of the adjacent walls prone to the spraying of sprinklers in bathrooms shall not be less than 2m（see Fig. 17.4.7）.

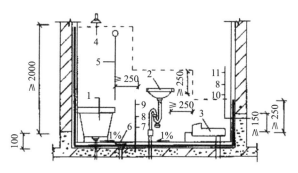

图 17.4.7 厕浴间墙面防水高度示意

Fig. 17.4.7 Schematic Diagram of Waterproof Height of Walls in Toilet Rooms and Bathrooms

1—浴缸；2—洗手池；3—蹲便器；4—喷淋头；5—浴帘；6—地漏；7—现浇混凝土楼板；8—防水层；9—地面饰面层；10—混凝土泛水；11—墙面饰面层

1-Bathtub; 2-Hand-washing tank; 3-Squatting pan; 4-Sprinkler; 5-Shower curtain; 6-Floor drain; 7-Cast-in-situ concrete floor slab; 8-Waterproof layer; 9-Floor finish layer; 10-Concrete flashing; 11-Finish layer of the wall surface

17.4.8 【室内防水规】3.2.4 条：有填充层的厨房、下沉式卫生间，宜在结构板面和地面饰面层下设置两道防水层。单道防水时，防水应设置在混凝土结构板面上，材料厚度参照水池防水设计选用。填充层应选用压缩变形小、吸水率低的轻质材料。填充层面应整浇不小于 40mm 厚的钢筋混凝土地面。排水沟应采用现浇钢筋混凝土结构，坡度不应小于 1%，沟内应设置防水层。

17.4.8 [Technical specification for interior waterproof of residential buildings] 3.2.4: For kitchens and sunken toilets with filling layers, two waterproof layers shall be arranged under the structural boards and floor finish layers. When the single waterproof layer is adopted, the waterproof layer shall be arranged on the concrete structural slab, and for the thickness of the materials, refer to the waterproof design of the water tank. For the filling layer, the lightweight materials with small compression deformation and low water absorption rate shall be selected. The reinforced concrete floor with the thickness not less than 40 mm shall be fully poured on the filling layer. The cast-in-situ reinforced concrete structures shall be adopted for drainage ditches, with the gradient of not less than 1%, and the waterproof layer shall be arranged in the ditches.

18
设备用房
Equipment Rooms

18.1　给水和排水

18.1　Water Supply and Drainage

1. 民用建筑给水和排水设计要求见表 18.1-1。

1. Design requirements for water supply and drainage in civil buildings see Tab.18.1-1.

【通则】8.1：民用建筑给水和排水设计要求　　　　　　表 18.1-1

[Code for design of civil buildings] 8.1:
Design Requirements for Water Supply and Drainage in Civil Building　　Tab.18.1-1

设计要求 Design requirements
1. 民用建筑给水排水设计应满足生活和消防等要求 1. The design for water supply and drainage in civil buildings shall meet the requirements for living and fire protection
2. 生活饮用水的水质，应符合国家现行有关生活饮用水卫生标准的规定 2. The water quality of domestic drinking water shall conform to the current national regulations relevant to the hygienic criteria of domestic drinking water
3. 生活饮用水水池（箱）应与其他用水的水池（箱）分开设置 3. The domestic drinking water pools (tanks) shall be set separately from other water pools (tanks)
4. 建筑物内的生活饮用水水池、水箱的池（箱）体应采用独立结构形式，不得利用建筑物的本体结构作为水池和水箱的壁板、底板及顶板。生活饮用水水池（箱）的材质、衬砌材料和内壁涂料不得影响水质 4. Independent structures shall be used as bodies of the domestic drinking water pools (tanks) in the building, and the structures of the building shall not be used as wall plates, bottom plates and top plates of the water pools (tanks). The materials, lining materials and the coatings of inner walls of domestic drinking water pools (tanks) shall not affect water quality
5. 埋地生活饮用水贮水池周围 10m 以内，不得有化粪池、污水处理构筑物、渗水井、垃圾堆放点等污染源，周围 2m 以内不得有污水管和污染物 5. There shall be no pollution sources of septic tanks, sewage treatment structures, leakage wells, garbage dumps, etc. within 10 m of the underground living drinking water reservoir, and no sewer pipes or pollutants within 2 m around
6. 排水管道不得布置在食堂、饮食业的主副食操作烹调备餐部位的上方，也不得穿越生活饮用水池部位的上方 6. The drainage pipes shall not be arranged above the dining room and staple and non-staple food operation, cooking and preparation positions of catering trade or run above domestic drinking water tanks
7. 排水立管不得穿越卧室、病房等对卫生、安静有较高要求的房间，并不宜靠近与卧室相邻的内墙 7. The drainage riser pipes shall not run through the rooms with higher sanitation and silence requirements such as bedrooms, wards, etc, or near the interior walls adjacent to bedrooms
8. 给水排水管不应穿越配变电房、档案室、电梯机房、通信机房、大中型计算机网络中心、音像库房等遇水会损坏设备和引发事故的房间内 8. Water supply and drainage pipes shall not penetrate the rooms such as electric transformer and distribution rooms, archives rooms, machine rooms of the elevators, communication machine rooms, large and medium-sized computer network centers, phonotape and videotape warehouses, etc, where the equipment can be damaged by water and where accidents can be caused by water
9. 给水泵房、排水泵房不得设置在有安静要求的房间上面、下面和毗邻的房间内；泵房内应设排水设施，地面应设防水层；泵房内应有隔振降噪设置。消防泵房应符合防火规范的有关规定 9. Water supply pump rooms and water drainage pump rooms shall not be arranged above and below the rooms with silence requirements and in the adjacent rooms; drainage facilities shall be arranged in the pump rooms, and the waterproof layers shall be made on the floor; vibration isolation and noise protection devices shall be arranged in the pump rooms. Fire pump rooms shall conform to the relevant provisions in the code for fire protection design of buildings

2. 给水排水设备用房见表18.1-2。

2. Schedule of water supply and drainage equipment rooms see Tab.18.1-2.

给水排水设备用房一览表　　　　表 18.1-2
Schedule of Water Supply and Drainage Equipment Rooms　　Tab.18.1-2

类别 Category	设备用房名称 Name of equipment room	面积（m²） Area (m²)
给水排水 Water supply and drainage	水泵房及水池 Water pump rooms and water tanks	190～300
	中水处理间及水池 Reclaimed water treatment rooms and water tanks	150～250
	热交换器 Heat exchanger	60～100
	泡沫灭火间 Foam fire-extinguishing rooms	10（设2个泡沫罐）/间 10 (two foam tanks shall be arranged) /room
	气体灭火间 Gas fire-extinguishing rooms	4～10

18.2 【通则】8.3 条：建筑电气

18.2 [Code for design of civil buildings] 8.3: Building Electricity

1. 民用建筑物内配变电所设计要求见表18.2-1。

1. Distribution and transformer substations in civil buildings see Tab.18.2-1.

民用建筑物内配变电所　　　　表 18.2-1
Distribution and Transformer Substations in Civil Buildings　　Tab.18.2-1

类别 Category	设计要求 Design requirements
配变电所位置 Locations of distribution and transformer substations	1. 宜接近用电负荷中心 1. The locations shall be appropriately close to the center of the power loads
	2. 应方便进出线 2. The locations shall facilitate incoming and outgoing lines
	3. 应方便设备吊装运输 3. The locations shall facilitate lifting and transport of equipment
	4. 不应设在厕所、浴室或其他经常积水场所的正下方，且不宜与上述场所相贴邻；装有可燃油电气设备的变配电室，不应设在人员密集场所的正上方、正下方、贴邻和疏散出口的两旁 4. The locations shall not be arranged right below and near toilets, bathrooms or other places frequently with accumulated water; the transformer and distribution rooms with fuel oil and electrical equipment shall not be arranged right above, right below and near crowded places and on both sides of evacuation exits
	5. 当配变电所的正上方、正下方为住宅、客房、办公室等场所时，配变电所应作屏蔽处理 5. When residential rooms, guest rooms, offices, etc, are right above and right below distribution and transformer substations, shielding treatment shall be made to the distribution and transformer substations

2. 可燃油油浸电力变压器设计要求见表18.2-2。

2. Design requirement for Combustible oil immersed power transform see Tab.18.2-2.

【通则】8.3.1.3 条：可燃油油浸电力变压器　　　　　　表 18.2-2

[Code for design of civil buildings] 8.3.1.3: Combustible Oil Immersed Power Transformer　Tab.18.2-2

类别 Category	设计要求 Design requirements
可燃油油浸电力变压器 Combustible oil immersed power transformer	1. 总容量不超过 1260kVA、单台容量不超过 630kVA 的变配电室可布置在建筑主体内首层或地下一层靠外墙部位 1. The transformer and distribution room with total capacity not exceeding 1260 kVA and single capacity not exceeding 630 kVA shall be arranged at the position near the exterior wall of the first floor or first floor underground in the main body of the building
	2. 应设置直接对外的安全出口，变压器室的门应为甲级防火门；外墙开口部位上方，应设置宽度不小于 1m 不燃烧体的防火挑檐 2. The emergency exit directly toward the outside shall be arranged, and grade A fire door shall be used as the door of transformer rooms; the fire-proof overhanging eave of the incombustible body with the width not less than 1m shall be arranged above the opening positions on exterior walls

3. 高压配电室设计要求见表18.2-3。

3. Design requirement for high-voltage distribution room see Tab.18.2-3.

【通则】8.3.1.3 条：高压配电室　　　　　　表 18.2-3

[Code for design of civil buildings] 8.3.1.3: High-voltage Distribution Room　Tab.18.2-3

类别 Category	设计要求 Design requirements
高压配电室 High-voltage distribution rooms	1. 不带可燃油的高、低压配电装置和非油浸的电力变压器，可设置在同一房间内 1. The high- and low-voltage power distribution devices without flammable oil and non-oil immersed power transformers can be arranged in the same room
	2. 高压配电室宜设置不能开启的距室外地坪不低于 1.80m 的自然采光窗，低压配电室可设能开启的不临街的自然采光窗 2. The high-voltage distribution rooms shall be appropriately equipped with unopenable natural lighting windows with the distance to the outdoor floor not less than 1.80m, and the low-voltage distribution rooms can be equipped with openable natural lighting windows not adjacent to streets
	3. 长度大于 7m 的配电室应在配电室的两端各设一个出口，长度大于 60m 时，应增加一个出口 3. One exit shall be arranged respectively at both ends of the power distribution rooms with the length greater than 7 m, and when the length is greater than 60m, another exit shall be arranged additionally
	4. 变压器室、配电室的进出口门应向外开启 4. The entrance and exit doors of transformer rooms and distribution rooms shall be opened outward
	5. 变压器室、配电室等应设置防雨雪和小动物从采光窗、通风窗、门、电缆沟等进入室内的设施 5. For transformer rooms, distribution rooms, etc, facilities preventing the entry of rain, snow and small animals from lighting windows, ventilating windows, doors, cable trenches, etc. shall be arranged
	6. 变配电室的电缆夹层、电缆沟和电缆室应采取防水、排水措施 6. Waterproof and drainage measures shall be adopted for cable interlayers, cable trenches and cable chambers of transformer and distribution rooms
	7. 变配电室不应有与其无关的管道和线路通过 7. Irrelevant pipelines and lines shall not pass through the transformer and distribution rooms

4. 配变电所防火门设计要求见表 18.2-4。

4. Design requirement for fire doors of distribution and transformer substation see Tab.18.2-4.

【通则】8.3.2 条：配变电所防火门　　　　　　　　　　表 18.2-4

[Code for design of civil buildings] 8.3.2: Fire Doors of Distribution and Transformer Substation　Tab.18.2-4

类别 Category	设计要求 Design requirements
配变电所防火门 Fire Doors in Distribution and Transformer Substation	1. 设在高层建筑内的配变电所，应采用耐火极限不低于 2.00h 的隔墙、耐火极限不低于 1.50h 的楼板和甲级防火门与其他部位隔开 1. The distribution and transformer substation in a high-rise building shall be separated from other positions by the partition walls with the fire endurance not less than two hours and floor slabs and grade A fire doors with the fire endurance not less than 1.50 hours
	2. 可燃油油浸变压器室通向配电室或变压器室之间的门应为甲级防火门 2. Grade A fire doors shall be used as the doors from the flammable oil immersed transformer room to the power distribution room or transformer room, etc.
	3. 配变电所内部相通的门，宜为丙级的防火门 3. Grade C fire doors shall be used as the doors in the distribution and transformer substation
	4. 配变电所直接通向室外的门，应为丙级防火门 4. Grade C fire doors shall be used as the doors of distribution and transformer substations directly to the outdoor

5. 柴油发电机房设计要求见表 18.2-5。

5. Design requirement for diesel generator rooms see Tab.18.2-5.

【防火规】5.4.13 条：柴油发电机房　　　　　　　　　　表 18.2-5

[Code for fire protection design of buildings] 5.4.13: Diesel Generator Rooms　Tab.18.2-5

类别 Category	设计要求 Design requirement
柴油发电机房 Diesel generator rooms	1. 宜布置在首层或地下一、二层 1. They shall be arranged on the first floor or first and second floors underground
	2. 不应布置在人员密集场所的上一层、下一层或贴邻 2. They shall not be arranged on floor above, below or adjacent to a crowded place
	3. 应采用耐火极限不低于 2.00h 的防火隔墙和 1.50h 的不燃性楼板与其他部位分隔，门应采用甲级防火门 3. They shall be separated from other positions with fireproof partition walls with fire resistance not lower than 2.00 hours and floor slabs with 1.50hour incombustibility, and Grade A fire doors shall be used
	4. 机房内设置储油间时，其总储存量不应大于 1m³，储油间应采用耐火极限不低于 3.00h 的防火隔墙与发电机间分隔；确需在防火隔墙上开门时，应设置甲级防火门 4. When the oil storage room is arranged in the machine room, the total storage capacity shall not be greater than one m^3; and the oil storage room shall be separated from the generator room with fire partitions with the fire endurance not less than 3.00 hours; Grade A fire doors shall be used when doors are required to be arranged in the fire partitions
	5. 应设置火灾报警装置 5. Fire alarm devices shall be arranged
	6. 应设置与柴油发电机容量和建筑规模相适应的灭火设施，当建筑内其他部位设置自动喷水灭火系统时，机房内应设置自动喷水灭火系统 6. The fire-extinguishing facilities appropriate to the capacity of diesel generator and the construction scale shall be arranged, when the automatic sprinkler systems are provided at other positions in the building, the automatic sprinkler system shall be arranged in the machine room

6. 电气用房面积见表 18.2-6。

6. Schedule of areas of electrical rooms see Tab.18.2-6.

电气用房面积一览表　　　　　　　　表 18.2-6
Schedule of Areas of Electrical Rooms　　Tab.18.2-6

类别 Category	设备用房名称 Name of equipment room	面积（m²） Area（m²）
电气用房 Electrical rooms	变、配电室（变压器室、高低压配电室） Transformer and distribution rooms（transformer rooms, and high- and low-voltage distribution rooms）	0.6% 总建筑面积，且 ≥ 200 0.6% of the total building area, and ≥ 200
	柴油发电机房 Diesel generator rooms	40 ~ 100（内含油箱间） 40-100（Including oil tank rooms）
	消防控制室 Fire control rooms	24 ~ 60
	数字自控室（BAS） Digital automatic control rooms（BAS）	30
	闭路电视室 Closed circuit television rooms	20
	电话机房 Telephone rooms	6
	电视前端室 Television front-end rooms	4 ~ 5
	电梯机房 Machine rooms for elevators	16 × 电梯数量 16 × quantity of elevators

18.3 智能化系统机房设计要求

18.3 Design Requirements for Intelligent System Room

【通则】8.3.4 条、【防火规】5.2.7/8.1.7 条：智能化系统机房 表 18.3
[Code for design of civil buildings] 8.3.4, [Code for fire protection design of buildings] 5.2.7/8.1.7: Intelligent System Room Tab.18.3

类别 Category	设计要求 Design requirements
机房分类 Category of the machine room	消防控制室、安防监控中心、电信机房、卫星接收及有线电视机房、计算机机房、建筑设备监控机房、有线广播及（厅堂）扩声机房等 Fire control room and security monitoring center, telecommunications room, satellite reception and cable TV room, computer room, construction equipment monitoring room, cable broadcasting room, (hall) amplification room, etc
机房设置 Arrangement of machine room	1. 可单独设置，也可合用设置 1. The machine room shall be set separately or shared with other rooms 2. 通风、空气调节机房和变配电室开向建筑内的门应采用甲级防火门，消防控制室和其他设备房开向建筑内的门应采用乙级防火门 2. Grade A fire doors shall be used as the doors to be opened inward in ventilation rooms, air conditioning rooms, and transformer and distribution rooms, and the grade B fire doors shall be adopted as the doors to be inward opened in fire control rooms and other equipment rooms 3. 消防控制室宜设置在建筑物内首层的靠外墙部位，亦可设置在建筑物的地下一层。乙级防火门设计要求见《建筑设计防火规范》6.2.7 条 3. The fire control room shall be arranged at the position on the first floor of the building near the exterior wall or on the first floor underground the building. grade B Fire Doors shall be adopted in accordance with the specifications in 6.2.7 of *Code for fire protection design of buildings* 4. 消防控制室、安防监控中心宜设在建筑物的首层或地下一层，且应采用耐火极限不低于 2.00h 或 3.00h 的隔墙和耐火极限不低于 1.50h 或 2.00h 的楼板与其他部位隔开，并应设直通室外的安全出口 4. The fire control rooms and security monitoring centers shall be appropriately arranged on the first floor or first floor underground of the building, be separated from other positions by partition walls with the fire endurance not less than two hours or three hours and floor slabs with the fire endurance not less than 1.50 hour or two hours, and be provided with safety exits directly to the outdoor 5. 智能化系统的机房宜铺设架空地板、网络地板或地面线槽；宜采用防静电、防尘材料；机房净高不宜小于 2.50m 5. For machine rooms of the intelligent systems, the overhead floors, network floors or on-ground trunking shall be appropriately paved; anti-static and dustproof materials shall be appropriately adopted; and the net height of the machine rooms shall be appropriately not less than 2.50m 6. 智能化系统的机房不应设在厕所、浴室或其他经常积水场所的正下方，且不宜与上述场所相贴邻 6. The machine rooms of the intelligent systems shall not be arranged right below and near toilets, bathrooms or other places frequently with accumulated water

18.4 【通则】8.3 条：暖通和空调

18.4 [Code for design of civil building] 8.3: Heating, Ventilation and Air Conditioning

1. 暖通和空调机房设计要求见表 18.4-1。

1. Design requirements for heating, ventilation and air conditioning machine rooms see Tab.18.4-1.

暖通和空调机房　　　　　　　表 18.4-1
Heating, Ventilation and Air Conditioning Machine Rooms　　Tab.18.4-1

类别 Category	设计要求 Design requirements
通风系统 Ventilation systems	1. 机械通风系统的进风口应设置在室外空气清新、洁净的位置 1. The air inlets of the mechanical ventilation system shall be arranged at the outdoor positions with fresh and clean air
	2. 废气排放不应设置在有人停留或通行的地带 2. The exhaust discharge ports shall not be arranged at the place where people stay or pass
	3. 机械通风系统的管道应选用不燃材料 3. For the pipes of the mechanical ventilation system, the incombustible materials shall be selected
	4. 通风机房不宜与有噪声限制的房间相邻布置 4. The ventilator rooms shall not be arranged adjacent to the rooms with noise limit requirements
	5. 通风空调机房应设置甲级防火门 5. For the ventilation and air conditioning rooms, Grade A fire doors shall be arranged
空气调节系统 Air conditioning system	1. 空气调节系统的民用建筑，其层高、吊顶高度应满足空调系统的需要 1. For civil buildings with the air conditioning system, the floor height and ceiling height shall meet the requirements of the air conditioning system
	2. 空气调节系统的风管管道应选用不燃材料 2. For air ducts of the air conditioning system, the incombustible materials shall be selected
	3. 空气调节机房不宜与有噪声限制的房间相邻 3. The air conditioning rooms shall not be arranged adjacent to the rooms with noise limit requirements
	4. 空气调节系统的新风采集口应设置在室外空气清新、洁净的位置 4. The fresh air collection ports of the air conditioning system shall be arranged at outdoor positions with fresh and clean air
	5. 通风空调机房应设置甲级防火门 5. For the ventilation and air conditioning rooms, Grade A fire doors shall be arranged

18.5 暖通空调设备用房一览表

18.5 Schedule of HVAC Equipment Rooms

暖通空调设备用房一览表　　　　　　表 18.5-1
Schedule of HVAC Equipment Rooms　　　Tab.18.5-1

类别 Category	设备用房名称 Name of Equipment Room	面积（m²） Area（m²）	位置及要求 Positions and Requirements
空调机房 Air conditioning machine rooms	锅炉房 Boiler rooms	100 ~ 150	独立设置或地下一层、首层，靠近外墙部位。应设置只对外安全出口。外墙上门窗开口部位的上方应设置≥1m宽的防火挑檐或≥1.2m的窗槛墙。应设防火、防爆墙和甲级防火门，其外墙、楼地面或屋面，应该有≥锅炉房占地面积10%的泄压面积 Boiler rooms shall be independently arranged at the locations near the exterior wall of the first floor underground and the first floor. External safety exits shall be provided. At the position above the door or window openings in the external wall, fireproof overhanging eaves with the width of 1m or window spandrels with height of 1.2m shall be arranged. The rooms shall be provided with fireproof and explosion-proof walls and Grade A fire doors, and there shall be pressure relief areas greater than or equal to 10% of the floor spaces of the boiler rooms on the walls, floor grounds or roofs
	空调机房 Air conditioning machine rooms	2% ~ 2.5% 空调面积 2% ~ 2.5% of the air-conditioning area	空调层、设备层。靠近负荷中心，应有一边靠外墙。不得开门直接与商场相通 Air conditioning floors and equipment floors. Near the load center, one side shall be against the exterior wall. The air conditioning machine rooms shall not be directly connected to the mall after door is opened
	地下室通风排烟机房 Ventilation and smoke exhaust machine rooms in the basement	0.25% ~ 0.3% 通风面积 0.25% ~ 0.3% of the ventilation area	地下室。排烟口应设置在下风向并远离楼梯出口 Basement. The smoke exhaust ports shall be arranged at the downwind direction, and shall be kept far away from the staircase exits

18.6 管道井

18.6 Pipeline Shafts

管道井 表 18.6
Pipeline Shafts Tab.18.6

类别 Category		设备用房名称 Name of equipment room	面积（m²） Area（m²）	位置及要求 Positions and requirements
管道井 Pipeline shafts	水 Water	管道水表井 Water meter and pipeline shaft	0.7 × 1.2	住宅建筑可设置在防烟楼梯间前室、合用前室内，但检查门应采用丙级防火门 For residential buildings, the shafts can be arranged in the anterooms of smokeproof staircases and share the anterooms, and for inspection doors, grade C fire doors shall be adopted
		管道排气井 Exhaust pipeline shaft	0.7 × 1.2	
	电气 Electrical	强电井 High-current shafts	1.0 × 1.5	
		弱电井 Low-current shaft	0.8 × 1.2	
	通风 Ventilation	排烟井 Smoke exhaust shaft	0.4 × 1.5	防烟楼梯间及前室、消防电梯与防烟楼梯合用前室、地下室通风排烟机房 Smokeproof staircases and anterooms, fire elevator and smoke prevention stairs, can share the anteroom or the ventilation and smoke exhaust machine room in the basement
		送风井 Air supply shaft	单用 0.4 × 2.0 Single use 0.4 × 2.0 合用 0.4 × 2.5 Shared 0.4 × 2.5	

注：1. 机房净高要求：锅炉房 6m，中水处理 5m，水泵房、电梯机房 3m，变配电室 4.5m，制冷机房 3.5～5m，发电机房 4m，空调机房 3.5～4.5m。
Notes:1. Requirements for the net height of the machine room: 6m for the boiler room, 5m for the reclaimed water treatment room, 3m for the water pump room and elevator machine room, 4.5m for the transformer and distribution room, 3.5～5m for the refrigerating machine room, 4m for the generator room, and 3.5～4.5m for the air conditioning machine room.
2. 管道占用净空：空调管道 300～700mm，常用 500mm；商住楼、综合楼给水排水管道 200mm。
2. Clearance occupied by pipeline: the clearance occupied by the air conditioner pipelines is 300～700mm, and 500mm is in common use; and the clearance occupied by the water supply and drainage pipelines of commercial and comprehensive buildings is 200mm.

19

电缆井、管道井、烟道、通风道和垃圾道
Cable Wells, Pipeline Shafts, Flues, Ventilation Channels and Refuse Chutes

19.1 【防火规】6.2.9 条：建筑内的电梯井等竖井应符合下列规定。

19.1 [Code for fire protection design of buildings] 6.2.9: Shafts such as elevator shafts in buildings shall conform to the following provisions.

1. 电梯井应独立设置，井内严禁敷设可燃气体和甲、乙、丙类液体管道，不应敷设与电梯无关的电缆、电线等。电梯井的井壁除设置电梯门、安全逃生门和通气孔洞外，不应设置其他开口。

1. Elevator shafts shall be independently arranged. It is prohibited to lay pipelines for combustible gases and liquids of Classes A, B and C, and cables, wires, etc, irrelevant with elevators shall not be laid. On the walls of elevator shafts, except that elevator doors, safety escape doors and ventilation holes can be arranged, other openings shall not be arranged.

2. 电缆井、管道井、排烟道、排气道、垃圾道等竖向井道，应分别独立设置。井壁的耐火极限不应低于 1.00h，井壁上的检查门应采用丙级防火门。

2. Vertical wells such as cable wells, pipeline wells, smoke exhaust passage, air exhaust passage, refuse chutes, etc. shall be independently arranged. The fire endurance of the shaft walls shall be not less than 1.00 hour, and for the inspection doors on the shaft walls, Grade C fire doors shall be adopted.

3. 建筑内的电缆井、管道井应在每层楼板处采用不低于楼板耐火极限的不燃材料或防火封堵材料封堵。

3. Incombustible materials or fireproof blocking materials with the fire endurance not less than that for floor slabs shall be used to seal the joint between the cable and pipeline shafts in the building and each floor.

建筑内的电缆井、管道井与房间、走道等相连通的孔隙应采用防火封堵材料封堵。

Holes connecting cable and pipeline shafts and rooms, corridors, etc. in the building shall be sealed with fireproof blocking materials.

4. 建筑内的垃圾道宜靠外墙设置，垃圾道的排气口应直接开向室外，垃圾斗应采用不燃材料制作，并应能自行关闭。

4. Refuse chutes in buildings shall be arranged close to exterior walls. The exhaust vents of refuse chutes shall be arranged directly to the outside, and the refuse buckets shall be made of incombustible materials and can be closed automatically.

5. 电梯层门的耐火极限不应低于 1.00h，并应符合现行国家标准《电梯层门耐火试验 完整性、隔热性和热通量测定法》GB/T 27903 规定的完整性和隔热性要求。

5. The fire endurance of elevator landing doors shall be not less than 1.00 hour and shall conform to the integrity and heat shielding requirements specified in the current national standard *Fire resistance test for lift landing doors - methods of measuring integrity, thermal insulation*

and heat flux GB /T 27903.

19.2 【通则】18.3条:烟道和通风道的断面、形状、尺寸和内壁应有利于排烟(气)通畅,防止产生阻滞、涡流、窜烟、漏气和倒灌等现象。

19.2 [Code for design of civil buildings] 18.3: The cross sections, shapes, dimensions and inner walls of flues and ventilation ducts shall facilitate smoke (gas) exhaust, to prevent phenomena of blocking, eddy current, smoke channeling, air leakage, backflow, etc.

19.3 【通则】18.4条:烟道和通风道应伸出屋面,伸出高度应有利烟气扩散,并应根据屋面形式、排出口周围遮挡物的高度、距离和积雪深度确定。平屋面伸出高度不得小于0.60m,且不得低于女儿墙的高度。坡屋面伸出高度应符合下列规定。

19.3 [Code for design of civil buildings] 18.4: The flues and ventilation ducts shall extend out of roofs, and the extension height shall facilitate fume diffusion, and shall be determined as per the roof forms, the height and distance of the obstructions around exhaust ports and the snow retention depth. The extension height out of flat roofs shall not be less than 0.60 m, and not less than the height of the parapet wall. The extension height out of slope roofs shall conform to the following provisions.

1. 烟道和通风道中心线距屋脊小于1.50m时,应高出屋脊0.60m;

1. When the distance between the centerlines of the flues and ventilation ducts and the roof ridges is less than 1.50m, the extension height shall be 0.60m higher than the roof ridge;

2. 烟道和通风道中心线距屋脊1.50~3.00m时,应高于屋脊,且伸出屋面高度不得小于0.60m;

2. When the distance between the centerline of the flue and ventilation duct and the roof ridge is 1.50-3.00m, the flue and ventilation duct shall be higher than the roof ridge, and the protrusion height out of the roof shall be less than 0.60m;

3. 烟道和通风道中心线距屋脊大于3m时,其顶部同屋脊的连线同水平线之间的夹角不应大于10°,且伸出屋面高度不得小于0.60m。

3. When the distance between the centerline of the flue and ventilation duct and the roof ridge is greater than 3m, the angle between the connecting line of the top and the ridge and the horizontal line shall be greater than 10°, and the protrusion height out of the roof shall be not less than 0.60m.

20

景观设计
Design of Landscape

20.1 一般规定

20.1 General Provisions

20.1.1 【09措施】2.1.1条：场地景观设计是场地总平面规划的重要组成部分。应因地制宜，充分利用自然地形、原有水系和植被，对原有生态环境进行保护。

20.1.1 [Measures 2009] 2.1.1: Site landscape design is an important part of general site plan planning. Original ecological environments shall be protected by adaptation to local conditions and full use of natural terrains, original water systems and vegetations.

20.1.2 【09措施】2.1.2条：各类民用建筑场地景观设计，应与建筑群体、区内道路、地下建筑物、构筑物、场地竖向布置、地下管线设计等进行综合考虑。

20.1.2 [Measures 2009] 2.1.2: For the landscape design of various civil building sites, comprehensive consideration shall be given to the design of building groups, roads in the area, underground buildings and structures, vertical layout of the site, design of underground pipelines, etc.

20.1.3 【09措施】2.1.3条：根据所处地区的气候、土壤类型、自然植被特点进行景观设计，植物配置设计应以适于本地生长的植被为主。

20.1.3 [Measures 2009] 2.1.3: The landscape design shall be made as per the climate, soil types and natural vegetation characteristics of the area, and in the plant configuration design, the vegetation suitable for local area will prevail.

20.1.4 【09措施】2.1.4条：场地景观种植设计，应采取绿色生态措施，通过植物起到防晒、防尘、降温、调节小气候、提高空气负氧离子浓度、减少二氧化碳量、降低噪声等作用。

20.1.4 [Measures 2009] 2.1.4: In the planting design for the site landscape, green ecological measures shall be adopted, to achieve the effect of sun protection, dust prevention, cooling, microclimatic modification, increase of negative oxygen ion concentration in the air, reduction of the carbon dioxide content, noise reduction, etc., through plants.

20.1.5 【09措施】2.1.5条：场地景观设计，应就地取材，选用可再生和可再利用的环保材料，应采取节能措施，积极利用可再生能源，如太阳能、风能等，并有效利用中水、雨水等资源。

20.1.5 [Measures 2009] 2.1.5: In the site landscape design, materials shall be obtained locally, the renewable and recyclable environmentally friendly materials shall be selected and energy saving measures shall be adopted. Renewable energy, such as solar energy, wind energy, and other energy, can be actively used, and the resources of reclaimed water, rainwater, and other water resources can be effectively utilized.

20.1.6 【09措施】2.1.6条：公共活动空间应有无障碍设施。

20.1.6 [Measures 2009] 2.1.6: In public space, barrier-free facilities shall be arranged.

20.1.7 【09措施】2.1.7条：地震烈度在6度以上（含6度）的地区，城市开放绿地必须结合绿地布局设置专用防灾、救灾设施和避难场地。

20.1.7 [Measures 2009] 2.1.7: In the areas with the seismic intensity above 6 degree inclusive. For the open urban greenland, the dedicated disaster prevention and disaster relief facilities and shelter sites must be arranged in combination of the layout of green space.

20.2 平面布局

20.2 Plane Layout

20.2.1 【09措施】2.2.1条：场地景观设计的平面布局应以场地总平面布局为依据，根据场地使用要求合理进行总体构思、景区划分、景点设置、出入口布置、竖向设计。处理好园路、铺装场地与绿化、水景的用地比例及相互关系，并结合活动需要布置各类景观小品。

20.2.1 [Measures 2009] 2.2.1: The plane layout of site landscape design shall be based on the total plane layout of the site. According to the application requirements of the site, overall conception, scenic area division, scenic spot setting, entrance and exit layout, and vertical design shall be reasonably carried out. The proportion of land use and correlation between the garden path, and pavement site and the greening and waterscape shall be well made. Various kinds of landscape accessories shall be laid out in combination of the activity needs.

20.2.2 【09措施】2.2.2条：总体构思是场地景观设计的关键，应结合建筑布局、建筑性质、使用特点、地域文化等综合考虑，控制景观设计的格调，达到功能性、文化性、艺术性的有机结合。

20.2.2 [Measures 2009] 2.2.2: The overall conception is the key to the site landscape design. Comprehensive consideration shall be given to the layout, properties, use characteristics, regional culture, and others of the buildings. The style of the landscape design shall be controlled to achieve organic combination of functionality, culture and artistry.

20.2.3 【09措施】2.2.3条：场地景观设计应在场地总平面布局的基础上进一步进行景区划分，确定各分区的规模及特色，并结合主次景区进行相应景点设置。

20.2.3 [Measures 2009] 2.2.3: For the site landscape design, the division of scenic spots shall be further made on the basis of the total plane layout of the site. The scale and characteristics of each subarea shall be determined, and the setting of corresponding scenic spots shall be carried out in combination of primary and secondary scenic spots.

20.2.4 【09措施】2.2.4条：出入口一般是场地景观设计的重点，应根据场地外部及内部的具体要求，确定主、次和专用出入口的位置，合理设置出入口内外广场、大门围墙、停车场、自行车存放、管理设施等，并注重景观效果。

20.2.4 [Measures 2009] 2.2.4: Usually entrances and exits are key points in the site landscape design.The positions of primary, secondary and dedicated entrances and exits shall be determined as per the specific requirements for the exterior and interior of the site. The squares inside and outside of the entrance, gate enclosure wall, parking area, bicycle storage and management facilities, etc, shall be reasonably arranged, and significance shall be paid to the landscape effect.

20.2.5 【09措施】2.2.5条：景观竖向设计有利于丰富场地的空间特征，应控制好以下内容：山顶、地形等高线，水底、常水位、最高水位、最低水位、驳岸顶部，园路主要转折点、交叉点和变坡点，各出入口内外地面、铺装场地、建构筑物地坪，地下工程管线及地下构筑物的埋深等。

20.2.5 [Measures 2009] 2.2.5: The vertical landscape design shall help enrich the space features of the site. The following contents shall be well controlled: topographic contours for mountain tops; water bottoms, normal water levels, maximum water levels, minimum water level and revetment tops; main turning points, intersections and slope changing points of garden paths; grounds at the inside and outside of entrances and exits, pavement sites, and the floors of buildings and structures; burying depth, of underground engineering pipelines and underground structures,etc.

20.2.6 【09措施】2.2.6条：园路、铺装场地设计，应根据场地规模、各分区活动内容、人员数量等需要，确定园路的路线、分类分级和铺装场地的位置和面积规模。绿地中园路的路网密度宜在200~380m/hm² 之间。

20.2.6 [Measures 2009] 2.2.6: In the site design of garden paths and pavement, the routes, classification and grading of garden paths and the position and area scale of the pavement site shall be determined in accordance with the site size, contents of activities in sub-areas, number of personnel, etc. The density of road network in the green land shall be 200 ~ 380m/hm².

20.2.7 【09措施】2.2.7条：各类场地中绿化用地的比例：

20.2.7 [Measures 2009] 2.2.7: Proportion of Greening Lands in Various Sites:

1.居住区内绿化用地占总景观用地的比例宜大于50%；

1. The proportion of the green land in the residential area to the total landscape land shall be more than 50%;

2. 一般公共建筑公共广场中集中成片绿地不宜小于广场总面积的25%;

2. The concentrated green areas in squares of public buildings shall be appropriately not less than 25% of the total area of the square;

3. 车站、码头、机场等集散广场中集中成片绿地不宜小于广场总面积的10%。

3. The concentrated green areas in the distribution squares such as stations, terminals, docks, airports, etc, shall be appropriately not less than 10% of the total area of the square.

20.2.8 【09措施】2.2.8条：居住区景观平面布局要点。

20.2.8 [Measures 2009] 2.2.8: Key Points for Plane Layout of Landscapes in Residential Areas.

1. 居住区景观设计应以创造轻松自然的环境氛围为主，应尽可能增加绿地面积并形成乔灌草地被花卉相结合的布置方式，形成地面、屋顶和垂直绿化相结合的立体景观效果；

1. In the landscape design of residential areas, the creation of a relaxed and natural environment shall prevail. The area of green lands shall be increased as much as possible, and the combined arrangement of arbors, grass, ground cover plants and flowers shall be adopted, to form the stereoscopic landscape effect uniting ground, roof and vertical greening;

2. 应充分重视老人和儿童活动的需要，合理设置老人、儿童活动场地；

2. Full attention shall be paid to the requirements for the activities of the aged and children, and the activity sites for the aged and children shall be reasonably arranged;

3. 园路系统应满足居民散步、游憩的需要；

3. The garden path system shall meet the requirements for walking and recreation of residents;

4. 集中活动场地应与住宅保持一定距离，或采取措施以避免噪声对居民造成影响。

4. A certain distance shall be kept between the places with intensive activities and the residential buildings, or appropriate measures can be adopted to avoid the influence of noise on residents.

20.2.9 【09措施】2.2.9条：公共建筑场地景观平面布局要点：

20.2.9 [Measures 2009] 2.2.9: Points of landscape plane layout for public building places:

1. 公共建筑场地景观设计应根据建筑属性及特点，确定设计构思；

1. The concept for site landscape design of public buildings shall be determined in accordance with the properties and characteristics of buildings;

2. 应充分考虑人流集散的需要，解决好人行与车行的关系；

2. The gathering and distribution requirements of people shall be fully considered, and the relationship between pedestrians and vehicles shall be solved appropriately;

3. 广场尺度及规模应与建筑相匹配，并注重组织视线关系；

3. The dimensions and scales of the square shall match the buildings, and attention shall be paid to the relationship of the sight lines;

4. 应注重细节设计，如铺装、井盖、水池、灯具、标识等。

4. Attention shall be paid to the detail design, such as pavements, manhole covers, water tanks, lighting fixtures, signs, etc.

20.3 竖向设计

20.3 Vertical Design

20.3.1 【09措施】2.3.1条：场地景观竖向设计应以场地总平面控制高程为依据。

20.3.1 [Measures 2009] 2.3.1: The vertical design for the site landscape shall be based on the total plane control elevation of the site.

20.3.2 【09措施】2.3.2条：场地景观竖向设计的山坡、谷底必须保持稳定。当土坡超过土壤自然安息角呈不稳定时，必须采用挡土墙、护坡等技术措施，防止水土流失或滑坡。

20.3.2 [Measures 2009] 2.3.2: The hillsides and valley bottoms in the vertical design for the site landscape must be kept stable. When the soil slopes exceed the natural repose angle of the soil and are not stable, technical measures of retaining walls, revetments, etc, must be adopted, to prevent soil erosion or landslide.

20.3.3 【09措施】2.3.3条：人工土山堆置高度应与堆置范围相适应，并应防止滑坡、沉降而破坏周边环境。

20.3.3 [Measures 2009] 2.3.3: The stacking height of the artificial soil hills shall be appropriate to the stacking range. Landslide and sedimentation shall be prevented from damaging the surrounding environment.

20.3.4 【09措施】2.3.4条：竖向设计除了创造一定的地形空间景观外，还应为植物种植设计、给水排水设计创造良好的条件，为植物生长和雨水排蓄创造必要条件。

20.3.4 [Measures 2009] 2.3.4: In vertical design, in addition that certain terrain landscapes are created, good conditions for the design of plant cultivation, water supply and drainage shall be provided, and necessary conditions for the growth of plants and storage and drainage of rainwater shall be produced.

20.3.5 【09措施】2.3.5条：竖向设计应合理利用和收集地面雨水，有效控制场地内不可渗透地表的面积，设置阻水措施，减缓径流速度、增强雨水下渗，并利用人工或自然水体蓄存雨水。

20.3.5 [Measures 2009] 2.3.5: In the vertical design, the rainwater on the ground shall be reasonably used and collected. The impervious surface area in the site shall be effectively controlled and water blocking measures shall be arranged, to lower the runoff speed and enhance rainwater infiltration, so that artificial and natural water bodies can be used for rainwater storage.

20.3.6 【09措施】2.3.6条：竖向设计应考虑软质地表的排水坡度，宜符合本规范表2.3.6的规定。

20.3.6 [Measures 2009] 2.3.6: In the vertical design, the drainage gradient of the soft ground surface shall be considered, and the design shall conform to the provisions in Tab. 2.3.6 of the code.

20.4 园路及铺装场地

20.4 Garden Paths and Pavement Sites

20.4.1 【09措施】2.4.1条：园路

20.4.1 [Measures 2009] 2.4.1: Garden paths

1. 各级园路应以总体设计为依据，确定路宽、曲线的线形以及路面结构。

1. For the garden paths of all levels, the road widths, curved line shapes and pavement structures shall be determined on the basis of the overall design.

2. 园路宽度宜符合表20.4.1的规定。

2. The widths of garden paths shall conform to the provisions in Tab. 20.4.1.

3. 园路线形设计应符合下列规定：

3. The geometric design for garden paths shall conform to the following provisions:

1）与地形、水体、植物、建筑物、铺装场地及其他设施结合，形成完整的平面布局；

1) To form the complete plane layout in combination of the terrain, water bodies, plants, buildings, pavement sites and other facilities;

2）创造连续展示景观的空间或欣赏前方景物的透视线；

2) To create the space for continuous display of the landscape and perspective line for appreciating the front scenery;

3）路的转折、衔接通顺，符合行人的行为规律。

3) The smooth turning and joining of the roads shall conform to the behavior rules of pedestrians.

4. 园路技术标准见表20.4.1。

4. Technical standards for garden paths in Tab.20.4.1.

园路宽度　　　　　　　　　　　　　　　　　　　　表 20.4.1

Widths of Garden Paths　　　　　　　　　　　　　　Tab. 20.4.1

园路级别 Grade of garden paths	场地面积（hm^2） Area of the site（hm^2）			
	< 2	2 ~ 10	10 ~ 50	> 50
主路 Main roads	2 ~ 3.5	2.5 ~ 4.5	3.5 ~ 5.0	5.0 ~ 7.0
支路 Branch roads	1.2 ~ 2.0	2.0 ~ 3.5	2.0 ~ 3.5	3.5 ~ 5.0
小路 Paths	0.9 ~ 1.2	0.9 ~ 2.0	1.2 ~ 2.0	1.2 ~ 3.0

注：1. 主路纵坡宜小于8%，横坡宜小于3%，粒料路面横坡宜小于4%，纵、横坡不得同时无坡度。
Notes:1.The gradient of longitudinal slopes of the main road shall be appropriately less than 8%; the gradient of transverse slopes shall be appropriately less than 3%; the gradient of transverse slopes of the aggregate road surface shall be appropriately less than 4%; and it is prohibited to arrange no gradient for longitudinal and transverse slopes at the same time.
　　山地场地的园路纵坡应小于12%，超过12%应做防滑处理。
　　The gradient of longitudinal slopes of the garden path at mountain places shall be less than 12%, and if the gradient exceeds 12%, slip preventive treatment shall be carried out.
2. 支路和小路，纵坡宜小于18%。纵坡超过15%路段，路面应做防滑处理；纵坡超过18%，宜按台阶、梯道设计，台阶踏步数不得少于2级，坡度大于58%的梯道应做防滑处理，宜设置护栏设施。
2.For branch roads and paths, the gradient of longitudinal slopes shall be appropriately less than 18%. The slip preventive treatment shall be carried out on the road surface of the section with the longitudinal slope above 15%. When the longitudinal slope exceeds 18%, the design shall be done as per the steps and stairway, the number of step treads shall not be less than two, slip preventive treatment measures shall be taken if the gradient is more than 58%, and railing facilities shall be provided.
3. 园路在地形险要的地段应设置安全防护设施。
3.Safety protection facilities shall be arranged in the dangerous part of a garden path.
4. 园路应根据不同功能要求确定其结构和饰面。宜使用天然砂石等透水透气材料，提高园路的自然生态功能，使雨水自然渗透。在北方地下水位较高地区，为防止冬季灰土冻胀，园路垫层不宜选用灰土，宜选用级配砂石。饰面、垫层等材料及构造均要透水，才能达到透水的效果。
4. The structures and finishes of garden paths shall be determined as per different functional requirements. Permeable and breathable materials of natural sand and stone, etc, shall be adopted, to improve the natural ecological functions of the garden paths, and make rain water naturally permeate. In areas with higher groundwater level in the north, in order to prevent frost heaving of lime soil in winter, lime soil shall not be adopted for the cushion layers of the garden paths, but graded sand and stone shall be adopted. The materials and structures of finishes, cushion layers, etc. shall be permeable, and the effect of water permeability can be achieved.
5. 园路应平整，路缘不得采用锐利的边石。
5.The garden paths shall be smooth, and sharp curbs shall not be adopted.
6. 场地出入口及主要园路宜便于通过残疾人使用的轮椅，其宽度及坡度的设计应符合有关规范规定。
6.The site entrances and exits and main garden paths shall facilitate the passage of wheelchairs for the disabled people, and the design for the width and slope shall conform to the provisions in the relevant code.

20.4.2 【09 措施】2.4.2 条：铺装场地

20.4.2 [Measures 2009] 2.4.2: Pavement sites

1. 应根据场地总平面布局的要求，确定各种铺装场地的类型和面积。铺装场地应根据集散、活动、演出、赏景、休憩等使用功能要求进行不同设计；

1. The types and area of various pavement sites shall be determined in accordance with the requirements in the total plane layout of the sites. Different design shall be carried out for the pavement sites in accordance with the function requirements for distribution, activities, performances, viewing, rest, etc.;

2. 铺装场地地面材料应考虑平整、耐磨、防滑，并需考虑儿童车、行李车等通过时的震动及噪声影响，材料尽量选用透水砖等环保产品；

2. The ground material for pavement sites shall be smooth, wear resistant and anti-slip, vibration and noise influences when children's cars, luggage carts, etc., pass through shall be considered, and the environmental protection materials of water permeable bricks, etc, shall be adopted as far as possible;

3. 安静休憩场地应利用地形或植物与喧闹区隔离；

3. The quiet sitting-out areas shall be separated from noisy areas through terrains or plants;

4. 演出场地应有方便观赏的适宜坡度和观众席位；

4. The performance places shall be equipped with the appropriate gradient and audience seats convenient for viewing;

5. 铺装场地应考虑各种景观小品及设施的配置，并应考虑夜景照明效果。

5. For pavement sites, consideration shall be given to the configuration of various landscape accessories and facilities, and the lighting effects for night scenes.

20.5 景观小品

20.5 Landscape Accessories

20.5.1 【09 措施】2.5.1 条：景观小品具有分割空间、观赏、休息、标志、使用等功能。各类小品造型设计应有特色，尺度适宜，与周围环境相协调。

20.5.1 [Measures 2009] 2.5.1: The landscape accessories have the functions of partitioning space, viewing, rest, sign, use, etc. The design for the shapes of various landscape accessories shall be provided with the characteristics and appropriate sizes, and shall be coordinated with the surroundings.

20.5.2.【09措施】2.5.2条：景观小品应与主体建筑设计风格协调一致，与绿化种植设计结合组织景观。

20.5.2. [Measures 2009] 2.5.2: The landscape accessories shall be coordinated with the design styles for the main building, and shall form landscapes in combination with the design of greening planting.

20.5.3【09措施】2.5.3条：应保证结构牢固安全。

20.5.3 [Measures 2009] 2.5.3: Structures shall be ensured to be firm and safe.

20.5.4【09措施】2.5.4条：应便于清洁和维护。

20.5.4 [Measures 2009] 2.5.4: Cleaning and maintenance shall be convenient.

20.5.5【09措施】2.5.5条：景观小品的类型及设计要求见表20.5.5。

20.5.5 [Measures 2009] 2.5.5: For the type and design requirements of featured landscapes, see Tab. 20.5.5.

景观小品的类型及设计要求　　　　　　　　　　　　表20.5.5
Type and Design Requirements of Featured Landscape　　　　Tab. 20.5.5

类型 Type		功能 Function	设计要求 Design requirements	配合关系 Mating relation
门 Door		分隔空间、限界标志、出入口 Separation of space, boundary mark and entrance and exit	庭园、园林内如月亮门、尺度宜人、可富有趣味性。限界标志门尺度适当加大，形体多样 Such as moon gates in courtyards and gardens, with appropriate dimensions, and rich interest. The dimensions of the boundary marking doors shall be appropriately increased, and the shapes of doors are diverse	与廊柱墙体结合 Combined with colonnade walls
墙 Walls	花墙 Tracery wall	分隔空间、造景、景观渗透 Space separation, landscaping and landscape infiltration	墙体利用预制混凝土花格、木制漏花窗或金属、玻璃花格，使墙体通透 The prefabricated concrete flower lattices, wood ornamental perforated windows or metal or glass flower lattices shall be used on the walls, to make the walls fully connected	—
	景墙 Landscape wall	观赏、遮挡、衬托背景 Viewing, sheltering and background setting	墙体表面绘制壁画、浮雕、刻字或在墙体作凹凸光影效果，墙体高度适当，符合人体视觉感受 Wall paintings, embossing and lettering shall be made on the surface of the walls, or concave and convex light and shadow effect shall be made on walls, the height of the walls shall be proper, and conform to the human visual sensation	与绿化、广场、水景等结合，也可以用挡土墙作景墙 Combined with the green belts, squares, waterscapes, etc., or with the retaining walls as landscape walls

续表

类型 Type		功能 Function	设计要求 Design requirements	配合关系 Mating relation
墙 Walls	围墙 Enclosure wall	维护、限界 Maintenance and limit	根据不同的位置达到维护、安全要求，高度在 1.8~2.2m The maintenance and safety requirements can be achieved in accordance with different positions, and the height is 1.8-2.2m	可设置花盆或垂直绿化相结合 The combination of flowerpots and vertical greening can be arranged
亭（伞） Pavilions (umbrellas)		休息、观赏、过渡 Rest, viewing and transition	体量适宜，与周围建筑协调一致 The scale shall be appropriate, and the accessories shall be consistent with the peripheral buildings	与坐凳结合 Combined with stools
膜结构建筑 Buildings with membrane structures		休息、观赏、过渡 Rest, viewing and transition	利用钢结构、模制作、顶部透光、可做成类似亭廊建筑 The buildings shall be fabricated with steel structures through moulding; the tops shall be pervious to light; and the buildings can be built into buildings similar to pavilions and galleries	可与入口结合，或作为独立式庭院建筑 In combination with entrances, or used as independent-type courtyard buildings
柱 Columns	门柱 Door columns	分隔不同空间，作为出入口 Separation of different spaces, and as entrances and exits	尺度适宜，有特色 The scale shall be appropriate, and the columns shall have their own characteristics	出入口与广场、水景结合 Entrances and exits, combined with squares and waterscapes
	廊柱 Colonnades	观赏、过渡、休息 Viewing, transition and rest	以柱组成的过渡空间，尺度宜人，作为观赏性尺度可稍大 The sizes of the transition spaces formed with columns are appropriate, and the sizes can be slightly larger for viewing	出入口与广场、水景结合 Entrances and exits, combined with squares and waterscapes
	装饰柱 Decorative columns	观赏、标志 Viewing and marking	观赏性强，可作浮雕柱、造型柱、（图腾）花柱，尺度与周围环境协调 The decorative columns are provided with high ornamental value and can be used as embossed columns, moulding columns and (totem) style columns, and the sizes shall be coordinated with the surroundings	可与出入口、广场、庭园等地景观协调 Coordinated with landscapes of the places of entrances and exits, squares, courtyards, etc
廊 Corridors		观赏、过渡、休息 Viewing, transition and rest	形式多样，有全封闭、开放、半敞开式、尺度宜人，单独过渡式仅作通廊，开敞、半开敞式仅供休息观赏 Various types, such as fully enclosed type, open type and semi-open type, appropriate sizes. The separate transition type is only used as corridors. The open and semi-open types are only for rest and viewing	—

续表

类型 Type	功能 Function	设计要求 Design requirements	配合关系 Mating relation
栅栏 Fences	标明界限，防止进入，球场防止球类飞出 Indicating range to prevent access and preventing balls flying out from the playing court	材料多为木、竹制、金属。金属网应考虑强度，防倾倒，球场宜采用网状，孔洞尺寸大小应考虑避免球飞出，限人高度≥1.80m，隔离植物0.50m左右，网球场挡球高度3~4m Materials can be wood, bamboo and metal. For metal meshes, strength shall be considered. To prevent inclination, the meshy type shall be adopted for playing courts. For the dimensions of holes, avoidance of flying out of the balls shall be considered. The height limit for people shall be greater than or equal to 1.80m. The height of isolating plants shall be about 0.50m, and the backstop height in the tennis court shall be 3~4m	—
桥 Bridges	分隔水面，联系交通、点缀风景 Water surface separation, traffic link and scenery embellishment	结构牢固、稳定，桥面防滑，应根据通航、通车、行人等的要求确定具体尺寸规模。水面较宽时，近2m范围内水深大于0.5m时，桥两侧需设栏杆，作用在栏杆上的竖向力和栏杆顶部的水平荷载均按1.0kN/m计算；非通行车辆的桥应有阻止车辆通行的措施，桥面人群荷载按3.5kN/m² 计算。通行车辆的桥按市政相关要求计算和设计，通过小游船的桥梁，其桥底雨水位之间净空高度不应小于1.5m The structure shall be firm and stable, with slip-resistant decks. The specific dimensions shall be determined as per the requirements for navigation, transport service, pedestrians, etc. When the water surface is wider, and the water depth in the range of 2m is greater than 0.5m, railings shall be arranged on both sides of the bridge. The vertical force acting on the railing and the horizontal load acting on the top of the railing shall be calculated as per 1.0kN/m. For the bridges not for the passage of vehicles, measures prevent the passage of vehicles shall be arranged, and the pedestrian load on the bridge deck shall be calculated as per 3.5kN/m². The bridges for the passage of vehicles, the calculation and design shall be carried out as per the relevant municipal requirements. For bridges for the passage of small amusement boats, the clearance height of the rainwater level at the bottom of the bridge shall be not less than 1.5m	可与廊、亭结合，廊桥需设观赏坐凳 The bridge can be combined with galleries and pavilions, and the viewing stool shall be arranged on the corridor bridge

续表

类型 Type	功能 Function	设计要求 Design requirements	配合关系 Mating relation
假山叠石 Rockery fold stone	观赏、登高、穿越、分隔空间 Viewing, climbing, penetration and space separation	可与自然山石堆砌或人造石浇筑，注意基础稳定，结构安全可靠，并从多角度考虑造型整体效果 The rockery fold stone can be stacked and built with natural rocks or poured with artificial rocks. Attention shall be paid to maintain the stable foundation and safe and reliable structures. The overall effect of modeling shall be considered from multiple angles	可与水景结合，形成跌水等效果 Combined with the waterscape for the effect of cascade, etc.
汀步 Stepping stone	临水，步行道路 Water-facing pedestrian walkways	结构牢固稳定，步距≤0.5m，水深不大于0.5m，应考虑防滑 The structure is firm and stable. The pace length is less than or equal to 0.5m. The water depth shall not be more than 0.5m, and anti-slip measures shall be taken	—
座椅（凳） Seats (stools)	休息用 For rest	稳定牢固，座高 0.35～0.45m，座板倾角 6°～7°，椅座面宽 0.40～0.60m，椅背与座板夹角 98°～105°，椅背高 0.35～0.65m，凳面 0.4m×0.4m The seats (stools) shall be stable and firm. The seat height is 0.35～0.45m. The inclination angle of the seat plates is 6°～7°. The width of the seat surface is 0.40～0.60m. The angle between the chair back and the seat plates is 98°～105°. The height of chair backs is 0.35～0.65m, and the dimensions of the chair surfaces are 0.4m×0.4m	广场景点、绿荫、路旁、游戏场，可与灯柱、种植容器结合制作 The seats in square attractions, greenery, roadsides and playgrounds shall be manufactured in combination with lamp posts and planting containers
桌 Tables	休息用 For rest	桌高 0.60～0.70m，面宽≥0.70～0.80m（四人用） Height of the tables: 0.60～0.70 m, and surface width ≥0.70～0.80m (for four people)	与凳椅结合 Combined with chairs
雕塑 Sculptures	观赏 Viewing	要有艺术性，与整体规划主题一致，尺度适宜 Artistry is required and consistent with the overall planning theme, and the scale is appropriate	—
台阶 Steps	高低差过渡 Transition between the high position and low position	踏步高≤0.15m，踏步宽度≥0.30m，踏步间平台宽度≥1.50m，应考虑轮椅坡道 Step height ≤0.15m, step width ≥0.30m, width of the platform between steps ≥1.50m, and the wheelchair ramps shall be considered	—

续表

类型 Type		功能 Function	设计要求 Design requirements	配合关系 Mating relation
指示标志 Indicator signs	位置标志 Position indication	表明基地总面积，各分区及设施配置 Indicating the total area of the base, all sub-areas and facility configuration	清晰、明确、尺度宜人，满足人近观要求 Signs shall be clear and definite. The dimensions shall be delightful, and the signs shall meet the requirement for close viewing	在基地入口处 At base entrances
	导向标志 Guidance signs	方向指示 Direction indication	醒目、位置适当 Signs shall be eye-catching, and the locations shall be proper	在基地道路交叉口和各类场地入口处 At intersections of the roads in bases and entrances of various places
	名称标志 Name signs	居住区名称，组团名称，游园名称、楼号及树木 Names of residential communities, group buildings, amusement parks, building numbers and trees	根据不同的名称要求，分类设置，同类标志风格一致 Signs shall be arranged in categories as per the different name requirements, and the style of the same kind of signs shall be consistent	各处入口或建筑物上，固定式或嵌入式。树木采用悬挂式 At various entrances, or on buildings, in a fixing or embedding mode. For trees, the suspension mode shall be adopted
	警示标志 Warning signs	限速、禁人、行动规范 Speed limiting, entry forbidding and conduct normalization	涉及游人安全处必须设置，警示内容明确 For tourist safety, signs must be arranged in these places and the warning contents shall be definite	—
植物容器 Plant containers		观赏、突出所需种植物 Viewing and highlighting the plants to be needed	各种材质容器必须设置泄水孔，单独设置或成从 The drainage holes shall be arranged for containers of various materials separately or together	可与台阶、道路及各种景观物结合 Combined with steps, roads and various landscapes
饮水台 Drinking water station		饮水用，可与洗手结合设置 For drinking water, arranged in combination with wash basins	水台水流向上如喷泉，台高0.80~1.00m，儿童饮水台高0.65m左右，洗手部分在饮水侧面偏低处高0.50m，上设置水箅或水盘 The water from the drinking fountain stands flows upwards like a fountain. The height of the stands shall be 0.80 ~ 1.00m. The height of drinking water stations for children shall be about 0.65m. The height of hand washing parts at lower places on drinking water sides shall be 0.50m, and water grilles or water trays shall be arranged on the washing parts	设在路旁、游乐场地或人流集中的景点 Arranged by roads, at playground or scenic spots with crowded people

续表

类型 Type	功能 Function	设计要求 Design requirements	配合关系 Mating relation
果皮箱 Garbage bins	垃圾收集 Refuse collection	垃圾分类设置，高度 0.60~0.80m 左右，以确保卫生，使用和清洁方便 The garbage bin shall be arranged in categories, with the height of about 0.60~0.80 m, to ensure sanitation and convenient use and cleaning	与椅凳结合，成组分布 Combined with benches and stools, and arranged in groups
旗杆 Flagpoles	标志 Signs	嵌入式、基座式，旗杆高度 5~6m 时杆间距约 1.5m，旗杆高度 7~8m 时杆间距约 1.8m，旗杆高度 9m 以上，杆间距 2m， The embedded type and foundation base type can be adopted. When the height of the flagpole is 5~6m, the spacing of poles shall be about 1.5m. When the height of the flagpole is 7~8m, the spacing of poles shall be about 1.8m. When the height of the flagpole is greater than 9m, the spacing of the poles shall be 2m	主入口或靠近主体建筑 At main entrances or near the main buildings
车挡 Vehicle stopper	作路障用，防止车辆通行 Used as barricade, to prevent vehicles from passing through	有移动式、固定式，金属揽柱高约 0.50~0.70m，间距 0.60m，混凝土墩成球，高 0.30~0.40m The movable and fixed types are adopted. When the height of metal bollards is about 0.50-0.70m, the spacing shall be 0.60m. For concrete pier balls, the height shall be 0.30-0.40m	—
照明灯 Lighting lamp	照明，同时可利用彩色灯，丰富景点夜间景观及色彩 For lighting, at the same time, colored lamps can be used, to enrich the landscapes and colors of scenic spots at night	灯具选择依不同景点设置，如草坪灯、建筑射灯、水下灯、地灯、指示灯、装饰灯等；庭院灯高度一般为 3~4m 左右，间距一般为 15~20m 左右；草坪灯高度一般为 0.3~1m 左右，间距一般为 5~8m 左右，并注意防眩光措施。安装在水池内、旱喷泉内的水下灯具必须防止触电等级为 III 类、保护等级为 IPX8 的加压水密型灯具，电压不得超过 12V The lighting fixtures, such as lawn lights, architectural spotlights, underwater lights, floor lights, indicator lights, decorative lights, etc. shall be arranged as per different scenic spots. The height of courtyard lamps is commonly about 3~4m. The spacing is commonly about 15~20m. The height of the lawn light is commonly about 0.3~1m. The spacing is commonly about 5~8m. The glare preventive measures shall be taken. The pressurized watertight-type lamps with Grade III electric shock prevention, IPX8 protection grade, and voltage within 12V must be adopted as the underwater lights mounted in the water tank and dry fountain	

20.6 【09措施】2.6.1条：水景

20.6 [Measures 2009] 2.6.1: Waterscape

20.6.1 自然水体

20.6.1 Natural water bodies

1. 水景设计应充分利用自然水体，创造临水空间和设施，并加强沿岸防护安全措施；

1. Natural water bodies shall be fully utilized in waterscape design, to create water-facing space and facilities, and enhance the safety measures for coastal protection;

2. 充分考虑当地防排洪的要求，确定最高洪水位、最高潮位、常水位及最低水位等关系，按当地有关规定进行处理；

2. Local requirements for flood protection and drainage shall be fully considered. The relationships between highest flood levels, highest tide levels, normal water levels and lowest water levels, etc. shall be ascertained. Local relevant provisions shall be followed;

3. 自然水体的进水口、排水口、溢水口和闸门的标高，应适应于水位、泄洪和清淤的要求；

3. The elevations of water inlets, water outlets, overflow holes and gates of natural water bodies shall adapt to the requirements of water levels, flood discharging and desilting;

4. 自然湖泊可供划船的最小面积为 7.5hm^2，最小水深为 0.7m。

4. The minimum area of a natural lake for boating is 7.5hm^2, and the minimum water depth is 0.7m.

20.6.2 人工水体

20.6.2 Artificial water bodies

1. 人工水体应充分注重不同地区、不同季节变化对水景的影响，合理确定水景规模，尽量降低维护成本，并应考虑流水声音、雾气等可能对人们形成的影响；

1. For artificial water bodies, full attention shall be paid to the effects of different seasonal changes in different regions on waterscapes. The scales of waterscapes shall be reasonably determined and the maintenance costs shall be reduced as much as possible. The influences of water sounds, fog, etc. on people shall be considered;

2. 人工水体供水宜采用循环水，可采用冷却水、中水处理水，减少用水量和能源消耗。应采用过滤、循环、净化、充氧等技术措施，保证水质符合卫生及观感要求；

2. The artificial water bodies shall be supplied with the circulating water, cooling water and treated reclaimed water, to reduce the water consumption and energy consumption. The technical measures of filtration, circulation, purification, oxygenation, etc., shall be adopted, to ensure that the water quality conform to the hygiene and perception requirements;

3. 人工水体的进水口、溢水口、排水坑、泵坑、过滤装置等宜设置在相对隐蔽的位置。

3. The water inlets, overflow openings, drainage pits, pump pits, filtration devices, etc. of artificial water bodies shall be arranged at the relatively concealed positions.

4.【公园规】4.3.2条：硬底人工水体的近岸2.0m范围内的水深，不得大于0.7m，达不到此要求的应设护栏。无护栏的园桥、汀步附近2.0m范围以内的水深不得大于0.5m。

4. [Code for design of park] 4.3.2: The depth of the hard-bottom artificial water body within the range of 2.0m near the bank shall not be greater than 0.7m, and if the requirements are not reached, railings shall be arranged. The depth of the water within the range of 2.0m near the garden bridge without railings and stepping stone shall not be greater than 0.5m.

5. 人工池体应采用防水及抗渗漏材料，并依据不同地区气候条件考虑防冻等特殊措施，刚性池体应根据要求设置伸缩缝。

5. Waterproof and anti-leakage materials shall be adopted for artificial ponds. Special anti-freezing measures and others shall be considered as per the climatic conditions in different regions. Expansion joints shall be arranged for the rigid pool body as per the requirements.

6. 人工溪流缓流坡度0.3%~0.5%，急流处3%左右。可涉入的溪流水深不应大于0.3m，底部宜石砌，便于清理。

6. The gradient of places with slow flow of an artificial stream is 0.3% ~ 0.5%, and the gradient at the rapid flow is about 3%. The water depth of accessible streams shall be not greater than 0.3 m, and the bottom shall be built with stone, to facilitate cleaning.

7. 室外游泳池应形成独立区域并设置管理及配套服务设施。游泳池深度应根据使用人群确定，儿童游泳池水深0.5~1.0m为宜，成人游泳池水深1.2~2m为宜。池底和池岸应防滑，池壁应平整光滑，池岸应做圆角处理，并应符合游泳池设计的相关规定。

7. For outdoor swimming pools, the independent area shall be formed, and the management and ancillary service facilities shall be arranged. The depth of swimming pools shall be determined as per the users. The water depth of children's swimming pool shall be 0.5 ~ 1.0m, and the water depth of adult swimming pools shall be 1.2 ~ 2m. The bottoms and banks of pools shall be skip-resistant. The pool walls shall be flat and smooth. For banks, fillet processing shall be made. The specifications relevant to the design of swimming pools shall be met.

8. 养鱼池深度因所养鱼种而异，一般池深0.8~1.0m，并需确保水质的措施。

8. The depth of fishponds shall differ due to different fish species. Usually, the pond depth

shall be 0.8 ~ 1.0m, and measures to guarantee water quality shall be arranged.

9. 水生植物种植池深度应满足不同植物的栽植要求，浮水植物（如睡莲）水深要求 0.5-1.0m，挺水植物（如荷花）水深要求 1.0m 左右。

9. The depth of aquatic planting pools shall meet the planting requirements of different plants. The water depth for floating plants (such as water lilies) is required to be 0.5 ~ 1.0m, and the depth of the water for emerging plants is required to be about 1.0m.

10. 喷泉可与水池结合，与游人接触的喷泉，不得使用再生水。

10. Fountain and water tanks can be combined, and the reclaimed water shall not be used for the fountain in contact with visitors.

11. 旱喷泉喷洒范围内不宜设置道路，以免喷洒时影响交通，地面铺装还需考虑防滑。旱喷泉内禁止直接使用电压超过 12V 的潜水泵。

11. No road can be arranged within the spraying range of the dry fountain to avoid affecting the traffic during spraying. Anti-slip measures shall be considered in floor pavement. The submersible pump with voltage above 12V shall not be directly used in the dry fountain.

20.6.3 驳岸及护坡

20.6.3 Revetment or Slope Protection

1. 素土驳岸岸顶至水底坡度小于 1：1 的应采用植被覆盖，坡度大于 1：1 的应有固土和防冲刷的技术措施。

1. Vegetation coverage shall be adopted when the gradient of plain soil revetments from the top to the water bottom is less than 1 ：1. When the gradient is more than 1 ：1, technical measures for soil fixation and erosion control shall be provided.

2. 人工砌筑或混凝土浇筑的驳岸及护坡，边坡一般为 1：1 或 1：1.5，并应有良好的透水构造，以防土壤自坡下流失，驳岸或护坡基础应在冰冻线以下，并应根据水体及土层冻胀对驳岸的影响提出相应技术措施。

2. The side slope of manually-built or concrete-poured revetment or slope protection is commonly 1 ：1 or 1 ：1.5. Proper permeable structures shall be provided to prevent the loss of soil from the lower part of the slope. The revetment or slope protection foundation shall be below the freezing line. Corresponding technical measures shall be proposed in accordance with influence of frost heaving of the water body and soil layer on revetment.

3. 水体岸边应有安全防护措施，并满足相关设计规范。

3. Safety protective measures shall be provided at the banks of a water body, and relevant design code shall be met.

20.7 【09措施】2.7条：种植设计

20.7 [Measures 2009] 2.7: Planting Design

20.7.1 植物配置设计原则：

20.7.1 Principles for plant configuration design:

1. 种植设计应根据当地光照、土壤、朝向等自然条件选择生长健壮、病虫害少、养护管理方便、对人体无害的植物材料；

1. In planting design, plants with good growth, less diseases and insect pests, easy cultivation and management and harmlessness to human bodies shall be selected as per the natural conditions such as local light, soil, orientation, etc;

2. 充分发挥植物材料的各种功能和观赏特点，乔、灌、草与地被、花卉等合理配置，常绿与落叶、速生与慢生相结合。提倡屋顶绿化和垂直绿化，形成多层次的复合结构，植物群落构图和谐、色彩季相丰富，具有地域特点；

2. Various functions and ornamental features of plant materials shall be fully exerted. Arbors, shrubs and grass shall be reasonably configured with the ground cover plants, flowers, etc. The evergreen plants shall be combined with the deciduous, fast-growing and slow growing plants. Roof greening and vertical greening shall be advocated, to form the multi-level composite structure, harmonious composition of the plant communities and rich colors and seasonal aspects, and the geographical characteristics;

3. 在统一的基调基础上，树种力求丰富、配置形式多样，并应根据植物生长速度、对近远期景观提出要求，必要时需采取过渡种植措施；

3. On the basis of unified tone, tree species shall be diverse as much as possible and the configuration forms shall be multiple. Requirements shall be proposed for the landscapes in the short term and long term as per the growth speed of the plants. If necessary, transitional planting measures shall be adopted;

4. 要注意种植的位置，与建筑、地下管线、高压线等设施的距离要符合要求，一般乔木距建筑物 5~8m，以免影响室内采光和通风；

4. Attention shall be paid to the places to plant. The distances between plants and facilities, or buildings, underground pipelines, high-voltage lines, and others shall conform to the requirements. The distance between common arbor trees and buildings shall be 5 ~ 8m to avoid affecting indoor daylighting and ventilation;

5. 居住区内儿童游乐区严禁配置有毒、有刺等易对儿童造成伤害的植物；

5. It is prohibited to arrange toxic, thorny and other plants that can cause injury for children in children's playground in residential areas;

6. 居住建筑朝阳面种植设计应避免植物对居室内阳光的遮挡。

6. During the design of planting on the sunny side of a residential building, plants shielding indoor sunlight shall be avoided.

20.7.2 道路绿带设计，行道树定植株距应以树种成年期冠幅为准，最小株距4m，树干中心至路缘石外侧最小距离宜为0.75m。

20.7.2 For the design of green belts along roads, the specified distance between street trees shall be subject to, the crown sizes of the trees in the adult stage. The minimum plant spacing is 4m, and the minimum distance between the trunk center and outside of the curbs shall be 0.75m.

20.7.3 广场植物配置，应考虑协调与四周建筑的关系，根据广场功能、规模和尺度，宜种植高大乔木，应考虑安全视距及人流通行要求，树木枝下净空应大于2.2m。

20.7.3 For the configuration of plants on squares, the coordination with the surrounding buildings shall be considered. Tall arbors shall be planted in accordance with functions, scales and sizes of the square. The safety sight distance and passing requirements of pedestrian flow shall be considered, and the clearance under the branches of trees shall be greater than 2.2m.

20.7.4 停车场周边宜种植乔木，停车场内宜结合停车间隔种植乔木，树木枝下净空应符合停车位高度要求，小型汽车高2.5m，中型车高3.5m，载货车高4.5m。

20.7.4 Arbor trees shall be planted around the parking area and in the parking area arbor trees can be planted in combination with the parking distance. The clearance under the branches of trees shall conform to the height requirements of the parking spaces. For compact vehicles, the height shall be 2.5m; for medium-duty vehicles, it shall be 3.5m, and for trucks, it shall be 4.5m.

20.7.5 树木与地下管线最小水平距离见表20.7.5。

20.7.5 For the minimum horizontal distance between trees and underground pipelines, see Tab. 20.7.5.

【公园规】附录2：树木与地下管线最小水平距离　　　　表20.7.5

[Code for design of park] Appendix II: Minimum Horizontal Distance between Trees and Underground Pipelines　　　Tab. 20.7.5

名称 Name	新植乔木 Newly-planted arbor trees	现状乔木 Current arbor trees	灌木或绿篱外缘 Outer edge of shrubs or hedges
电力电缆 Power cables	1.50	3.50	0.50

续表

名称 Name	新植乔木 Newly-planted arbor trees	现状乔木 Current arbor trees	灌木或绿篱外缘 Outer edge of shrubs or hedges
通信电缆 Communication cables	1.50	3.5	0.50
给水管 Water supply pipes	1.50	2.0	—
排水管 Drainage pipes	1.50	3.0	—
排水盲沟 Blind drainage ditches	1.00	—	—
消防笼头 Fire faucets	1.20	2.0	1.20
煤气管道（低中压） Gas pipelines (low- and-medium pressure)	1.20	3.0	1.00
热力管 Heating pipe	2.00	5.0	2.00

注：乔木与地下管线的距离是指乔木树干基部的外缘与管线外缘的净距离。灌木或绿篱与地下管线的距离是指地表处分蘖枝干中最外的枝干基部的外缘与管线外缘的净距。

Note: The distance between the arbor tree and underground pipelines means the clear distance between the outer edge of the trunk base of the arbor tree and the outer edge of the pipeline. The distance between shrubs or hedges and underground pipelines means the clear distance between the basal outer edge of the outermost branch of the tillering branch on the ground surface and the outer edge of the pipeline.

20.7.6 树木与地面建筑物、构筑物外缘最小水平距离见表20.7.6。

20.7.6 For the minimum horizontal distance between trees and the outer edges of the ground buildings and structures, see Tab. 20.7.6.

【公园规】附录3：树木与地面建筑物、构筑物外缘最小水平距离　　表20.7.6

[Code for design of park] Appendix III: Minimum Horizontal Distance between Trees and the Outer Edges of the Ground buildings and Structures　　Tab. 20.7.6

名　称 Name	新植乔木 Newly-planted arbor trees	现状乔木 Current arbor trees	灌木或绿篱外缘 Outer edge of shrubs or hedges
测量水准点 Measurement reference point	2.0	2.0	1.0
地上杆柱 Column and pole	2.0	2.0	—
挡土墙 Retaining wall	1.0	3.0	0.5
楼房 Storied buildings	5.0	5.0	1.5

续表

名称 Name	新植乔木 Newly-planted arbor trees	现状乔木 Current arbor trees	灌木或绿篱外缘 Outer edge of shrubs or hedges
平房 Single-storey building	2.0	5.0	—
围墙（高度小于2.0m） Enclosure walls (the height is less than 2.0 m)	1.0	2.0	0.75
排水明沟 Open drainage ditches	1.0	1.0	0.5

20.7.7 树木与架空电力线路导线的最小垂直距离见表20.7.7。

20.7.7 For the minimum vertical distance between trees and overhead power lines and conductors, see Tab. 20.7.7.

【城市道路绿化规】6.1.2：树木与架空电力线路导线的最小垂直距离　　表20.7.7

[Code for design of urban road greening] 6.1.2: Minimum Vertical Distance between Trees and Overhead Power Lines and Conductors　　Tab. 20.7.7

电压（kV） Voltage (kV)	1~10	34~110	154~220	330
最小垂直距离（m） Minimum vertical distance (m)	1.5	3.0	3.5	4.5

20.7.8 单行整形绿篱的生长空间距离见表7.7.8。

20.7.8 Tab. 7.7.8 for Growth Spacing between Single-Row Shaped Hedges.

【城市道路绿化规】6.3.1条：单行绿篱空间距离（m）　　表20.7.8

[Code for design of urban road greening] 6.3.1: Space Distance between Single-Row Hedges (m)　　Tab. 20.7.8

类型 Type	地上空间高度 Space height over the ground	地上空间宽度 Space width over the ground
树墙 Espalier	>1.6	>1.5
高绿篱 High hedge	1.2~1.6	1.2~2.0
中绿篱 Medium hedge	0.5~1.2	0.8~1.5
矮绿篱 Short hedge	0.5	0.3~0.5

注：双行种植时，其宽度增加0.3~0.5m。
Notes: When the hedges are planted in double rows, the width shall be increased by 0.3~0.5m.

20.7.9 古树名木的保护。

20.7.9 Protection of Ancient and Precious Trees.

1. 古树，指树龄在100年以上的树木；名木，指国内外稀有的以及具有历史价值和纪念意义及重要科研价值的树木。

1. Ancient trees mean the trees over 100 years old. Precious trees refer to trees that are rare at home and abroad and have historical value, commemorative significance and important value for scientific research.

2. 古树名木分为一级和二级。凡是树龄在300年以上，或特别珍贵稀有，具有重要历史价值和纪念意义、重要科研价值的古树名木为一级；其余为二级。

2. Ancient and precious trees can be divided into grade I and grade II. Grade I trees cover all the ancient and precious trees which are more than 300 years old, or are especially precious and rare, with important historical value, commemorative significance and important value for scientific research; others are Grade II.

3. 新建、改建、扩建的建设工程影响古树生长的，建设单位必须提出避让和保护措施。

3. If the new construction, reconstruction and extension projects have influences on the growth of ancient trees, the construction organizations must adopt measures to avoid and protect the ancient trees.

4. 古树名木的保护必须符合下列要求：

4. The protection of ancient and precious trees must conform to the following requirements:

1）古树名木必须原地保留；

1）The ancient trees and precious trees must be preserved at the original place;

2）距古树名木树冠垂直投影5m的范围内严禁堆放物料、挖坑取土、兴建临时设施建筑；

2）Within the range of 5m from the vertical projection of the crown of the ancient and precious tree, storage of materials, digging for earth and construction of temporary facilities or buildings shall be strictly prohibited;

3）保护范围内不得损坏表土层和改变地表高程，除保护及加固设施外，不得设置建筑物、构筑物及架（埋）设各种过境管线，不得栽植缠绕古树名木的藤本植物；

3）Within the protection scope, the topsoil shall not be damaged and the ground elevation shall not be changed. Except the protection and reinforcement facilities, buildings, structures and racks, and various transit pipelines shall not be arranged (embedded); and vines that can entwine ancient and precious trees shall not be planted;

4）保护范围附近，不得设置造成古树名木处于阴影下的高大物体和排泄危及古树的有害水、气的设施；

4）Any tall objects that can keep ancient and precious trees in shadow and harmful water and gas drainage facilities that endanger ancient and precious trees shall not be arranged near the protection range;

5）采取有效的工程技术措施和创造良好的生长环境，维护其正常生长。

5）Effective technical measures for the engineering shall be adopted, and good growth environment shall be created, to maintain the normal growth.

20.7.10 种植屋面。

20.7.10 Planting roofs.

1. 种植屋面设计应包括下列内容：

1. The design of planting roofs shall include the following contents:

1）计算建筑屋面结构荷载；

1）The structural load of the building roof shall be calculated;

2）因地制宜设计屋面构造系统；

2）The design of the structural system of the roof shall be made in accordance with the specific conditions of the local area;

3）设计排水系统；

3）The drainage system shall be designed;

4）选择耐根穿刺防水材料和普通防水材料；

4）The root-penetration-resistant waterproof materials and ordinary waterproof materials shall be selected;

5）确定保温隔热方式，选择保温隔热材料；

5）To determine the insulation and heat shielding modes, and select the insulation and heat shielding materials;

6）选择种植土类型；

6）To select the types of planting soil;

7）选择植物种类，制定配置方案；

7）To select plant types, and establish the configuration program;

8）设计并绘制细部构造图。

8）To design and draw the detail structural drawings.

2. 植被层应根据屋面大小、坡度、建筑高度、受光条件、绿化布局、观赏效果、防风安全、水肥供给和后期管理等因素选择，并应符合下列要求：

2. The vegetation layer shall be selected on the basic of the factors like roof size, gradient, building height, light condition, greening layout, ornamental effect, wind protection, water and fertilizer supply, later-stage management, etc. The following requirements shall be conformed to:

1）不宜选用根系穿刺性强的植物；

1) Plants of high root penetration are appropriately not to be selected;

2）不宜选用速生乔木、灌木植物；

2) Fast-growing arbor trees and shrubs are appropriately not to be selected;

3）高层建筑屋面和坡屋面宜种植地被植物；

3) For roofs of high-rise building and slope roofs, groundcover plants shall be adopted;

4）乔木、大灌木高度不宜大于7.5m，距离边墙不宜小于2m。

4) The height of arbor trees and large shrubs shall be appropriately not greater than 7.5m, and the distance from the side wall shall be appropriately not less than 2m.

3. 根据气候特点、屋面形式，宜选择适合当地种植的植物种类。

3. The species of plants suitable for growing locally shall be appropriately selected as per the climate characteristics and roof forms.

4. 植物荷重设计应按植物在屋面环境下生长10年后的荷重估算，初栽植物的荷重应符合表20.7.10-1的规定。

4. In the load design of plants, the load shall be estimated as per the load of plants growing for 10 years in the roof environment. The load of the newly-grown plants shall conform to the provisions in Tab. 20.7.10-1.

【种植规程】4.6.4条：初栽植物种植荷载　　　　表20.7.10-1

[Technical specification for planting roof engineering] 4.6.4:Load of Newly grown Plants　Tab. 20.7.10-1

植物类型 Plant type	小乔木（带土球） Small arbor (with soil ball)	大灌木 Large shrub	小乔木 Small arbor	地被植物 Groundcover plants
植物高度或面积 Plant height or area	2.0～2.5m	1.5～2.0m	1.0～1.5m	1.0m^2
植物荷载（kN/株） Plant load (kN/plant)	0.8～1.2	0.6～0.8	0.3～0.6	0.15～0.3kN/m^2
种植荷载（kN/株） Planting load (kN/plant)	2.5～3.0	1.5～2.5	1.0～1.5	0.5～1.0

注：种植荷载应包括种植区构造层自然状态下的整体荷载。选择植物应考虑植物生长产生的活荷载变化，一般情况下，树高增加2倍，其重量增加8倍，需10年时间。
Note: the planting load shall include the overall load of the tectonic layers of the planting area in natural state. The plants shall be selected in consideration of changes in the live load generated due to plant growth, in normal cases, the tree height is increased by 2 times, and the weight is increased by 8 times for 10 years.

5. 建筑屋面种植宜选用改良土或无机复合种植土，地下建筑顶板种植宜选用田园土。种植土的厚度应根据植物种类按表20.7.10-2选用。

5. For planting on the building roof, the improved soil or inorganic composite planting soil shall be appropriately selected. For planting on the roof of the underground building, the pastoral soil shall be appropriately selected. The thickness of the planting soil shall be selected in accordance with plant species and Tab. 20.7.10-2.

【09 措施】表 2.7.10-2：种植土　　　　　　　　　　表 20.7.10-2
[Measures 2009] Tab. 2.7.10-2: Planting Soil　　　Tab. 20.7.10-2

种植土类型 Type of planting soil	种植土厚度（mm） Thickness of planting soil (mm)			
	大乔木 Large arbor	大灌木 Large shrub	小灌木 Small shrub	地被植物 Ground cover plants
田园土 Pastoral soil	800 ~ 900	500 ~ 300	300 ~ 400	100 ~ 200
改良土 Improved soil	600 ~ 800	300 ~ 400	300 ~ 400	100 ~ 150
无机复合种植图 Inorganic composite planting soil	600 ~ 800	300 ~ 400	300 ~ 400	100 ~ 150

注：此表摘自《种屋面工程技术规程》JGJ 155—2007。
Notes: the table is extracted from *Technical Specification for Planting Roof Engineering* JGJ 155—2007.

6. 屋面种植乔木、大灌木时，宜局部增加种植土的厚度。

6. When arbor trees and large shrubs are planted on roofs, the thickness of the planting soil shall be partially increased.

7. 花园式屋面种植的布局应与屋面结构相适应；乔木类植物和亭台、水池、假山等荷载较大的设施，应设在承重墙或柱的位置。

7. The layout for planting on the garden-style roof shall be appropriate to the roof structure. The heavy-load facilities like arbor trees, pavilions, water tanks, rockeries, etc. shall be arranged at the positions of bearing walls or columns.

8. 种植屋面宜设置雨水收集系统，并应根据种植形式的不同，确定水落口数量和落水管直径。

8. On planting roofs, the rainwater collection system shall be appropriately arranged, and the quantity of downspout ports and the diameter of downpipes shall be determined as per the different planting forms.

9. 种植屋面为平屋面时，其坡度宜为 1% ~ 2%。单向坡长小于 9m 的屋面可用材料找坡，单向坡长大于 9m 的屋面宜结构找坡。天沟、檐沟坡度不应小于 1%。

9. When planting roofs are flat roofs, the gradient shall be 1% to 2%. For the roofs with the

length of the one-way slope less than 9 m, sloping can be made by means of the material, and for the roofs with the length of the one-way slope greater than 9 m, sloping shall be made by means of the structure. The gradient of gutters and trenches shall be not less than 1%.

10. 种植屋面配套设施应符合下列规定：

10. The ancillary facilities for planting roofs shall conform to the following provisions:

1）水管、电缆线等设施，应铺设在防水层之上；

1）The facilities of water pipe, cable, etc, shall be laid on the waterproof layers;

2）屋面周边应有安全防护设施；

2）On the periphery of the roof, safety protective facilities shall be provided;

3）花园式种植屋面宜有照明设施；

3）On the garden-type planting roofs, lighting facilities shall be appropriately provided;

4）灌溉可采用滴灌、喷灌和渗灌设施；

4）For irrigation, facilities for trickle irrigation, sprinkling irrigation and infiltrating irrigation can be adopted;

5）新移植的植物宜采用遮阳、抗风、防寒和防倒伏支撑等设施。

5）For newly-transplanted plants, shading, wind-resistant, cold-resistant, anti-falling and supporting facilities shall be appropriately adopted.

20.7.11 种植屋面基本构造见图 20.7.11。

20.7.11 For the basic structure of planting roofs, see Fig. 20.7.11.

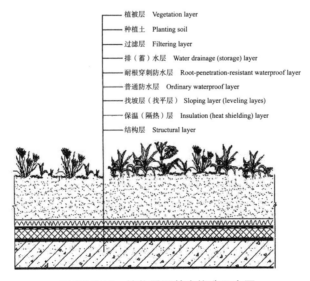

图 20.7.11　植物屋面基本构造示意图

Fig. 20.7.11　Schematic Diagram for Basic Structure of Planting Roofs

20.7.12 种植屋面排（蓄）水层及过滤层材料。

20.7.12 Materials for water drainage（storage）layers and water drainage（storage）layers on planting roofs.

1. 排（蓄）水层可选用下列材料，并注意在年降水量小于蒸发量的地区，宜选用蓄水功能强的排水板；在坡屋面上做种植时，种植土厚度小于150mm时，不宜设排水层。

1. For water drainage（storage）layers, the following materials can be selected, but for the area with the annual precipitation amount less than the evaporation amount, the drainage boards with strong water storage function shall be appropriately selected. When planting is carried out on slope roofs, and the thickness of the planting soil is less than 150 mm, drainage layers shall not be arranged.

1）凹凸型排（蓄）水板，其主要物理性能应符合表20.7.12-1的要求。

1）For the concave-convex water drainage（storage）boards, the main physical performances shall conform to the requirements in Tab. 20.7.12-1.

【种植规程】4.5.1-1：凹凸型排（蓄）水板主要物理性能　　　　表20.7.12-1

[Technical specification for planting roof engineering] 4.5.1-1 : Main Physical Performances of Concave-Convex Water Drainage（Storage）Boards　　Tab. 20.7.12-1

项目 Project	单位面积质量（g/m²） Mass per Area（g/m²）	凹凸高度（mm） Height of Concave-Convex Area（mm）	抗压强度（kN/m²） Compressive Strength（kN/m²）	抗拉强度（N/50mm） Tensile strength（N/50 mm）	断裂延伸率（%） Elongation（%）at break
性能要求 Performance requirements	≥ 50	≥ 90	85 ~ 90	≥ 380	稳定 Stable

2）网状交织排（蓄）水板主要物理性能符合表20.7.12-2。

2）The main physical performances of the mesh interwoven water drainage（storage）boards shall conform to Tab. 20.7.12-2.

网状交织排（蓄）水板主要物理性能　　　　表20.7.12-2

Main Physical Performances of the Mesh Interwoven Water Drainage（Storage）Board　　Tab.20.7.12-2

项目 Project	抗压强度（kN/m²） Compressive Strength（kN/m²）	表面开孔率（%） Porosity of the surface（%）	孔隙率（%） Porosity（%）	通水量（cm³/s） Discharge capacity（cm³/s）	耐酸碱性 Resistance to acid and base
性能要求 Performance requirements	≥ 50	≥ 95	85 ~ 90	≥ 380	稳定 Stable

3）陶粒，其粒径不应小于25mm，堆积密度不宜大于500kg/m³，铺设厚度宜为100~150mm。

3) Ceramsites, the diameter of particles shall be not less than 25mm. Stacking density shall be not greater than 500 kg/m³. The pavement thickness shall be 100 ~ 150mm.

2. 过滤层宜采用单位面积质量为200~400g/m²的材料，过滤层材料的搭接宽度不应小于150mm，过滤层应沿种植土周边向上铺设，并与种植土高度一致。

2. For the filtering layer, materials with the mass per area of 200 ~ 400g/m² shall be used. The overlapping width of the material of the filtering layer shall be not less than 150 mm. The filtering layer shall be laid upwards along the periphery of the planting soil and be consistent with the height of the planting soil.

20.7.13 种植屋面种植土。

20.7.13 Planting Soils on Planting Roofs.

1. 种植土可选用田园土、改良土或无机复合种植土，其湿密度应符合表20.7.13-1的规定。

1. For the planting soil, pastoral soil, improved soil or inorganic composite planting soil can be used, and the wet density shall conform to the provisions in Tab. 20.7.13-1.

【09措施】表 2.7.13-1: 种植土湿密度　　　　　　表 20.7.13-1

[Measures 2009] Tab. 2.7.13-1: Wet Density of Planting Soils　　Tab. 20.7.13-1

类别 Category	湿密度（kg/m³） Wet density（kg/m³）
田园土 Pastoral soil	1500 ~ 1800
改良土 Improved soil	750 ~ 1300
无机复合种植土 Inorganic composite planting soil	450 ~ 600

2. 常用种植土配置见表20.7.13-2。

2. The configuration of common planting soils shall conform to the provisions in Tab.20.7.13-2.

【09措施】表 2.7.13-2: 种植土常用配置　　　　　表 20.7.13-2

[Measures 2009] Tab. 2.7.13-2: Configuration of Common planting soils　　Tab. 20.7.13-2

主要配比材料 Main proportioning materials	配制比例 Preparation ratio	湿密度（kg/m³） Wet density（kg/m³）
田园土：轻质骨料 Pastoral soil: lightweight aggregate	1 : 1	1200

续表

主要配比材料 Main proportioning materials	配制比例 Preparation ratio	湿密度（kg/m³） Wet density（kg/m³）
腐殖土：蛭石：沙土 Humus soil: vermiculite: sandy soil	7：2：1	780～1000
田园土：草炭：蛭石和肥料 Pastoral soil: turf: vermiculite and fertilizer	4：3：1	1100～1300
田园土：草炭：松针土：珍珠岩 Pastoral soil: turf: pine needle mulch: perlite	1：1：1：1	780～1100

3. 种植土物理性能和种植土理化指标应符合表20.7.13-3和表20.7.13-4。

3. The physical performances and physicochemical indexes of the planting soil shall conform to Tab. 20.7.13-3 and Tab. 20.7.13-4.

【09措施】表2.7.13-3：种植土物理性能　　　　表20.7.13-3

[Measures 2009] Tab. 2.7.13-3: Physical Performances of the Planting Soil　Tab. 20.7.13-3

项目 Project	湿密度（kg/m³） Wet Density（kg/m³）	导热系数[W/(m·K)] Coefficient of Thermal Conductivity [W/(m·K)]	内部孔隙度(%) Internal Porosity（%）	有效水分（%） Available Moisture（%）	排水速率（mm/h） Drainage Rate（mm/h）
田园土 Pastoral soil	1500～1800	0.5	5	25	42
改良土 Improved soil	750～1300	0.35	20	37	58
无机复合种植土 Inorganic composite planting soil	450～650	0.046	30	45200	

【09措施】表2.7.13-4：种植土物理化指标　　　　表20.7.13-4

[Measures 2009] Tab. 2.7.13-4: Physical Performances of the Planting Soil　Tab. 20.7.13-4

项目 Project	非毛管孔隙度（%） Non-capillary porosity（%）	pH	含盐量（%） Salinity（%）	含氮量（g/kg） Nitrogen content（g/kg）	含磷量（g/kg） Phosphorus content（g/kg）	含钾量（g/kg） Potassium content（g/kg）
理化指标 Physicochemical indexes	＞10	7.5～8.5	＜0.12	＞1.0	＞0.6	＞17

20.7.14　地下建筑顶板种植设计

20.7.14　Design for planting on roofs of underground buildings

1. 地下建筑顶板种植设计应满足本节第20.7.10条中相关要求，并符合下列规定：

1. The design for planting on the roofs of underground buildings shall meet the relevant

requirements in Clause 20.7.10 of this section, and conform to the following provisions:

1）地下建筑顶板种植土与周界地面相连时，可不设排水层；

1) When the planting soil on the roofs of underground buildings is connected with the surrounding ground, drainage layers may not be arranged;

2）地下建筑顶板高于周界地面时，应设找坡层和排水层。

2) When the roofs of underground buildings are higher than the surrounding ground, sloping layers and drainage layers shall be arranged.